Bulldozer

Bulldozer

Demolition and Clearance of the Postwar Landscape

FRANCESCA RUSSELLO AMMON

Yale UNIVERSITY PRESS/NEW HAVEN & LONDON

Published with assistance from the Ronald and Betty Miller Turner Publication Fund, from the income of the Frederick John Kingsbury Memorial Fund, and from the foundation established in memory of Philip Hamilton McMillan of the Class of 1894, Yale College.

Illustrations in this book were funded in part or in whole by a grant from the SAH/Mellon Author Awards of the Society of Architectural Historians.

Yale University Press books may be purchased in quantity for educational, business, or promotional use. For information, please e-mail sales.press@yale.edu (U.S. office) or sales@yaleup.co.uk (U.K. office).

Designed by Mary Valencia.
Set in Century type by Tseng Information Systems, Inc.
Printed in the United States of America.

Library of Congress Control Number: 2015950794
ISBN 978-0-300-20068-3 (cloth : alk. paper)

A catalogue record for this book is available from the British Library.

This paper meets the requirements of ANSI/NISO Z39.48-1992 (Permanence of Paper).

10 9 8 7 6 5 4 3 2 1

Frontispiece: Lowell Georgia/Denver Post/Getty Images (modified; see p. 201).

For Peter, Isabella, and Oliver

Contents

Acknowledgments ix

Introduction
A Culture of Clearance 1

Part One *Bulldozers at War*

1 "A Dirt Moving War"
Engineers and Seabees as World War II Heroes 21

2 Prime Movers
Equipment Manufacturers Prepare for Postwar Prosperity 61

Part Two *Bulldozers at Work*

3 Grading Groves and Moving Mountains
Suburban Land Clearance in Orange County, California 97

4 "Armies of Bulldozers Smashing Down Acres of Slums"
Urban Renewal Demolition in New Haven, Connecticut 140

5 "The Intricate Blending of Brains and Brawn"
Engineering the Postwar Highway Boom 182

Part Three *Bulldozers of the Mind*

6 Unearthing "Benny the Bulldozer"
Children's Books and Tonka Trucks 221

7 Bulldozers as Paintbrushes
Earthworks and Building Cuts in Conceptual Art 251

Conclusion
Toward a Culture of Conservation 287

Notes 307
Index 369

Acknowledgments

This book is the product of years of research, writing, and conversation. It is my pleasure to acknowledge the many people who have enriched my thinking on this topic and the multiple institutions that have afforded me access to this book's wide-ranging source material. I also offer a quiet nod of thanks to the varied social spaces—from libraries and archives to conference rooms, coffee shops, and kitchen tables—that provided both scenery and soundtrack for the solitude of writing.

I have benefited immeasurably from the support of valued mentors. Dolores Hayden initially observed that a study of demolition might appeal to my interests. Not only was this a wonderful idea, but Dolores has proven to be an ideal advisor. I have appreciated her encouragement, advice, and example, and I thank her immensely for the countless investments she has made in me throughout this process. Jean-Christophe Agnew shared his enthusiasm for cultural history and inspired me with his expansive ability to see connections across space and time. Seth Fein, Matthew Jacobson, and Laura Wexler also read drafts of parts of this project and shared feedback that has greatly improved the shape of the book. A fellowship from the Miller Center for Public Affairs introduced me to Edmund Russell, who offered valuable suggestions on early chapters as well.

I have appreciated the opportunities to present and discuss parts of this project at meetings of the Society for American City and Regional Planning History (SACRPH), Urban History Association, International Planning History Society, American Studies Association, Business History Conference, and Policy History Conference, and at symposia and workshops organized by Hagley Museum and Library, the Program on the Study of Capitalism at Harvard, the Agrarian Studies and American Studies programs at Yale, and the *Under Construction* conference at Stuttgart University. Among the many scholars I have met throughout these academic communities, Lizabeth Cohen, Alison Isenberg, Mark Rose, Zachary Schrag, and Lawrence Vale have been particularly generous in nurturing my work. Thanks also to Thomas Andrews, Brian Balogh, Kevin Boyle, Richardson Dilworth, Philip Ethington, Richard Harris, Roger Horowitz, Carl Kramer, Pamela Laird, William Rankin, Adam Rome, Elihu Rubin, Aaron Shkuda, Steven Tolliday, and Shirley Wajda for their helpful comments on papers and presentations along the way. Without having read a word of this project, David P. Billington provided the example that first motivated me to begin studying the built environment. I thank him for his multi-dimensional approach to studying the urban environment, and for the inspiration of his teaching and scholarship.

Other colleagues have provided equally vital support. My writing group prompted me to start writing before I thought I was ready. Of course, that was precisely when it was time to begin. I thank Helen Curry, Julia Guarneri, Sara Hudson, Kathryn Gin Lum, Catherine McNeur, and Robin Morris for many spirited potluck dinners and generous feedback. Happily, Catherine remained a diligent reader and a ceaseless source of encouragement and advice well after our writing group had dispersed. I am also grateful for exchanges of work with Alice Moore, who especially pushed me to make the most of my visual material. Andy Horowitz shared his deep knowledge of New Haven and remained unflagging in his enthusiasm for this project. Dara Orenstein has been a valued fellow-traveler and gracious Washington, D.C., host. Daniel Amsterdam, Brian Goldstein, Jennifer Hock, Lauren Pearlman, and Sara Stevens offered further support, alongside the inspiring example of their own scholarship. Although Roxanne Willis and I crossed paths only briefly in New Haven, I was fortunate to reconnect with her near this project's end. Roxanne's editorial advice on multiple chapters significantly sharpened my ideas and my prose.

Several organizations have generously funded research, travel, and writing time. I thank Yale University and its Lamar Center for Frontiers and Borders; the Mrs. Giles Whiting Foundation; the Miller Center for Public Affairs and the Ambrose Monell Foundation; the Business History Conference; the Center for the History of Business, Technology, and Society, Hagley Museum and Library; and the Lemelson Center for the Study of Invention and Innovation. I had the privilege of spending a year at the American Academy of Arts and Sciences. This fellowship provided the space, time, and community necessary for me to push this project forward at a critical stage. I thank my American Academy colleagues for their insights, suggestions, and Thirsty Scholar companionship. Thanks to Melinda Baldwin, Hillary Chute, Matthew Karp, Christopher Loss, Nikki Skillman, and Peter Wirzbicki. Thanks also to Mary Maples Dunn and Patricia Spacks for their nurturing leadership of this group.

My colleagues at the University of Pennsylvania have warmly welcomed this historian into the School of Design. Domenic Vitiello and Aaron Wunsch have been wonderful co-instructors, showing me the joys of teaching collaboratively. I thank other fellow faculty members in City and Regional Planning and Historic Preservation—including Stefan Al, Eugenie Birch, Thomas Daniels, Erick Guerra, Amy Hillier, John Landis, Randall Mason, Frank Matero, Megan Ryerson, and Laura Wolf-Powers—for making PennDesign a comfortable home. Penn's Humanities + Urbanism + Design Initiative has offered another valuable sounding board for my work. My students have also enriched my thinking on this topic. In particular, I thank Amber Woodburn for her thoughtful feedback on Chapter 5.

I heartily acknowledge the wonderful research centers (and their staffs) whose collections underpin this project. I spent many months wading through demolition files at Yale University's Manuscripts and Archives, where I particularly appreciated the assistance of Michael Frost, Stephen Ross, and Claryn Spies. I am also grateful to Lisa Marine and Lee Grady at the Wisconsin Historical Society, Alison Oswald and Eric Hintz at the Lemelson Center, Jim Roan at the National Museum of American History Library, Nicole Thaxton at Caterpillar Inc. Archives, Tanya Brun at Hagley Library, and numerous staff members at the New York Public Library (both the Main Branch and the Science, Industry, and Business Library) and the Library of Congress (including its American Folklife Center). Robert Cullen patiently provided access to the American Association of State Highway and Transportation Officials' rich oral history collection. Michael Taylor welcomed me to explore the files at the headquarters of the National Demolition Association. I also thank William Keller and the staff at Penn's Fine Arts Library. Finally, I acknowledge the Interlibrary Loan staffs at Yale University, Harvard University, and the University of Pennsylvania, without whose assistance a large portion of the research for this project would not have been possible.

At Yale University Press, Jean Thomson Black has shared my enthusiasm for this project's many dimensions since we first spoke. I am grateful for her editorial assistance in bringing the manuscript to fruition. I also thank Samantha Ostrowski and Sara Hoover for their patience in guiding me through the many details along the way. Toward the end of this process, Susan Laity lent her critical copyediting eye to the project, providing perceptive suggestions for which I am incredibly grateful. I also thank my anonymous reviewers for thoughtful and thorough feedback that significantly improved the manuscript. A modified version of Chapter 6 appeared in the April 2012 issue of *Technology and Culture*. I thank the Johns Hopkins University Press for permission to reuse that work, and I also gratefully acknowledge helpful feedback I received from Suzanne Moon and three anonymous reviewers during that article's revision process.

Visual culture has always been critical to my conception of this project. It is with great appreciation that I acknowledge the many organizations that have permitted me to reproduce their images here. These organizations' names appear in the image credits. But several individuals particularly helped make the difficult process of reproducing twentieth-century images as smooth as possible. I especially thank Noel Sturgeon (Theodore Sturgeon Trust); Jonathan Braun (Ernest Braun Photography); Larry McCallister (Paramount); Ted Ryan and Tom Barber (Coca-Cola); Matthew Zebrowski and Tanja Hrast (UCLA); Marisa Bourgoin (Smithsonian Archives of American Art); Domenic Iacono (Syracuse University Art Galleries); Nick Lesley (Electronic Arts Intermix); and Philip Tan (James Cohan Gallery). I also thank the many photographers, authors, illustrators, artists, servicemen, and toy makers—and the guardians of their estates—who graciously granted me permission to reproduce or quote from their work. The Andrew W. Mellon Foundation, Society of Architectural Historians, Yale University Press, and University of Pennsylvania provided financial support that offset the costs associated with this visual program. Madeleine Helmer proved a deft research assistant, drafting multiple illustrations and resourcefully acquiring several images and permissions. I am also grateful to Elise Fariello for locating one outstanding image in the final hours.

Family and friends outside the academic world have offered just the right balance of interest in my work, coupled with even more appreciated distraction. I thank my parents, Chris and Toni Russello, for first supporting me on my academic journey so long ago. I thank Regina Lazarus, Andy Lazarus, Elena Russello, Amanda Russello, Sandy Ammon, Helmut Ammon, Kristin Ammon Shimano, Rich Shimano, Christopher Ammon, and Pamela Sifuentes Ammon for the camaraderie that only close family can provide. The warmth of many friendships has also sustained me across this project's duration. Among these, I especially thank the Blackstone-Schaenen family for welcoming our family so completely to Philadelphia.

Most enthusiastically of all, I recognize the little family that has grown up alongside this project. You make me laugh, dance, sing, and smile; for that I am forever grateful. Thank you to Isabella and Oliver for your enthusiasm and curiosity about the book that Mommy has been writing. Thank you to Peter for so many things: for your sharp editorial advice and unflagging willingness to read every chapter just one more time; for your company as we worked side by side at our computers over so many evenings; for your cheerful shouldering of our family obligations during this project's multiple crunch times; and, for always being such a wonderful and understanding husband and father. I dedicate this book to all three of you, with much love.

Bulldozer

Introduction

A Culture of Clearance

In 1951, Stephen Meader published a young adult novel about an entrepreneurial youth and a bulldozer. In his story, simply titled *Bulldozer,* friends Bill Crane and Ducky Davis stumble upon a Caterpillar D2 tractor submerged in a lake. They soon also discover a bulldozer blade, complete with hydraulic controls, hidden among the weeds on shore. After rescuing the equipment, Ducky repairs the tractor, drawing upon the experience he had acquired during his previous three years in the Army Corps of Engineers. Although Bill is unsure if the waterlogged tractor will function properly, Ducky has no doubts. "Those babies are hard to kill," he declares. "Why, I've seen 'em hauled ashore on islands in the Pacific where they'd been sunk in salt water ever since the war. They'd have coral an' barnacles growing on 'em, but in a week or two the boys would have 'em running again!"

Ducky's confidence proves well founded, as Bill is soon deploying the machine for his earthmoving business. With his "Cat" and a growing equipment supply, he clears orchards, digs cellars, topples buildings, plows fields, skids logs, fights forest fires, and even bravely rescues his girlfriend, Betty Barlow, when she gets caught in a blizzard. Bill's accomplishments earn him continued contracts and local fame. More important, Bill's work helps transform him from a youth into a man—both in his own eyes and in the eyes of the community. Yet Bill consistently deflects credit for his accomplishments to his machine. As he notes toward the novel's end, "It was the bulldozer—not me. If they want to put up a monument to the D2, I'm all for it."[1]

In this story, the American bulldozer opens a path to both personal and professional improvement. One reviewer described the novel as a "literary come-on for mechanically minded boys," and "the story of an enterprising young man who carves a career for himself with a bulldozer." Drawing upon his years as an ad man for the Caterpillar Tractor Company, Meader created a text that is, in some respects, an extended advertisement for the manufacturer. As Ducky pulls out replacement nozzles while repairing the newfound machine, he proclaims his love for the brand. "That's the beauty of 'Cat' equipment," he says. "Everything about it's simple. Just about foolproof. An' when you do have to replace a part you know it'll fit." Through Bill and Ducky's experiences, Meader educates readers about construction equipment operation. He also describes the positive economics of the earthmoving business. Bill learns the hourly rates of earth removal for a machine like the D2. Ducky advises him on pay grades for equipment operators. And bids submitted by Bill and others reveal the fees a contractor can charge for this work, by acreage and by time.[2]

Meader's story reached a wide and enthusiastic audience. *Bulldozer* was the twenty-fifth of the more than forty young adult novels he wrote in his nearly half-century career. Harcourt Brace followed its original publication with a school edition in 1955 and a paperback shortly afterward. When Stephen King was asked to name five books that were significant to him as a high school student, he listed *Bulldozer* alongside *The Grapes of Wrath.* He even mentions it in his own novel *It.* Today, hundreds of libraries continue to carry the title, and some enthusiasts recently put all of Meader's novels back into print. Contemporary young readers can now purchase the books that may have once delighted their parents and grandparents.[3]

Despite the book's endurance, it was a product of its time. How else to account for a bulldozer as the central figure in this tale? Or for the raw enthusiasm that protagonist Bill Crane voices for the machine? When Ducky tells Bill of the many Caterpillar D7 and D8 bulldozers on which he worked while in the Engineers, his friend responds with envy. "Maybe I'd better try a hitch in the Army myself," he replies. "I'd rather drive one o' those babies than eat!" But Bill does not have to join the military to fulfill his dream. The domestic American landscape affords him plenty of comparable opportunities. The novel's closing scene features Bill driving his bulldozer home after rescuing Betty in the blizzard. Despite the frigid air, he has "summer warmth in his heart," buoyed by his feelings of optimism for the future, both with Betty and in his career. As the novel concludes, "He looked ahead. More equipment in the years to come. Big tractors, scrapers, motor graders. Bigger contracts—highways—dams—airfields—there was no limit to what they could do with those tough machines."[4]

The bulldozer was ubiquitous throughout both the material world and popular culture in the decades following World War II. The machine appeared not only in novels like Meader's but also in films, children's books, artworks, and journalistic photographs and prose. These representations reflected the increasing presence of the machines on rural, suburban, and urban construction sites. In both its physical and cultural manifestations, the bulldozer dramatically transformed the postwar American landscape. To clear space for new suburban construction, interstate highways, and urban renewal development, wrecking companies demolished buildings and earthmoving contractors leveled land at an unprecedented pace and scale. The bulldozer was the iconic instrument of and symbol behind this transformation. Yet its history has thus far remained hidden in plain sight. Elevating the story of the bulldozer can change our understanding of these transformative years. Although the postwar decades are understood as an era of rapid American construction, they were equally significant for their implementation and celebration of large-scale destruction. Using the bulldozer as my primary object of inquiry, in this study I ask how the nation came to embrace and implement widespread destruction as a means of achieving progress.

The landscape has long absorbed the process of creative destruction: capitalist society's tendency to destroy while creating anew. Since the early settlement of the nation, taming the natural environment has been integral to the establishment of America's "second nature." Early settlers cleared forests for farmland, carved up the plains, and drove canals and railroads through the land. These acts of clearance often involved simple tools like the ax in the hands of hired and slave labor. In the nineteenth century, more advanced

Robert Moses, head of the NYC Mayor's Committee on Slum Clearance, sits at the controls of a Caterpillar bulldozer on a building demolition site in 1959. (Meyer Liebowitz/The New York Times/Redux)

In this 1962 image for *Look* magazine, a bulldozer and operator move dirt as part of land-clearance efforts for tract housing development in California. (Photograph by Ernest Braun)

technologies, such as steam shovels and hydraulic cannons, facilitated clearance for mining, forest harvesting, and dam building in rural lands. In cities from Boston to Seattle, these new tools also leveled hillsides for urban development. At the same time, cities continued eviction and clearance practices that dated to at least medieval times. In the nineteenth and twentieth centuries, immigrant tenements and shantytowns fell, making way for new parks, infrastructure, and other public and commercial uses.[5]

The postwar United States stands out for undertaking large-scale clearance on a nationwide scale. Over the span of just a few decades, the quest for progress drove the destruction of cities, suburbs, and rural landscapes across the country. According to the U.S. Census of Housing, roughly 7.5 million dwelling units were demolished between 1950 and 1980—to say nothing of nonresidential losses. These demolitions reached their zenith in the 1960s, when wrecking crews tore down one out of every seventeen dwelling units nationwide. A sizable portion of this destruction occurred in the name of urban renewal, which had displaced nearly a million families and over a hundred thousand businesses by 1980. Interstate highways drove additional demolition, peaking during the 1960s at a rate of twenty-seven thousand dwelling units destroyed per year (nearly twice that figure if we include all federally funded roadway construction). Clearance for the interstates also displaced a total of 42 billion cubic yards of earth. The development of suburban housing, shopping, and industry added further acreage to the toll.[6]

A historically specific "culture of clearance"—the war-inflected ideology, technology, policy, and practice of large-scale destruction—made such rapid, dramatic, and distinctive transformation possible. The bulldozer's use as a World War II weapon advanced the machine's technology, trained its operators, rehearsed its applications, and glorified its exploits. Postwar federal policies subsidized the domestic clearance of buildings and land while also providing tax incentives for greenfield development. Equipment manufacturers engineered the technology to make clearance work fast and cost-effective, and growing numbers of wrecking and excavating companies maneuvered the machines upon the land. Finally, popular representations of clearance celebrated this work culturally. As engineers and operators physically cleared the landscape, books, films, photographs, artworks, and other cultural texts performed some of the vital ideological tasks of rallying the public in favor of remaking postwar property uses and rights.

The story of postwar clearance begins when bulldozers went to war. World War II established the groundwork for building demolition and land clearance on the postwar home front. In producing massive quantities of tractors, scrapers, and bulldozers for wartime application, manufacturers like Caterpillar, International Harvester, and Allis-Chalmers honed earthmoving technology. The resulting equipment was bigger, stronger, and more easily maneuverable. While these were not among the wholly new technologies instigated by the war, the conflict proved crucial to advancing and popularizing existing categories of machines. Both advertisements and enthusiastic journalistic coverage showcased the machines for potential future customers. Wartime needs mobilized army engineers and members of the newly formed Naval Construction Battalions (or Seabees). As they trained in the machines' maintenance and operation, they rehearsed large-scale clearance practices. The Seabees were a new division of the navy, introduced during World War II both to build and to fight. More than 325,000 men served in the Seabees during the war—a number roughly equivalent to the entire membership of the present-day navy.[7]

When Seabees like Bill Levitt, the builder behind Levittown, Long Island, came home, they brought their new and refined skills with them. War valorized earthmoving machines and their operators by associating them with masculinity, heroism, and patriotism. Building on a long history of American environmental conquest, wartime engineers and others reinforced a militarized view of the landscape as an enemy to be conquered for new construction. Meanwhile, through images of urban ruin and reconstruction in Europe, the ravages of war lent moral and visual valence to destruction on a dramatic scale. Finally, wartime partnerships forged between manufacturers and the state helped yield postwar policies that fueled future commercial demand. In these ways, World War II made the subsequent transformation of the domestic landscape both physically and psychologically possible.

Once the war ended, bulldozers went to work on the home front, creating and transforming urban, suburban, and rural construction sites. Federal legislation provided the economic stimulus for large-scale clearance projects across the country. The G.I. Bill of 1944 built upon existing federal mortgage subsidies to provide returning servicemen with zero-down-payment loans for new housing, stimulating residential demand. Title I of the Housing Act of 1949 covered two-thirds of the cost of acquiring and clearing "slums" for urban redevelopment. The Federal-Aid Highway Act of 1956 provided 90 percent federal subsidies for highway construction, spurring the removal of massive quantities of earth, rock, and physical infrastructure. Meanwhile, the introduction of accelerated depreciation (from forty years to seven) into the 1954 Internal Revenue Code made it more profitable to build new construction on greenfield sites than to reuse existing locations.

As these policies were implemented on the ground, sites emblematizing rapid progress gave way to the damaging processes of physical destruction. What followed all too often was the "slow violence" of clearance and its relatively unspectacular physical, social, and environmental injustice. Development need not always occur in this way. While expansion and reuse of existing infrastructure may often be possible, some clearance will continue to be necessary both to manage existing landscapes and to support growth. Yet the scale at which that clearance occurs and the process through which it is realized dramatically shape the work's resultant impacts. The postwar period serves as a prime example of the consequences of large-scale clearance when implemented to excess.[8]

In Orange County, California, for example, a growing number of earthmoving contractors—operating increasingly powerful heavy equipment—bulldozed orchards and leveled hills for suburban development. During the late 1950s, one orange tree fell every fifty-five seconds across the state. Meanwhile, Les McCoy & Sons, a typical contracting firm working for a single day on a single project in Anaheim Hills, moved enough dirt to cover an entire football field more than two stories high. Over time, the boosterish embrace of new "crops" of housing collided with the costs of declining agricultural land and open space, the erasure of local history, development-induced landslides, and the growth of sprawl.[9]

At the same time, in cities like New Haven, Connecticut, wrecking companies operating figurative "armies of bulldozers" performed the dirty and disruptive labor of demolition for urban renewal. Former World War II army engineers filled leadership roles at Gil Wyner Company, one of the city's typical demolition contractors. In the 1960s, these companies tore down one out of every six dwelling units across the city. In urban renewal projects throughout the country, deceptively speedy, triumphal performances of wrecking gave way to decades of costly neighborhood invasion. This work all too often yielded parking lots, vacant land, and the enduring trauma of lost community. Minority residents bore a disproportionate share of these burdens.[10]

Finally, across the nation, vast federal funding and the relatively unfettered implementation of an engineering mentality trampled wide ribbons of buildings and land for interstate highways. Each of the more than forty-six thousand miles of interstate highways consumed roughly forty acres of land and the homes of twenty individuals. Although these losses resulted from processes that mirrored those for urban renewal demolition and suburban land clearance, the destruction wrought by interstate highways also differed from them. The federal government subsidized a greater portion of construction, thereby encouraging states to clear new routes, rather than make use of existing ones, in order to maximize their receipts. The technocratic expertise of engineers, not the designs of municipal planners or private developers, primarily routed highways through diverse geographies. Particularly in the program's early days, highway displacees received less compensation than if they had been relocated for the era's other major federal building program, urban renewal.[11]

Alongside the bulldozer's transformation of the physical landscape came the machine's invasion of the popular imagination. These bulldozers of the mind—including the representations in Meader's young adult novel—both reflected and reshaped the built environment. In addition to piecemeal appearances of the machine in varied cultural forms, the rise of heavy construction equipment spurred the creation of new subgenres. In the realm of children's literature, a body of "bulldozer books" made clearance seem natural and heroic. The books' characters were friendly equipment operators and their mechanical counterparts: machines with names like Benny and Buster Bulldozer. These works promoted clearance as technological progress and put a friendly, masculine, patriotic face on otherwise violent acts. They also made clearance look like fun. Along with Tonka trucks and related postwar toys, the bulldozer books were child-sized counterparts to pro-clearance journalism that appeared in popular magazines like *Life* and *Look*.

At the opposite end of the cultural spectrum, and aimed at quite a different demographic, the bulldozer's reach extended into high art during the 1960s and 1970s. Robert Smithson, Michael Heizer, Walter De Maria, and Gordon Matta-Clark exposed the celebratory culture of clearance through their "anarchitecture" and earthworks. These artists offered subtle critiques, reappropriated heavy equipment and wrecking sites, and at times even engaged in social and environmental reclamation. In conjunction with other depictions of clearance, such as those in photography and film, these artistic forms track the rise and fall of the culture of clearance throughout the postwar period. This collection of sculptural, visual, and literary works reflects the significance of culture and its representations to the structure of the physical world.

By the mid-1970s, the violence encoded in both the application of the bulldozer and its representations came to estrange—if not discredit—the culture of clearance. The destruction stimulated criticism, protest, and, ultimately, legislation in support of preservation, environmentalism, and increased citizen participation in the planning of space. During the early postwar years, the Cold War atmosphere of modernization, militarization, and mass consumption had helped create an environment conducive to a culture of clearance. By the end of the era, however, the context was changing. Riots and arson wrought uncontrolled destruction in cities. Suburbs sprawled outside them. Foreign manufacturers like Komatsu began to erode American dominance of the construction equipment trade. Construction laborers aligned themselves with a different, and more contested, kind of war on people and land in Vietnam. Soon, Americans were more commonly battling against—rather than with—the once heralded machine. The movements that these protests spawned helped to slow, though certainly not to stop, the progress of the American bulldozer.

Why the Bulldozer?

Although the insinuation of the bulldozer into an increasing number of academic texts beginning around the early 1960s might suggest that this scholarly terrain is well covered, the term has served more as a symbol of protest than as a subject of inquiry. Authors name the machine as the enemy of minority urban residents, elite preservationists, and suburban environmentalists, but they do not typically investigate its initial appeal or physical process. More recent historical scholarship on the postwar period—on the separate subjects of urban renewal, suburbs, and highways—functions similarly. These works typically move directly from plans to built projects. Scholars skip over the intermediary clearance process on the ground and direct attention toward the new built environments or the protest movements that resulted. In illuminating the bulldozer's negative legacy, these works leave open the question of why the machine rose to prominence and how it was transformed from hero to villain. Although historians of technology have made the greatest headway into this subject, adulatory corporate histories and encyclopedic "buff books" dominate the field. In the twenty-first century, interest in demolition has focused on other time periods, or on spectacular implosions of monumental edifices, rather than on the slower dismantling of multitudes of anonymous vernacular buildings that dominated the postwar period.[12]

Bulldozer: Demolition and Clearance of the Postwar Landscape is the first scholarly history of the bulldozer, focusing on the machine and its material applications, the companies that profited from its manufacture and use, and the cultural meanings it produced in practice. I combine urban and environmental history with the histories of technology and business, marrying the traditional archival methods of social and political history with critical readings of visual and material sources. For my research I mined diverse archives, including naval cruise books, Army Corps of Engineers histories, trade journals, business records, municipal documents, census data, oral histories, magazines, newspapers, children's books, popular films, art installations, and the built environment itself. These sources expose the bulldozer—and related instruments of material destruction—as one of the most significant tools of postwar planning. In this book I expand the range of actors, industries, technologies, and vernacular practices that form our understanding of the shifting contours of the American landscape. Architects, planners, politicians, and activists certainly shape our built world, but so too do the military, engineers, construction workers, and the artists who create popular cultural media.

More than just a history of the bulldozer as object, this is also a history of clearance as process. While it is an oversimplification to single out the bulldozer from among a bevy of construction equipment, this machine particularly helps us distinguish clearance within the broader—and often obscuring—construction narrative. Clearance is the first step in many construction projects. It encompasses the leveling of buildings, natural growth, and the earth itself to ready a site for new building (although new construction need not always follow). In the postwar period, clearance often happened at a large—and largely destructive—scale. Illuminating this precursor to new building exposes the violence embedded in the apparent progress of property development. Further, tracing the history of a process helps excavate not only why but also *how* the postwar landscape developed as it did. Both the means and meanings of that process shine light on the shifting building practices and attitudes that emerged during that era. While postwar construction often spurred opposition on aesthetic grounds, it was the clearance process that undergirded many deeper social and environmental critiques. Particularly as the act of construction often physically obscures the clearance phase, a conscious isolation of this critical moment unearths a process that is normally erased and holds it up for closer inspection.

By reconnecting the postwar landscape with its wartime past, I reperiodize American urban and suburban history and place it in a global context, locating the technological, economic, and cultural foundations of modern, mass clearance in mid-century theaters of war. Allied forces deployed more than a hundred thousand tractors, many equipped with bulldozer blades, during World War II. They also commanded about twenty thousand large scrapers—large-bowled vehicles for moving masses of earth long distances—and a similar number of cranes and shovels. On the European front, these machines cleared rubble-strewn cities and crushed barriers to forward advancement. In the Pacific, they leveled jungles, coral mountains, coconut fields, and native housing in advance of roadway, airfield, and military-camp construction. When these machines, methods, and metaphors of war moved from international battlefields to the American home front, they made possible the large-scale landscape destruction that underpinned postwar development. Foreign battlefronts effectively served as proving grounds for domestic construction sites.[13]

By reconnecting World War II with the decades that followed, I expose the militaristic undertones of postwar domestic urbanism. War was not an interruption to ongoing discussions about housing and highways but rather a critical catalyst that would direct the future shape of the landscape. To see the war in this light shifts the lens by which we view the developments of the postwar period and unveils the aggression embedded in the processes of property acquisition, clearance, and construction. More broadly, it demonstrates the ways in which exceptional wartime innovations and practices become normalized in domestic life. As historians have shown, because of World War II, access to work for women and minorities expanded, the acceptable reach of the state was extended, industry furthered its influence on policy making, and design innovations focused on meeting military needs. Wartime needs also advanced potent technologies, including the atomic bomb, chemical weapons, and air power. The bulldozer stands alongside these. Such wartime-to-peacetime technological transfer is as relevant to World War II as it is to more recent conflicts in Afghanistan and Iraq. As with today's military drones, individual bulldozers typically did not transfer from one environment to the other, but the larger technology and ideology did.[14]

In addition to the military, business interests have profoundly shaped the landscape. While the influence of property designers, owners, realtors, and developers is well known, *Bulldozer* adds the oft-overlooked construction industry to that list. Construction equipment manufacturers and contractors made clearance both physically and economically possible. Their self-promotion in advertisements, journalism, and even children's literature shaped the perception of their work as patriotic and progressive. Some companies also promoted legislation that made clearance politically viable. In their pursuit of postwar profits, private-sector industries facilitated the ascendance of the culture of clearance. By exposing the processes and consequences of capitalism's creative destruction, this book adds buildings and land to the list of goods devoured by postwar mass consumption.[15]

Although the bulldozer physically carved apart countless landscapes, the machine conceptually unites the histories of postwar cities, suburbs, and rural regions. By incorporating an object-oriented, machine-bound focus from science and technology studies into urban planning history, I minimize the geographic divisions that define much of the existing work in urban history. My concentration on the bulldozer demonstrates how processes of postwar urban and suburban development—so often portrayed as cause and effect—also proceeded in a parallel manner. Further, that focus integrates the oft-overlooked rural landscape into the story. Common technological, cultural, and political economic catalysts made the machine a powerful agent for change in diverse geographies. Planners and builders across the landscape shared a common quest for the "blank slates" that were seemingly necessary for new, large-scale development. Advanced technologies and practices made these achievable goals. Thus, urban demolition and suburban and rural earthmoving practices were not isolated pursuits. Rather, they were part of a broader American culture that viewed large-scale destruction as endemic to progress. The postwar bulldozer knew no municipal bounds in its consumption of physical and social space.

Instead, the more subtle geographies of race, class, and gender directed the course of clearance, inflicting damaging consequences on varied grounds. Across the country, African Americans, Hispanics, and members of other typically poor minorities were disproportionately the victims of demolition. Nationwide, 60 percent of residents relocated by urban renewal were nonwhite. Highway clearance functioned similarly. In places like Atlanta, blacks constituted 95 percent of all residents displaced by highways and urban renewal, despite making up only about a third to one-half of the total population. These citizens lost their homes and communities as cities sought more affluent—and usually white—residents and consumers to take their place. Relocation also proved a challenge. Federal assistance was inconsistent and unreliable. Ninety percent of the destroyed housing was never replaced. Due in part to real estate market segregation, African Americans purchased only 2–3 percent of the new owner-occupied housing units in the suburbs. The dispossession of these citizens and their unequal access to home ownership furthered patterns of spatial and economic inequality that have enduring legacies today.[16]

Racial minorities rarely reaped the employment opportunities of postwar construction work. During World War II, African Americans serving in the Army Corps of Engineers often performed manual work with picks and shovels, while white men more typically gained skills in heavy equipment operation. After the war, contractors and trade unions continued to segregate the most skilled positions. Minorities received few of the ensuing construction jobs, and when they did get them, these jobs were typically the lowest-paying, nonunionized, manual-labor positions. White men operating construction equipment did double violence as they destroyed the postwar neighborhoods of so many poor—and often black—residents. As African American workers rose up against this discrimination during the 1960s and 1970s, both in opposition to the bulldozer and in support of affirmative action employment, they brought the Civil Rights movement and Black Power to the construction site.

Sex and gender also shaped postwar landscape destruction. Women of all races were even more limited than minority males in their postwar construction employment opportunities. Prior to passage of the Fair Housing Act of 1968, women were also denied access to mortgage financing that could help them with relocation or suburban homeownership. Clearance also implicitly targeted sexual minorities, including lesbians and gay men, through demolition of the neighborhoods and hotels that housed many of these nontraditional households. Low-income single mothers suffered a similar fate.

Gendered language influenced popular perceptions of clearance equipment and practices. From the deployment of bulldozers during the war to their implementation on postwar American terrain, heavy equipment operators often cultivated an image of heroic masculinity. In photography and fiction, the bulldozer operator became a new version of the American cowboy. He was strong, handsome, and white; and he controlled a mighty beast. Blighted urban neighborhoods and untrammeled rural landscapes were the new frontier, and he conquered them with reckless (but not "wreckless") abandon. Even some artists who wielded heavy construction tools in their artwork adopted this cowboy persona. The bulldozer operator, harnessing powerful, American-made technology, offered a patriotic antidote to Cold War anxieties about masculinity. Subsequently, opponents of the machine also employed gendered vocabulary to reframe the image of the bulldozer. In their portrayals, the bulldozer became the attacker and the rapist of feminized environments.[17]

As these metaphors suggest, both critics and boosters turned to culture—an intangible set of ideologies and values, as well as their symbolic representations—to make sense of the rapidly changing world around them. Whereas urban historians have often privileged the political and social ideas embedded in texts, here I also advance the case for the power of visual and material culture—and the ideologies embedded within them—to shape the physical world. Parsing the meaning and influence of these multimedia sources re-

In May 1959, a photographer snaps a picture of a bulldozer razing the house of the Arechiga family, occupants of the last home left standing at Chavez Ravine, in Los Angeles. Through a combination of building demolition and large-scale earthmoving on the grounds of this largely Mexican-American community, land originally slated for urban renewal housing development eventually became home to Dodger Stadium. (Los Angeles Times Photographic Archive, Department of Special Collections, Charles E. Young Research Library, UCLA)

quires close reading of the documents but also consideration of them within their broader social, political, and cultural contexts. Only then do they reveal themselves not as objective illustrations but as carefully framed arguments.

Cultural materials both represent history and participate in it. A photograph of the bulldozing of the last home left standing on the Chavez Ravine urban renewal project visually illustrates this point. Chavez Ravine was a pastoral hillside community located a mile from downtown Los Angeles. By the mid-twentieth century, it was home to primarily Mexican-American families living in modest houses. In 1950 the city targeted the area for the development of public housing, and clearance soon followed. As the bulldozer pictured in this image pauses next to the half-toppled home of an evicted family, a photographer—pictured in the lower right-hand corner—snaps the scene.

His presence explicitly illustrates the role of image-makers in shaping representations of history. The demise of this particular home is a consequence of both physical and social geography. This image reveals space as a site of historical action and a determinant in the course of history. The built environment and its representations serve as both source and subject of this book's story.[18]

Through a three-part investigation of the bulldozer as instrument and icon, in *Bulldozer* I demonstrate the central place of clearance in the making of postwar American growth. The story begins with men, machines, and mythologies at war. World War II is not a prequel to the central narrative of postwar development; it is when and where postwar development experienced its most important rehearsal. Reconnecting these two moments is one of my central goals. I then follow the men, machines, and methods into the postwar era, where hospitable policies and influential businesses set them to work on the domestic home front. Through the widespread implementation of *destruction,* they simultaneously shaped the *construction* of the American urban, suburban, and rural landscape. By excavating the process behind postwar building demolition and land clearance in varied geographies I expose the sheer messiness of this work in practice, especially as popular faith in technological expertise gave way to more complex and contested understandings. Finally, I argue that the realization and reshaping of postwar clearance relied on changes in both material practices and cultural values. These values take representational form in photographs, films, novels, and other media. But they most dramatically reveal themselves in two new creative subgenres—children's bulldozer books and built environment–based conceptual art—that developed in direct response to real-world clearance. By marrying these three spheres of war, work, and cultural imagination, I reveal the bulldozer's widespread influence and appeal, while also uncovering the mechanisms behind its eventual self-destruction.

A Brief Prehistory of the Postwar Bulldozer

The term *bulldoze* originated in violence. During the Reconstruction Era, a bulldoze (or bulldose) referred to a severe "dose" of whipping or flogging, as applied to a "Negro" in the southern states. In the early twentieth century, *bulldozing* also meant coercion through violence or intimidation, with or without physical assault. Sometimes the objectives of this coercion were political, as when powerful individuals or mobs forcefully attempted to shape public opinion and voting patterns. The "bulldozer" was the one who did the bulldozing. Or *bulldozer* could also refer to a type of large pistol that a person might brandish during these coercive acts.[19]

It was not until the first quarter of the twentieth century that the term was applied to mechanized implements designed to force the earth. In 1917, the Russell Grader Manufacturing Company's catalogue included a "bull dozer," a heavy, horizontal metal blade designed for moving dirt. (Although the term *bulldozer* continues to apply only to the blade itself, common usage subsumes both tractor and blade under the title.) This early bulldozer's power came from the team of mules pushing it. Even more impressive were its capabilities when attached at the front of a steam engine tractor. Wheeled tractors came first, but crawler models soon followed. In 1904, Benjamin Holt introduced a self-laying track system of the treadmill type, which could propel a steam engine forward without getting mired in soft or wet soil. Holt's objective was to distribute the pressure of heavy farm implements while moving them across the delta soil of Central California valleys.[20]

Since the crawler tractor's movement resembled that of a caterpillar, Holt registered it under the name Caterpillar Tractor. In 1925, Holt's company combined with that of another California farm implement maker, C. L. Best, to become the Caterpillar Tractor Company. Today, Caterpillar is still one of the best-known manufacturers of these machines. Although there is some dispute as to whether Caterpillar or LaPlant-Choate first placed a blade in front of a crawler tractor—thereby creating the implement referred to as a bulldozer—the modern machine emerged sometime in the mid-1920s. The gradual replacement of tractor steam engines with gasoline engines, beginning in 1908, and then with diesel, starting in 1931, increased the bulldozer's power.[21]

The bulldozer belongs to a long lineage of agricultural, industrial, and construction equipment, of which it is just the most versatile and best-known example. The machine is a descendant of the steel and iron plows that colonial settlers and western expansionists used to grade land for subsistence and development. In industrial applications, the bulldozer is the more powerful and mobile successor to the steam shovel, which helped move earth for canals and railroad construction beginning in the nineteenth century. Among more modern equipment, the bulldozer is a close relative of the scraper, a wheeled machine with a large bowl and cutting blade for digging, loading, hauling, and dumping the earth. The original nineteenth-century scraper was a metal scoop drawn by horses or mules and raised from the rear to deposit its dirt load. Later operators replaced the animals with crawler tractors, increasing the tool's power and capacity.

Advancements in cable controls, diesel engines, bowl size, and hydraulic lifts during the first half of the twentieth century improved the functionality of heavy construction equipment. LeTourneau produced the first high-speed motor scraper in 1938, and manufacturers like Euclid and Caterpillar soon followed with their own designs. At the same time, front-end loaders—or shovel dozers—added new capabilities to basic dozer machines. Considered "the most advanced development of the bulldozer," the loader paired a frame element with a bucket that could be pushed, raised, lowered, or dumped. (In their imprecise application of construction equipment terminology, lay observers often lump the loader under the term *bulldozer* as well.) By the start of World War II, then, the bulldozer, scraper, and related machines were versatile pieces of equipment capable of moving large quantities of material cheaply and efficiently. They were primed to meet substantial battlefront clearance demands and, soon afterward, to return home to wage war on the American landscape.[22]

Part One

Bulldozers at War

(*Overleaf*): An Engineer guard accompanies a Caterpillar Diesel D4 Tractor and Bulldozer on Munda Airfield, New Georgia Island. (*Military Engineer* 36, no. 223, May 1944. Photo by U.S. Signal Corps. Printed with permission of the Society of American Military Engineers.)

Chapter 1

"A Dirt Moving War"

Engineers and Seabees as World War II Heroes

In 1944, Colonel K. S. Andersson of the U.S. Army Corps of Engineers penned "The Bulldozer—An Appreciation." The essay was an ode to a piece of technology then proving critical to Allied success in World War II. "Of all the weapons of war," he wrote, "the bulldozer stands first. Airplanes and tanks may be more romantic, appeal more to the public imagination, but the Army's advance depends on the unromantic, unsung hero who drives the 'cat.'" That same year, a *Chicago Tribune* writer predicted, "When the axis powers are defeated and the time comes for distributing credit for the allied victory everywhere that it is due, an appreciable share should go to a machine that is constructed fundamentally to take no direct part in the fighting. This machine, tho not designed to play the role of weapon, bears the belligerent name of bulldozer." Major General Eugene Reybold, chief of the army's Engineers, added, "Victory seems to favor the side with the greater ability to move dirt." An advertisement for International Harvester, one of the manufacturers of these machines, took his assertion a step farther, proclaiming the entire conflict "a dirt moving war." As Reybold later concluded, "By the war's end, it was evident that American construction capacity was the one factor of Ameri-

THIS is a dirt-moving war . . . *a tractor war.* Already the history of World War II is brimful of heroic jobs done by crawler tractors, equipped with bullgraders, bulldozers and scrapers. Those tractors will continue to smash their way through jungle and swamp, over mountain and plain, to Victory.

As a two-star general of the Army Engineers puts it: "Victory seems to favor the side with *the greatest ability to move dirt.*"

Munda . . . Rendova . . . the Solomons . . . Kiska . . . Sicily . . . Salerno . . . everywhere our fighting forces go, you'll find these armored giants building roads, smoothing shell-torn landing fields, pulling heavy guns, handling aircraft bombs.

The Armed Forces have first call on International TracTracTors today. That's why so few new ones are available for civilian use. The new TracTracTors you need so much today, to replace badly worn equipment, are more urgently needed on the fighting fronts.

Many of your old Internationals have a lot of work-hours left in them. Keep those tractors well serviced. Work closely with your International Industrial dealer. He has the skilled service men, the well-equipped shop, and the stock of International Parts to help keep your TracTracTors plugging on the home front, backing up the military TracTracTors on the battle front.

INTERNATIONAL HARVESTER COMPANY
180 North Michigan Avenue — Chicago 1, Illinois

INTERNATIONAL POWER

can strength which our enemies consistently underestimated . . . [the one] for which they had no basis for comparison. They had seen nothing like it."[1]

While military construction men, equipment makers, and boosterish wartime journalists were not unbiased witnesses, battlefront evidence corroborates their accolades of the bulldozer. During its first six months engaged in World War II, the Corps of Engineers moved nearly half a billion cubic yards of earth. This exceeded the volume handled in digging the entire Panama Canal. Little wonder that Douglas MacArthur, supreme allied commander of the South West Pacific Area and general of the army, labeled the clash "an Engineer's war." In addition, over the length of the entire war, the Engineers' naval counterparts, the Seabees, cleared land and built structures for more than 400 bases, 100 air strips, 235,000 roads, 700 acres of warehouses, housing for 1.5 million men, and storage tanks for 100 million gallons of gasoline. The contributions of these men and machines extended beyond construction alone. They cleared bomb-damaged cities of rubble, rescued landing craft that were tossed at sea, and even directly unleashed their power upon the enemy.[2]

Scholars have typically viewed the war as an interruption to the housing and highway work of the construction trades, but in fact the industry participated vitally in the conflict. During World War II an army of construction workers, including 325,000 Seabees, rehearsed large-scale land clearance practices on foreign terrain. When these men returned home, they brought their new and refined skills with them. Their war work also elevated the cultural standing of bulldozer operators and their equipment. Contrary to Andersson's characterization of the "unsung" Caterpillar driver, diverse media—including trade journals, popular journalism, novels, photographs, and films—publicized his accomplishments. The glorified portrayal of the masculine operator and his patriotic machine rendered them hero and weapon. And the war shaped ideology. Building upon a long history of Ameri-

---

(Opposite) This 1944 International Harvester advertisement, which appeared in *Construction Methods* magazine, depicts tractors armed with bulldozer blades as they move earth on a Pacific Island battleground. The accompanying text incorporates General Reybold's equation of victory with "*the greatest ability to move dirt*" and proclaims the entire conflict "a dirt moving war." (Courtesy of Navistar, Inc.)

can environmental conquest, the conflict reinforced a militarized view of the natural and built landscape as enemies to be conquered for new construction. The largely overlooked environmental scars left by the war, coupled with relatively positive representations of urban ruins, made large-scale demolition seem both possible and palatable. Postwar Americans later redeployed wartime machines, methods, and metaphors on the domestic landscape.[3]

"We Build! We Fight!" "We Clear the Way"

The U.S. military has long incorporated engineering and construction into its mandate. The army deployed engineers during the Revolutionary War and formally established the Corps of Engineers in 1802. Later that century, the corps added civil projects to its responsibilities, constructing coastal fortifications and mapping the American West. In World War I, corps members utilized picks and shovels on international road maintenance projects and the digging of trenches. Where this work was mechanized, it was limited in scale. By the mid-1930s, the corps began a formal reexamination of its methods, motivated by recent improvements in commercial vehicles, experience gained on large-scale New Deal development projects, and the expectation that future wars would require equipment with increased mobility on land and in the air. While the corps had already begun adopting some larger construction machines, the country's entry into World War II stimulated expansion of its mechanized equipment stores and the pursuit of projects of greater volume, scale, and significance. The corps mantra, summed up on World War II recruiting posters, was "We clear the way."[4]

The navy prepared for World War II by building up its coastal and Pacific bases. American companies, including Turner Construction and Morrison-Knudsen, relocated their workers and hired local laborers to execute this work. Once the United States entered the war, these private contractors proved inappropriate for the expanded task. Navy leaders worried that civilian crews lacked the tight coordination, fast tempo, and strict discipline necessary to function productively under fire. In the event of an attack, they would be physically unprepared to fight back and legally prohibited from doing so. If captured, they would be treated as guerrillas, rather than as military prisoners of war, thereby at risk of summary execution. Finally, if an act of war terminated their labor, these civilians' employers would be under no obligation to continue compensating them.[5]

When the Japanese attacked American-occupied Pacific Islands in the hours and days following Pearl Harbor, these theoretical concerns became lived nightmares. The experience at Wake Island is particularly well known. When enemy planes arrived, the marines there were unable to protect themselves and the more than eleven hundred Morrison-Knudsen civilian crew members temporarily residing there. After the war ended, only about seven hundred of these contractors returned home. Nearly fifty were killed while attempting to defend the island, the Japanese executed almost a hundred more, and the remainder died in prisoner work camps in China and Japan. Throughout the war's long duration, their families received only limited and unreliable updates on their conditions and minimal financial support.[6]

Disasters like this, combined with the vast construction demands anticipated to support an air-based war, prompted the creation of a new, construction-oriented military branch. During World War I, the navy spent less than two million dollars building shore bases—none of which ever came under enemy fire. By contrast, World War II required hundreds of new bases, including some located in the heart of battle. Admiral Benjamin Moreel, chief of the Bureau of Yards and Docks, spearheaded the proposal for a new group of soldiers to meet these demands. Their motto would be "We build! We fight!" On January 5, 1942, the Naval Construction Battalions (NCBs) were born. The first battalion shipped out later that month. The launch of the Seabees was the first time the military had absorbed construction industry personnel wholesale, rather than developing its own engineering crews. Seabee ranks grew quickly, reaching a wartime peak of approximately two hundred battalions consisting of more than 250,000 men. The majority deployed to the Pacific theater.[7]

The Seabees' earliest members arrived with valuable construction experience. The first officers came from the navy's existing Civil Engineer Corps. Lieutenant Commander James R. Ritter, of the 107th NCB, was typical of this group. He had earned a degree in civil engineering at the University of Texas before going to work for Shell Oil Company, the Texas State Highway Department, and the City of Houston (as a planner). Civilian construction men populated the ranks that officers like Ritter oversaw. They represented roughly sixty different trades, from carpenters to electricians and steam shovel operators. Given their substantial experience, their demographics could differ sharply from those of a typical military enlistee. The yearbook-style cruise books produced by each of the battalions illuminate their profiles. As the cruise book of the 27th NCB noted, "We were boys just out of trade or high schools. We were men in the first flush of strength and skill. We were men of 25, 30 and 40 years. We were men of 50 years. . . . Most of us were flabby and soft, except for our calloused hands. . . . There were men in the outfit who could have served in the first World War. They couldn't march five miles now with seventy-pound packs on their backs, nor stand guard on a vessel plowing through wintry seas at night. They couldn't fight, but they could work. And here was the place to show what work they could do."[8]

The navy recruited these civilians, 80 percent of whom were union men, directly from engineering societies and industry organizations. From oral histories conducted with veterans we can gain insight into these men's motivations. Some reported joining the Seabees for the benefits of longer initial leave and reduced boot camp length. Others viewed the Construction Battalions as a less dangerous form of military service, particularly relative to the infantry. Further, the Seabees offered a way for men to serve at an advanced age. Finally, there was the appeal of higher rank and pay, commensurate with their level of construction experience. With bulldozer and crane operators earning twice the entry-level Apprentice Seamen pay grade, and foremen receiving even higher petty officer rankings, Seabees were among the highest-paid participants in the war. Military service extended the construction knowledge of even these skilled men. As one such veteran Seabee reflected, "I was always ready to learn something. So I got filled with it for three years. And when I come out, I was a hell of a lot smarter than I was when I went in."[9]

Not all entering Seabees were experienced construction men. From the beginning, the navy recruited younger soldiers to perform the heavy lifting. Over time, as demand for experienced laborers began to exceed supply, and when the distribution of enlistees' skills did not match military needs, more young Seabees arrived who had no previous training. Some of these men self-selected the Seabees because of their interest in construction or because of the work's relative perceived lack of danger. Others found themselves assigned there. William Schlumpf, of Evansville, Indiana, for example, entered the Seabees right out of high school. As he later recalled, "I went out for my physical and they said my eyes weren't good enough to be in the Air Force, so they were going to put me in the Medical Corps in the Army. And I said, 'I don't want to go in the Army.' And the fellow said, 'Well, good. We'll just put you in the Seabees.' And that's where I went." Schlumpf worked his way up to become a diesel mechanic in the heavy equipment repair shop.[10]

Construction training became increasingly critical to these raw recruits' success. In the Seabee organization's initial days, new arrivals trained for six weeks before shipping out to "Island X." At training centers in Virginia and Rhode Island, they learned combat skills along with specialized construction trades. Later, urgent needs for field personnel moved construction training entirely into the field. Through practical experience and detailed manuals provided by construction equipment manufacturers, Seabees progressed to what one journalist referred to as "mastery of the bulldozer." As he elaborated, "Overseas, and under fire, [a Seabee] may take it apart, put it back together, and make it do tricks." Such proficiency proved essential for Seabees and Engineers against their dual wartime enemies: the Axis and the environment.[11]

Rehearsals of Clearance Practices

The U.S. military compiled a stunning construction record during World War II. Over two-and-a-half years of service, the 11th NCB alone erected enough infrastructure, according to its cruise book, "to furnish housing, factories, warehouses, power, light, sewers, water, telephone, airfields and water front facilities including lighterage and garbage barges for a city of 11,000 people." Before any of this construction went up, Seabees took down everything that stood in their way. In the course of building 73 miles of roads, 19 docks, nearly 300 Quonset structures, and 154 timber-frame buildings, plus tank farms and pipe lines, another battalion excavated 790,000 cubic yards of earth and coral and felled enough trees to produce a million board feet of native lumber. In the words of the 87th battalion's cruise book, the soldiers were performing "destruction for construction." This was a battle as challenging and real as military combat. In the contest of "Seabees and Bulldozers vs. the

"Seabees and Bulldozers vs. the Jungle." Seabees battle the trees and earth of Pacific jungles as they build roadways on "Island X." (US NCB, 4th, *Lil' Short-Runner Presents the Fourth U.S. Naval Construction Battalion Penguin, 1944–45* [Baton Rouge: Army and Navy Pictorial Publishers, 1945], 36)

Jungle," yet another cruise book noted, the men "pushed down forests gladly . . . slamming through the so-called impenetrable forests." As the cruise book of the 27th NCB summed up, "Time and time again we did the unexpected task, performed the miracle of making the impossible a completed fact."[12]

Not only Seabees themselves but also the authors of more popular accounts depicted this work in a triumphal light. A 1945 *Life* magazine article described the high-speed conversion of the "little island" of Guam into a "mighty base" with 360 miles of highway and the longest paved runway in the world. As the magazine author recounted, "The island shook to the pounding rhythm of rock crushers and heavy engines and echoed with the sound of tractor treads crunching through coral. . . . Cats and Macks and bulldozers puffed and backed and hacked, shaving away the jungle growth. Guam became alive and bustling with roads and road builders." Ironically, the price of this figurative emergence of life was the death of much of the natural landscape. Still, the author concluded, "Guam has been changed, irrevocably and permanently"—and, he seems to imply, for the better. Similarly, in 1950, the historian Samuel Eliot Morison enumerated the "many enemies besides [the] Japanese" encountered by the 6th NCB in Guadalcanal: "The soil, for instance, was an elastic and unstable muck which stimulated the toughest former road boss to new heights of profanity. Daily tropical rains taxed drainage ditches and kept the mud from drying. Equipment was long in arriving." "Yet," he concluded, "no challenge remained unanswered by the Seabees."[13]

This wartime work trained army engineers and Seabees in large-scale land-clearance practices, which they implemented with ingenuity, speed, and skill. On the island of Los Negros, one lieutenant devised a system for toppling about four thousand coconut trees in just thirty-seven hours, using only two tractors connected by a cable. On the Solomon Islands, another battalion learned cutting, filling, and compaction as the men dug thirty-three oil tank building pads out of steep hills and canyons separated by distances of up to two hundred feet. Even level land could require substantial amounts of earthmoving (or coral moving). The creation of an airstrip measuring 150 by 6,500 feet, for example, displaced a quarter million yards of material. Another airstrip for B-29 bombers required the connection of two islands in order to create a sufficiently long runway. Rather than building a bridge, Seabees combined nonstop labor with mechanical might to fill the waterway with earth. As Seabee Adolph Pisani recalled, "You take a group of 4–500 people there, working 24 hours a day, you know, you can move a lot a, lot of ground. And then we had Caterpillars there, D-8 Caterpillars there, running around. We had shiploads of 'em come in there. One of 'em would break down, they didn't bother fixing it. Bulldoze it off in the ocean, get it out of the way, so another one could go to work on it." In constructing a series of such airstrips across the Pacific, Seabees completed a virtual "Road to Tokyo" for Allied air power. The building of a more conventional roadway, the Ledo Road (later called the Stilwell Road) in Burma, proved to be one of the greatest earthmoving tasks of all. In clearing just the first 270 miles of this more than 1,000-mile roadway, corps engineers and Chinese laborers moved enough earth "to build a dirt wall three feet thick and ten feet high from San Francisco to New York."[14]

Such vast clearance practices sometimes unearthed archaeological finds. In Guam, mechanical earthmoving and exploded shells exposed ancient

Through feats of massive earthmoving and compaction, Seabees cut flat building pads out of steep hills, such as on this wartime oil tank project on Florida Island in the Solomon Islands. (US NCB, 27th, *Danger; Fighting Men at Work: A Work-a-Day Tale of How the Job Was Actually Done by the 27th Seabees,* ed. Willard G. Triest [Baton Rouge: Army and Navy Pictorial Publishers, 1945], 64)

relics, leading an archaeologist resident on the island to urge restraint by army and navy bulldozers whenever they uncovered ruins. Earthmoving activity temporarily ceased in England when wartime digging revealed Roman artifacts that established the city of Canterbury as being much older than had previously been thought. In stark contrast, more recent building debris functioned as sources of fill rather than archives of history. When insufficient earth was available to fill marshy areas slated for English aerodrome construction, Engineers turned to the plentiful supply of "blitz brick" instead. Over time, this stone and brick rubble from the country's bombed-out buildings combined with cinders to produce a stony soil that offered a strong foundation for the building of concrete slabs above.[15]

One of the most epic World War II construction projects was the Alaska, or Alcan, Highway begun in early 1942. The roadway would serve as a supply line and instrument of defense as it connected the western United States with Alaska by way of Canada. Consistent with the army's own mythologizing of this massive undertaking, one contemporary trade journal called construction of the highway "one of the most amazing battles man has ever waged with the frozen tundra and won." Forging the most strategic path for the roadway, unadjusted for topography, posed a physical challenge. Men labored in harsh blizzards and freezing temperatures, and on soil varying in character between hardened ground and soggy tundra. Operators inexperienced with diesel engines damaged machinery, and replacement parts proved scarce in the distant location. Yet the collective efforts of army engineers and civilian contractors successfully slashed out a 1,390-mile pioneer road in just eight months. This expansive accomplishment covered a distance equal in length to that between Bangor, Maine, and Jacksonville, Florida.[16]

Along with the unique obstacles imposed by Alaskan geography and climate, the Alcan Highway project also prepared workers for more routine road-building tasks. Operators at the helm of Caterpillar D4 and D8 bulldozers learned practices that subsequently would be put to use on domestic highway projects. First, the lead bulldozer would topple trees along the established center line. Two or more of these machines would follow and "mop out" the hundred-foot roadway width, while succeeding bulldozers pushed fallen trees and brush to the side. Ten to twelve of these machines could clear two to three miles of roadbed every twenty-four hours. Contractors working thirty to fifty miles behind them widened the path into a functioning trail road, following the contours of the land when possible in order to avoid heavy grading. In 1943, with the enhancement of the Alcan Highway, road-building crews executed deep cuts and fills. They used bulldozers and shovels to remove soft muskeg material, while scrapers and dozers leveled the overall route.[17]

With the urgency of war, this "construction proceeded," as the environmental historian Roxanne Willis put it, "in the most destructive manner possible." Troops felled large swaths of trees, both to make way for the roadway and to provide logs to create "corduroy" atop earth that fluctuated seasonally between frozen and mushy ground. Accidental fires increased the loss of trees and other growth. Excess supplies and equipment left behind despoiled the environment, and the occasional clean-up consisted largely of burial by bulldozing under the earth. Meanwhile, the large-scale earthmoving unsettled the topography, in some cases causing unstable terrain. The hasty manner in which the roadway was constructed resulted in a crooked route, disconnected from the all-important Pacific Coast, that would be of limited postwar civilian and commercial value. Thus, discussions of a second, more utilitarian highway road—with its own environmental costs—were already under way before the war's end.[18]

The completion of the Alcan Highway highlighted not only physical but also social challenges associated with clearance. At the outset of the war, the army used the high rates of illiteracy among African American soldiers as justification for consigning them to service, rather than combat, units. The Corps of Engineers was chief among these assignments, for both educated and uneducated blacks. This placement particularly rankled among men who felt their education and skills were being wasted on labor-intensive roles. More than a third of all engineers who worked on the Alcan Highway project were black. They served in four African American units: the 93rd, 95th, 97th, and 388th. These units had trained in Alabama, Florida, and Georgia, and many of their members hailed from the South as well. They were ill-prepared for the Alaska climate. The military further handicapped them by supplying inferior clothing, shelter, and equipment. The army allocated more heavy machinery to white troops, leaving most African Americans with shovels and wheelbarrows for tasks typically requiring more sweat than skill. These decisions deprived the men of the opportunity to maximize their technical skills. In late October 1942, the encounter of a southbound bulldozer operated by an African American engineer with that of a white engineer traveling north marked the completion of the pioneer trail. A photographer captured the image of the two smiling men shaking hands atop their machines. The image and associated story circulated widely, symbolizing racial equality on the construction site. This incident was, however, hardly representative of African Americans' wartime experience. Nor did it foretell the character of the postwar years. African Americans on the postwar home front would be relegated to the low-paid roles of manual labor, when they participated in the construction trades at all.[19]

The war deprived African American construction men not only of maximum training but also of due recognition for their work. (The NAACP, among other organizations, lobbied for the few military awards of any kind bestowed upon black soldiers.) Moreover, the popular media—from *Life* photo-essays to advertising campaigns and Hollywood films—rarely covered black soldiers. This even though African Americans made up one out of every six Engineers on the European front, for example, by the middle of 1945. On some projects, such as the building of the Ledo Road in Burma, they even constituted a majority (65 percent). Yet the army initially planned that white troops would lead the convoy at the ceremonial road opening, thereby garnering all the newsreel and newspaper coverage. Only at the last minute did troop dissent spur the airlifting of a group of black soldiers to the convoy assembly point, where they were supplied with trucks to drive in the caravan. Although an African American newspaper, the *New York Amsterdam News,* reported on the "combat glamour" of the black soldiers engaged in road building, the largely white popular press ignored them. A *National Geographic* profile of the road in 1945 was representative of mainstream coverage, with the author describing African Americans on the project in subtly condescending ways. The one black soldier he quotes tells a joke, and the only photograph in which blacks appear shows a crew of men standing around uselessly as a few soldiers inflate a pontoon for a bridge. The caption notes that the "Negro soldiers . . . did yeoman service building the Ledo Road. At night they filled the air with plantation songs and 'boogie-woogie.'" By contrast, the author ascribes the real credit for the road to the white engineers, for they were the men who "operate[d] the lead bulldozers in clearing the trail for the Negro troops to follow." Although the *New York Amsterdam News* told of having seen African Americans, "tall, lean and hard, operating bulldozers, road graders, tractors, carry-alls, quickway cranes, at work cutting paths in the hitherto impenetrable depths of this age-old jungle," a much more limited audience read these words.[20]

White engineers and Seabees were less likely to be serving alongside black soldiers than they were to be confiscating the property of various racial minorities. The military acquired land for defense development through what amounted to eminent domain takings. Countless farms and homes fell as a result. In Hawaii, for example, the navy hired the civilian contractor Morrison-Knudsen to clear a low ridge, known as Red Hill, for construction of a work camp and underground fuel-tank storage facility. The company sent out bulldozers and scrapers to crush sugarcane and pineapple plants and then burned the refuse on-site. Next those same machines plowed roadways down the hillside. Although the navy did not own the property, this posed no practical problem. As one observer noted, "The pineapple people had no choice. Eventually you must sell at the Navy price." As the navy continued to expand its Hawaiian outposts, the 4th NCB arrived in early 1944 to work on neighboring Moanalua Ridge. The battalion's cruise book recorded: "The cane was dozed like so much hay, loaded into trucks by clam, and hauled away." In tandem, the military moved some houses off the ridge, although structural relocation was less common than outright destruction. After several weeks, the entire area was cleared for new building.[21]

Claiming and clearing property proved even easier in the South Pacific. The homes of native islanders were often of simpler construction, and colonial powers could deprive locals of property rights to the land they occupied. In a road-building project on Florida Island in the Solomons, for example, Seabees encountered a village that had housed four to five hundred residents before the community abandoned it during a Japanese attack. The Seabees decided to demolish the homes: "We also demolished a teakwood and mahogany church. We didn't like to, but we had no choice. We had to have that sand for the road—for the road and the Marines. Before we left that site we had dug up a whole peninsula of sand. The place was two feet under water before we had enough." The 87th NCB's cruise book captured the visual image of this kind of destruction in a photograph of a bulldozer demolishing a "humble" native home in the path of roadway construction. After the battalion's carpenters had salvaged lumber and other souvenirs from the structure, the cruise book notes, "'Big John' Wines smashed his 'dozer' into the dwelling." This incident was characteristic of Allied evictions of villagers across the Pacific. Mid-war expansion of American bases in eastern Fiji, for example, required the acquisition of a five-square-mile zone of land that included Fijian Indian tenants, five villages, and significant sugarcane acreage. The military also evacuated nearly the entire population of Nissan (Green) Island, bulldozed homes at Ulithi (in Micronesia) to build a naval base, and cleared an occupied portion of Vanuatu to make way for recreation and target-practice facilities.[22]

While the Americans, unlike the Japanese, generally compensated locals for their losses, the payment did not always make up for the hardships imposed. The U.S. military generally attempted to ease a portion of the burden by restoring some old villages, building new ones, and offering temporary

A Seabee and bulldozer demolish a native home for future military construction. The battalion's cruise book labels this "destruction for construction." (US NCB, 87th, *The Earthmover: A Chronicle of the 87th Seabee Battalion in World War II* [Baton Rouge: Army and Navy Pictorial Publishers, 1946], 260)

housing in the interim. Yet one battalion cruise book's description of the relocation process as a "herding" of residents into "compounds" raises questions about the quality of the experience. Challenges of a different sort befell the large numbers of displaced Indian tenants in Fiji. Although the United States compensated them for relocation costs, the value of destroyed improvements, and the loss of productive land, the Indians struggled to find new property to lease. Many became impoverished as a result.[23]

Such necessary compliance with their displacement, combined with typical American wartime depictions of natives in the role of "loyal islander," might suggest a lack of dissent. Yet protest did exist. On the island of Los Negros, an islander confronted members of the 46th NCB as they surveyed his land for an airfield. According to the battalion's cruise book, "A native of no small proportions physically, demanded that work stop immediately at the chosen location. He would not hear to the idea of selling the land, saying repeatedly, 'Me no sell, me no sell.' The particular tract contained his betel-nut trees. . . . Lt. Maddux realized the work must go on with no interference so he hit upon the idea of a trade of an American item for the trees. Soon a bargain was reached and the native, following the downing of the trees, was cutting off the nuts with a brand new 'GI' knife." As this story suggests, the military used physical force and creative bargaining to bulldoze whatever stood in its way. Resentment could follow. As a *Life* magazine photo-essay noted of army land clearance in the Caribbean, "Bermudians especially deplore destruction of their famed, unique Bermuda cedars." Similarly, after clearance on Rennell Island in the Solomons resulted in an airstrip that was never used, one displaced resident composed a song of frustration. The lyrics included this complaint: "Dig up my coconuts, remove my house/kill my forest trees. . . . Would there were a path to tread upon,/as I would retaliate."[24]

Wartime destruction encompassed the domestic landscape as well. The military located more than a third of its facilities in the South. Development of the associated bases, ordnance plants, and airfields often began with the clearance of farmland. In Hinesville, Georgia, for example, the army purchased 360,000 acres for an anti-aircraft firing range, displacing 713 families. Black farm families, who had only recently acquired much of this property through hard-fought battles, felt this displacement particularly harshly. Black sharecroppers suffered as well, losing their means of employment at the same time that high military demand for land made it difficult to rent farmland elsewhere. Although the Farm Security Administration assisted with relocation by purchasing nearby land capable of accommodating a quarter of the displaced population, its efforts proved insufficient to halt what the historian Bruce Schulman has termed "a sort of agricultural enclosure movement across the South."[25]

Vast quantities of flat farmland in interior Nebraska also attracted military development. The navy bought and cleared forty-eight thousand acres in order to erect the nation's largest ammunition depot in the town of Hasting. While much of this property was farmland, it also contained some homes. Army air corpsman Don Greery recalled that in Kearney, the site of additional military facilities, the encroachment of bases onto rural lands rarely, if ever, provoked resentment. His recollection probably reflects the welcoming reception locals often gave new military arrivals, rather than the emotions of those actually displaced from their land. By contrast, George Leiser, then a teenage resident of property on which the Cornhusker Ordnance Plant was built, right outside Grand Island, remembered the army giving his family just five days to move out of their home in 1942. The compensation they received for the house they had built three years earlier failed to ease his mother's disappointment over having to leave. Leiser and his brother spent the next several months sleeping on mattresses on the floor of a nearby school until their family found new, permanent accommodation.[26]

In all these ways, in both domestic and international locales, war offered countless opportunities for honing the practices of demolition and clearance. Engineers and Seabees claimed large-scale parcels of private property and displaced populations. They hacked through jungles, leveled farms, gouged hillsides, and demolished housing. When the wrecking was through, they disposed of the refuse in order to create the clean slates necessary for new construction. As an advertisement for Bucyrus-Erie tractor equipment observed during the war, "The fighting fronts are training courses, too—perilous, exacting, efficient. The earth-moving problems encountered are staggering, but the speed with which victories have been won attests to the skill these men have acquired." A contemporaneous article in the trade journal *Excavating Engineer* further predicted, "The equipment and men who operate it at war will be the same equipment and men available for this postwar work. Men in the excavating industry who are progressively planning for the postwar era have a wonderful pool of experienced manpower from which to make up their organizations. Also, excavating equipment will have gone through the greatest test in the world and we will better know all of its capabilities." These wartime experiences demonstrated the capabilities of equipment, the skill of operators, and the effectiveness of clearance practices on some of the most challenging international proving grounds. In making commonplace the collateral social inequities this work entailed, they also helped establish the viability of these practices on postwar terrain.[27]

Bulldozers as Weapons

Allied forces landed on the beaches of Mono, in the Treasury Islands, on October 27, 1943. Japanese gunfire met the arrival of navy vessels, from which bulldozers soon descended. Seabees quickly put these machines to use, digging out sites for weapons and knocking down jungle trees and brush to create an island roadway. Aurelio Tassone, a twenty-eight-year-old former state road bulldozer operator from Milford, Massachusetts, sat at the controls of one of these machines. About an hour into his work, gunfire erupted from a nearby Japanese pillbox. This coconut log–covered bunker concealed soldiers and weapons. One of the nearby officers, Lieutenant Charles Turnbull, interrupted Private Tassone to point out the disturbance. He ordered him, "Hit that damn thing! Spread it all over the beach!" Tassone responded by redirecting his bulldozer and roaring toward the enemy. When Tassone raised the bulldozer blade as a makeshift shield, the bullets of Japanese snipers pinged off the metal. Still he pressed on, with the lieutenant striding behind him, firearm in hand, to protect the rear. Just as the bulldozer was about to hit the target, Tassone dropped the blade down as hard as he could, nearly stalling the motor. As he later recounted, "The blade bit through the obstructions as if they were snowdrifts. The gun mount toppled over and chunks of logs and Jap bodies flew up in the air. Everybody and everything was crushed and buried underneath that rip-roaring machine." The battalion cruise book recorded the conclusion of the event: "Methodically, as if he were smoothing a rough spot in a road, Tassone bladed earth over the wreckage. His mission was accomplished." A postmortem investigation revealed that the bulldozer had buried twelve bodies and a large, new, twin-mount 37-mm gun inside the mass grave.[28]

Allied participants in this event earned high praise for their deeds. "For conspicuous gallantry and intrepidity in action," President Roosevelt awarded Tassone one of thirty-three Silver Stars given to Seabees during the war. While later reflecting that he was "tickled" to receive this honor, Tassone also suggested that his act of bravery was really no big deal at all. "To tell the truth," he continued, "I was surprised at the fuss afterward." A different account of his reaction to the commendation depicts him even diminish-

"'Earthmover' Artist Depicts Famed Pillbox Episode." This cartoon depiction, from the 87th NCB's cruise book, portrays Private Aurelio Tassone and "that famous bulldozer" as they laid waste to a Japanese pillbox. (US NCB, 87th, *The Earthmover,* 59)

ing his own valor that day in favor of that of Lieutenant Turnbull, "the most courageous man he's ever known." "I had the bulldozer blade for protection," Tassone reportedly said. "The Lieutenant had nothing but guts." This observation of bravery also applied to many of Tassone's fellow Seabee equipment operators. In the absence of liftable bulldozer blades, the typical equipment operator working on or near the battlefront often found himself exposed to the hazards of enemy fire. Although the selective introduction of armored cabs, beginning in late 1943, offered some protection, it never matched the shield of a tank. In total, nearly three hundred Seabees died in action during the course of the war. Another five hundred lost their lives to other causes, including disease, fire, and the general hazards of construction work. With an overall Seabee casualty rate one-tenth that of average U.S. military men, the units offered a generally safer way to serve. But stories like Tassone's showcased the individual valor behind those statistics. An article in the *Saturday Evening Post* about the Mono Island incident directed the credit for that valor beyond the human protagonists alone. Under a photograph of the lieutenant and private, standing in front of a battered machine, the caption reads, "Lt. Turnbull, Tassone, and that famous bulldozer."[29]

This legendary event—and the award it precipitated—painted a valiant picture of military construction men and their machines. Although Seabees practiced varied trades using a range of tools, it was the bulldozer and bulldozer operator who commanded the most attention. In this exemplary story, both man and machine appear powerful, heroic, and patriotic. The bulldozer functions as a tool and a weapon, clearing land for future construction and demolishing enemy foes. Japanese coconut-log pillboxes had long served as "death traps" for Allied soldiers. Simply finding the well-buried hideouts could be a challenge. Once located, they proved difficult to eradicate with offshore shelling. Close-range attacks by foot soldiers armed with flamethrowers and grenades were the dangerous but necessary alternative. Such obstacles proved no match for the bulldozer.[30]

The media picked up Tassone's story and elevated it beyond local battalion lore, enhancing the reputations of bulldozers and Seabees everywhere. During an Abbott and Costello radio broadcast, a Camel cigarette ad briefly recounted Tassone's exploits and recognized him as "Yank of the Week." R. J. Reynolds Tobacco then sent three hundred thousand Camel cigarettes to navy men in the Pacific in his honor. Tassone's name also appeared in the *New York Times, Time,* and *Popular Mechanics.* The propagandistic depiction of the event in *Engineering News-Record* was particularly graphic, noting that the "crunching of heavy timber and screams of trapped Japs told Tassone his job was done." Other retellings further embellished the episode. *Southern Lumberman* had Tassone charging straight off the landing craft to attack the Japanese gunners. *Automotive War Production* ended its account with his rescuing a setter dog that was hidden in the bunker and then receiving the canine as part of his reward. While the frequent reporting of this incident did not make Aurelio Tassone a household name, the broad outlines of his story spread far and wide. The rehearsal of this singular event increased public awareness of the Seabees and their bulldozers, while shaping perceptions of both as symbolic of American might.[31]

Such repetition across media and time made exploits of bulldozer brawn seem as commonplace as Tassone had suggested. So did the ensuing similar stories. An ad the following year for Buckeye equipment attachments, for example, featured an excerpt of a letter from a soldier who had reported observing a bulldozer charging two Japanese pillboxes in New Guinea. Shades of Tassone also appeared in a wartime children's chapter book about Bill Scott, a fact-based but fictional "typical American boy who decides to cast his lot with the famed 'Fighting Seabees.'" The tale ends with a faux newspaper clipping about Scott's exploits in "[wiping] out Jap machine-gun nest as Seabees storm island in Solomons." The details of this incident match those of Tassone's, right down to Scott's receipt of the Silver Star. Scott's story spreads equally widely in the fictional world of the book. As the text concludes, "No matter *where* you live, you probably saw an account of it; for it was on the front pages of newspapers all over the country."[32]

Tassone also identified what he perceived as a more tangible outgrowth of his real-life deed: the development of the tankdozer. This military machine, introduced by American forces in late 1943, was purpose-built for combat against enemy obstacles. It typically consisted of a specially designed blade mounted onto an M4 Medium Tank. Tassone felt that no one could "want a sweller reward" than this physical legacy of his actions. In reality, the tankdozer's story began two years earlier, when army engineers followed the British in commissioning research into tank-mounted bulldozer blades. It was perhaps the logical resolution of the bulldozer and tank's shared twentieth-century roots in Benjamin Holt's track-mounted tractor.[33]

The D-Day beach landing at Normandy in June 1944 was one of the tankdozer's most memorable appearances. With the machine's capabilities in mind, the Allies strategically planned their attack for maximum benefits of daylight and low tide. They hoped to use tankdozers to clear the mine-laden steel structures the Germans had placed in the beachfront's intertidal zones. Only six of sixteen tankdozers from the 37th Engineer Combat Battalion made it to shore in working condition. Yet operators put these surviving vehicles to productive use. They ran them across the beach as they dragged capsized vehicles out of the surf. Then they set off to fill anti-tank traps, clear mines, and open roads. Their ability to topple hedgerows was particularly useful. These fences consisted of a four-foot-thick base of dirt topped by vines and trees. Norman farmers constructed hedgerows around their plots of land, leaving narrow, sunken lanes in between. After equipping tankdozers with special teeth, Engineers in Normandy powered through these roadblocks to improve access and visibility. The absence of armored cabs exposed these operators, as "geysers of earth and shells bursting all around [them] rained down." For their efforts, Privates Vinton Walsh Dove and William J. Shoemaker each received the Distinguished Service Cross, with citations highlighting their respective "extraordinary heroism" and "intrepid actions, personal bravery, and zealous devotion to duty." Shoemaker formerly had worked in construction in Pittsburgh, Pennsylvania, and the International Union of Operating Engineers published a story about his accomplishments in its monthly magazine. The article lauded not only the union man but also his mechanical "fighting partner, the 'Hellcat,' on which he rode roughshod over the Germans."[34]

Tankdozers and other retrofitted bulldozers followed Tassone's example even more directly when Engineers used them to bury German equipment and bunkers. In some cases, when enemy troops refused to leave their dugouts, Allied bulldozers even buried soldiers alive. According to the memoir of one former soldier, Allied forces encountered a dug-in Panzer Maus, a thickly armored enemy tank weighing 185 tons, on the German front. More impregnable than a pillbox, the machine had repeatedly deflected direct hits before the Engineers developed an alternative plan. The new weapon of attack would be a Caterpillar D7 bulldozer, modified with an armored cab made of one-inch-thick boiler plate. After the infantry took out the Maus's periscope and blinded its operators, the bulldozer moved in from the rear. It dropped piles of earth on the Maus in rapid succession until "nothing remained of the pride of the German Panzer force except a bent antenna, woefully pointing from the tomb of its crew to the heavens above." The courageous bulldozer operator, described in keeping with his machine as "a bull of a man," won a Distinguished Service Cross that day.[35]

Dozer-equipped vehicles offered even more varied utility in applications across the globe. On the beaches of Italy, Axis fire ignited vast stores of ammunition nearly every night. Although shovels and dirt initially helped extinguish the blazes, bulldozer blades and full-fledged tankdozer kits dramatically enhanced firefighting operations. Elsewhere in the country, bulldozer operators ventured into the waves to push landing craft into deeper, calmer waters. They also rescued heavy trucks that were mired in mud. At Okinawa

Shima in Japan, Seabee Roy Ellett and his bulldozer cleared a native building that had caught fire and collapsed into a supply road. During the same conflict, Quentin Carroll used his bulldozer to quell another blaze. Once he had pushed the flaming buildings back and covered them with earth, military movement could recommence through an important intersection. The two Seabees earned recommendations for the Bronze Star for their deeds.[36]

Not all the accomplishments of these men and machines were so glorious. Beginning with Pearl Harbor itself, bulldozers and shovels also engaged in the more somber duties of digging out trenches for burying the dead. The same was true of the bulldozing of enemy dead throughout the Pacific fighting. Earthmoving machines played a similar role at the war's end, when the Allies liberated concentration camps like Belsen. Some Seabees and Engineers also died in the course of duty. Among the ranks of the more than twelve hundred men who made up the 105th NCB, for example, six fell during wartime service. Their deaths were attributed to natural causes, drowning, Japanese fire, a truck accident, and being hit by a fallen tree while at work on a bulldozer. On the Alaska Highway, far removed from enemy battle, other construction men died because of their dangerous work environments. On one occasion, a storm capsized the makeshift raft being used to ferry a Caterpillar tractor and other equipment across a lake. Of the seventeen Engineers plunged into the frigid waters, only five survived.[37]

Most wartime media accounts focused on the men and machines' happier contributions. They showed trained operators making critical weapons out of basic construction equipment. In addition to completing large-scale clearance work, that equipment rescued bogged-down machines, facilitated water landings, filled tank traps, buried enemy tanks, and flushed out snipers. Stories of these exploits transformed workaday construction men and equipment into heroes and weapons. When they returned home from the war, they arrived there on a pedestal.

The Heroic Image of the "Bulldozer Man"

Alongside stories of heroic bulldozers, popular wartime media also glorified the "Bulldozer Man." A caption accompanying a photograph of Aurelio Tassone in his battalion's cruise book even labeled him as such. But this specificity was unusual. Typically imagery from the period helped create a new American hero without naming the Bulldozer Man explicitly. The ability to operate heavy construction equipment was only this figure's opening qualification. In addition, he was strong, brave, and patriotic. He encapsulated all these traits within an exterior appearance of white, wholesome, all-American masculinity. These may have been traits with which the Bulldozer Man was born, but battlefield experience nurtured them, and direct association with equally heroicized—and sometimes feminized—machines amplified them further. Three of the major representations of the Bulldozer Man from this period, found within the genres of film, photo-essay, and literary fiction, indicate the kinds of popular depictions of bulldozer operators that prevailed as the country entered the postwar period.

Wartime Hollywood produced a host of patriotic films, including *The Fighting Seabees,* a 1944 motion picture about a group of construction workers who enlist in a newly formed construction-oriented battalion of the U.S. Navy. This dramatized version of the actual Seabees' formation and service interweaves wartime battles with Hollywood romance. The most memorable depiction of the soldiers' contributions to the war occurs in the climactic battle scene. The Japanese have just attacked a Pacific island and are closing in on the outnumbered Americans. Losing in hand-to-hand combat, the Seabees mount their equipment to change the terms of battle. A crane operator lifts a Japanese infantryman inside the jaws of a clamshell, shoots him while he's suspended in the air, and then drops the slain soldier on the ground. Another Seabee drives his bulldozer over a hillside and pushes a feeble Japanese tank off a cliff. The enemy vehicle appears to be a cheap box of sheet metal in comparison to the powerful American machine. Finally, battalion commander Wedge Donovan, played by John Wayne, takes matters into his own hands. He boards a bulldozer and speeds toward Allied oil tanks. Although Japanese snipers strike him en route, the bulldozer continues on, plowing into one of the containers. The ensuing explosion sends the enemy fleeing. Donovan dies in the episode—a plot development relatively unusual for both Wayne and real-world Seabees—but his actions prove decisive in saving the island for the Allies.[38]

Essentially a Western recast on World War II Pacific island shores, *The Fighting Seabees* replaced the classic masculine American icon of the cowboy

This scene from the film *The Fighting Seabees,* 1944, depicts John Wayne and his construction crew manning a bulldozer during a World War II Pacific island battle. Like real-world Seabees, these fictional men have inscribed their vehicle with a female name, Natasha.

with that of the bulldozer operator. Casting the era's silver screen symbol of manliness in the lead role of Wedge Donovan helped cement this association. The bulldozer served as a stand-in for his trusty horse on the dusty battlefield. The duo demonstrate courage and strength as they advance together; when circumstances require, the machine even picks up the slack left by its operator. One can also read the bulldozer, like many a horse, as a female companion. A Seabee recalled a fellow battalion member who thought of his vehicle as the "greatest little wife he ever had. He says 'Waddle'—that's what he calls her—plowed under six Japs, then for weeks always snorted every time she saw one." The female personification of bulldozers mirrored the practice of military pilots, who frequently inscribed female names on their airplanes' noses alongside female art. The name "Natasha" written on a bulldozer in *The Fighting Seabees* demonstrates how military construction men also applied this practice to their machines. In the relative absence of significant female companionship during the war, equipment operators became Bulldozer Men as they tended to feminized mechanical mates.[39]

The widespread marketing of this motion picture probably bolstered the image of real-world Seabees, some of whom served as extras in the film. With advertising copy that called the film's title subjects "the American Navy's new supermen," *The Fighting Seabees* might have even improved recruiting. One former Seabee remembered the motivating impact of the film when he joined the World War II military at age seventeen: "They showed John Wayne and the fighting CBs and that made me gung ho, everybody was crazy for the CBs so I enlisted in the CBs immediately." The navy used the film in basic training through at least the Vietnam era, and its reach endures as it is promoted in exhibits at the Seabee Museum at Port Hueneme, California, today.[40]

Life offered the still-image analogue to *The Fighting Seabees.* Photographer J. R. Eyerman's framing of the real-world bulldozer operator Glenn Selby, in a 1945 photo-essay, constructs the all-American ideal of the Bulldozer Man. Selby's is one of just two full-page portraits included in the piece. The other depicts Admiral Chester Nimitz, commander-in-chief of the Pacific Ocean Areas. While the space afforded Nimitz matches his military significance, that given an otherwise unknown bulldozer operator elevates his stature and that of all his fellow Seabees. Eyerman photographed Selby from below—just as he did Nimitz—forcing viewers to look up to him, visually and metaphorically. At the end of his lens, the subject appears framed by an idyllic backdrop of palm trees and clear sky. Selby gazes back directly, flashing a gleaming smile. He is young, handsome, muscular, and confident. As the shadow from his arm partially conceals his bare chest, he exhibits his brawn with modesty. While no caption reveals Selby's thoughts, he may well have expressed the excited surprise articulated by another young equipment operator regarding his own time spent on Guadalcanal: "Here I'm eighteen years old, and what do they put me on? A T9 bulldozer. . . . An eighteen-year-old kid operating one of those monsters, clearing boundaries and that sort of stuff." While the photograph of Selby is static, it still suggests action. His hands and feet perched upon the controls of a powerful bulldozer hint that equipment operation is a technical skill, rather than an effort in brute force. By zooming in and cropping the machine on all sides, Eyerman offers a limited view of the tractor's treads. He reveals just enough to confirm the vehicle's identity, but

In 1945, *Life* magazine photographer J. R. Eyerman captured the visual exemplification of the ideals of the all-American "Bulldozer Man"—strong, smiling, wholesome, and masculine—in this image of Seabee Glenn Selby seated at the controls of a bulldozer in Guam. (J. R. Eyerman/The LIFE Picture Collection/Getty Images)

not so much as to overwhelm with its scale. This framing also keeps the focus on Selby, making the man, rather than the machine, the true object of adoring view.[41]

Outtakes from this photo shoot reveal the choices directing this image's selection. Eyerman photographed at least three different military equipment operators, in multiple poses, during his time on Guam. An image of the Engineer Robert Sellers operating his own bulldozer foregrounds the machine in a way that distances the viewer from the operator. In addition, the grit of the

Outtakes from Eyerman's photo shoot depict (*top left*) Engineer Robert H. Sellers, (*bottom left*) Seabee Glenn Selby, and an unidentified Seabee. By contrast with the published image, these photographs offer less idealized representations of the Bulldozer Man. (J. R. Eyerman/The LIFE Picture Collection/Getty Images)

ground, the tattered assemblage of trees behind him, and the cigarette dangling from his mouth all detract from the innocence and picturesque potential of the scene. The "U.S. Army" stenciled on the bulldozer's side also limits the potential for generalization, and the overall tone is much darker and more aggressive than the published image of Selby. Closer-range images of Sellers and an unnamed Seabee at the helm of their respective heavy equipment similarly do not idealize the operators to the same extent as the published photo. While these photographs combine close-up, friendly visages with commanding control in a tropical paradise, the men exhibit a less generically all-American appearance. The camera's flat angle puts them on less of a pedestal. Even more important, the bulldozer is less identifiable. Finally, a second image of Selby himself also failed to make the cut. In this close-up, the machine type is less obvious, Selby's countenance appears more serious than friendly, and his position is more sedentary. Instead of offering a smiling face amid a halo of palm fronds, this photographic subject eyes us more warily, and perhaps even a bit alluringly.

Other period depictions of Seabees in *Life* further illuminate the wholesome version of masculinity, in which sex is sanitized and made safe, that popular imagery ascribed to the Bulldozer Man. A brief photo-essay in the magazine's September 10, 1945, issue portrays Seabees showing a good time to some members of the WAVES (Women Accepted for Volunteer Emergency Service) on the occasion of the women's group's third birthday. The WAVES, created during World War II, were the women's division of the U.S. Naval Reserve. During the women's visit to Seabee Camp Parks, the men treat a group of them to a lift in the bucket of a crane. When one of the women rips her pants in the process, a Seabee lends her his shirt. In addition to positioning earthmoving equipment in a lighter context, this photo-essay also underscores the playful yet gentlemanly character of the equipment operators. They use their specialized tools to entertain the women, and then valiantly come to their guests' rescue when trouble arises. The Seabees' exemplification of hardy, wholesome fun probably endeared them to Coca-Cola as well, which featured sketched versions of some of the men in a *Life* advertisement that same year. With bare, muscular torsos and warm smiles, this group of Seabees and their bulldozer share refreshment while fraternizing with a group of stylized Pacific islanders. Their combination of sociable manliness and modernism presumably explains Coke's utilization of them to embody its marketing message of a "friendly way of life." This ad suggests that the image of the Bulldozer Man not only inspired, it sold.[42]

While Seabees largely disappeared from the pages of *Life* after the war, the writer James Michener extended popular accounts of their wartime exploits into the early postwar period. Michener drew upon his World War II service as a naval historian in his Pulitzer Prize–winning *Tales of the South Pacific* (1946). Seabees and bulldozers feature prominently. In the chapter "The Airstrip of Konora," for example, Seabees build an airstrip across

In this playful yet sexualized depiction of home front military construction machinery, Seabees give members of the WAVES a ride in the bucket of a crane. (Nat Farbman/The LIFE Images Collection/Getty Images)

(*Opposite*) Coca-Cola draws upon positive public perceptions of the Seabees as friendly, masculine, and modern, in this 1945 *Life* advertisement set on a Pacific island. (Copyright 1945 the Coca-Cola Company, courtesy of the Coca-Cola Company)

Now you're talking . . . Have a Coca-Cola

. . . or tuning in refreshment on the Admiralty Isles

When battle-seasoned Seabees pile ashore in the Admiralty's, the world's longest refreshment counter is there to serve them at the P. X. Up they come tired and thirsty, and *Have a Coke* is the phrase that says *That's for me*—meaning friendly relaxation and refreshment. Coca-Cola is a bit of America that has travelled 'round the globe, catching up with our fighting men in so many far away places—reminding them of home—bringing them *the pause that refreshes*—the happy symbol of a friendly way of life.

* * *

Our fighting men meet up with Coca-Cola many places overseas, where it's bottled on the spot. Coca-Cola has been a globe-trotter "since way back when".

You naturally hear Coca-Cola called by its friendly abbreviation "Coke". Both mean the quality product of The Coca-Cola Company.

a chasm using excavated coral. Michener's story of their triumph over environmental adversity combines details about the engineering process with veneration of the protagonists. The lead character, Commander Hoag, is a Connecticut contractor determined to complete the seemingly impossible assignment. His tale ends abruptly on the morning of the first bomber's arrival. When a hidden Japanese soldier attacks Hoag with a grenade, his death upon the completed runway bed of excavated coral cements the contractor's heroism.[43]

Although adulation abounds in the text, *Tales of the South Pacific* also reveals a darker side of military clearance. In one episode, the narrator travels to Norfolk Island to erect an airstrip. His project will require bulldozing a majestic avenue of pine trees growing on the only viable site. As locals plead with him to forgo his plan in order to save their "living cathedral" of trees, he starts to reconsider. Days pass before military imperatives finally impel him to proceed as planned. He provides the locals with a few hours to take some last pictures of themselves in front of the aged trunks and then orders bulldozers and explosives to level the area. On hearing an explosion later that night, he discovers that two native women have coordinated with an American military man to blow up a bulldozer in retaliation. Although the machine's worn-out condition makes their act more symbolic than substantive, it remains notable as an unusually overt protest against the wartime bulldozer. This atypical storyline incorporates oppositional environmental perspectives, challenging the positive image of the Bulldozer Man. In Michener's narrator, singularity of destructive purpose uncharacteristically cedes to introspective consideration of the price of progress achieved by his machine.[44]

Ruins and the Environmental Legacies of War

In addition to the enduring imagery of Bulldozer Men and machines, World War II also left the public with scenes of vast destruction. As the environmental historian Chris Pearson has noted, wartime geographies became "both sites and victims of military conflict." Memories of the ruins and their transformation underlay the ideology of the postwar future, as hundreds of American cities proactively destroyed their own domestic environments in the years that followed. The messages in popular and professional media about these ruins, however, depended upon the type of environment being discussed. Wartime and postwar visual culture celebrated the ruins of the urban *built* environment as sources of rebirth. Rather than discourage postwar demolition of home front cities, journalists and planners used the opportunities offered by these ruins as arguments for making more. By contrast, the popular media largely ignored the destruction of the *natural* environment, bolstering the appeal of clearance work by allowing destruction without consequences. Recognition of the need for environmentally conscious development became a lesson that the broader American public would have to learn over many years through firsthand experience.[45]

Wartime bombs and firestorms had laid waste to the global built environment. Military raids in Germany destroyed nearly half the average large city's residential housing. They left 72 million cubic yards of stone and rubble in Berlin, 47 million in Hamburg, and 32 million each in Dresden and Cologne. The Pacific corollary was the annihilation of Japanese cities. Their buildings' typically flammable wooden construction, coupled with the preventive effects of state-imposed firebreaks, significantly reduced the volume of resultant building refuse. Yet the physical impact was still substantial, with firebombs incinerating two-thirds of Tokyo alone. The more modest built environments of Pacific islands also fell. In Papua and northeastern New Guinea, for example, Allied bombings were largely responsible for the destruction of twenty thousand native dwellings.[46]

From another perspective, however, the resulting ruinscapes offered positive opportunities for clean-up and renewal. As the historian Jörg Friedrich has noted of Germany, voluntary urban demolition exceeded the direct physical casualties of war as "residents ultimately tore down more stone than the bomb did. They could not stand to look at those fateful shells." Urban images depicted not only ruins but also cranes, shovels, and clamshells toppling damaged buildings and disposing of the remains. In this way, visual culture from the wartime and postwar periods symbolically repositioned these previously destructive machines as engines of ordering and cleaning. A 1945 advertisement for Wickwire Spencer Steel Company plays up such associations for marketing purposes. The ad's portrayal of hard-hatted soldiers using a crane and bucket to dig through rubble brings to mind a wrecking crew demolishing a site for new construction. The bold lettering of the advertisement makes

Friend of the "*Cleaner Uppers*"

As the invasion forces go forward shell-holes must be filled—tottering walls brought down and the rubble of destruction removed. And friendly, dependable Wickwire Rope is an old hand at helping the boys who clear the roads and keep supplies moving.

Long-lasting Wickwire Rope is a world traveller these days—as Liberty ship rigging, and in reels for our fighters who are doing the jobs that lead to victory. There are jobs here at home that are vital to success and they also call for strong wire rope. If you need Wickwire Rope it's available on priority, but it pays to make what you now have last as long as possible.

We will be happy to help you with your wire rope problems and also to send you copies of "Know Your Ropes" which offers helpful advice on proper selection, application and usage. Write Wickwire Spencer Steel Co., 500 Fifth Avenue, New York 18, N. Y.

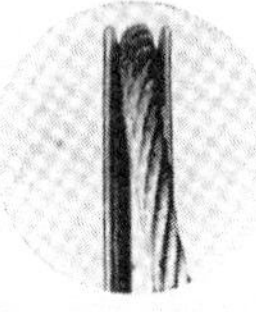

TO AVOID SHOULDER WEAR

Exact alignment of sheaves is important. Our Free book "Know Your Ropes" tells why and shows how to do it. This and 39 other wire rope life savers; 78 "right and wrong" illustrations and 20 diagrams and charts are all contained in this 82 page book.

SEND FOR YOUR FREE COPY

Send your wire rope questions to:

500 FIFTH AVENUE, NEW YORK 18, N. Y.

Abilene · Buffalo · Chattanooga · Chicago · Detroit · Houston · Los Angeles · Philadelphia · San Francisco · Tulsa · Worcester clear, however, that these workers and machines are not destroyers, but "cleaner uppers."[47]

The widespread circulation of images of urban ruins and their removal not only bolstered the reputations of the machines of demolition, they also naturalized large-scale destruction on a scale previously unimaginable. Photographs, films, and firsthand accounts of ruins helped familiarize wide audiences with wrecking and rubble. In their frequent portrayal of daily life going on amid the ruins, these representations further suggested that large-scale destruction was not only possible but people could live with it as well. Blitz photographs, in particular, employed this trope. They depict plucky London shopkeepers posting signs proclaiming "business as usual" under conditions that seem anything but usual, postal workers delivering mail to addresses that no longer exist, and passersby casually poring over the literary remains of devastated bookstores. Rather than ignore the damage, Londoners productively utilized the destruction's aftereffects. They planted gardens in the lots of demolished buildings, and gave band concerts in the outdoor amphitheaters of newly created open space. Others saw the picturesque potential in wartime ruins; *Architectural Review* even proposed that some bombed churches remain in their damaged state. While such romantic readings contradict the depiction of ruins as sites of new construction, they offer another example of the redemptive potential some contemporary viewers found in wartime destruction.[48]

Popular culture recognized positive potential even in German ruins. Despite the relatively dark context of Billy Wilder's film *A Foreign Affair* (1948), about German citizens and U.S. military men living in postwar Berlin, several aspects of the film's ruin depictions suggest resiliency. Footage shows young mothers animatedly pushing baby carriages past piles of rubble and children happily playing baseball on cleared city lots. The ruins even become new urban landmarks; one woman invites a guest to her home with the directions, "Let's go up to my apartment. It's only a few ruins from here." The signature song performed by Marlene Dietrich's character verbalizes the re-

---

(Opposite) This Wickwire Spencer Steel advertisement for wire rope appeared in a 1944 issue of *Southern Lumberman*. It associates the work of power shovels and other heavy equipment engaged in wartime urban rubble removal with cleaning, rather than destroying.

demptive potential of this forlorn landscape: "A brand new spring is to begin out of the ruins of Berlin." An image in Edward Steichen's famous *Family of Man* exhibit at the Museum of Modern Art (1955) further supports this positive reading. The photograph shows a backpack-wearing German boy walking down steps in a tattered urban landscape as if it were a normal living environment. Steichen derived his familiarity with bombing imagery from his own wartime experiences: he had learned aerial photography as commander of the photographic division of the U.S. Army Expeditionary Forces during World War I; later, during World War II, he directed the U.S. Naval Photographic Institute, overseeing combat images.[49]

The drive to rebuild real-world landscapes also helped conceptualize ruins as new beginnings, rather than ends. Planners widely embraced the opportunity to implement modernist planning theories on war-ravaged cityscapes. Le Corbusier and the International Congresses of Modern Architecture (CIAM) had advanced the concepts of the International Style and Functional City in the 1920s and 1930s, well before the war. They proposed replacing dense, small-scale urbanism with superblocks of open space punctuated by residential towers and bordered by multilane highways. A key impediment to realizing this functionally segregationist approach to city planning was the clearance of existing buildings necessary to initiate redevelopment. In cities from Europe to Japan, however, the war eliminated this obstacle, opening the door to new, large-scale construction.[50]

While some, like the German novelist W. G. Sebald, lamented the missed opportunity to process the physical loss, architects and planners led the public in looking immediately to a rebuilt future. This practical exploitation of wartime destruction was not a surprise outcome of the war but an explicit opportunity identified by designers both during and after the conflict. As the architect Frank Lloyd Wright had observed in 1940, "I will not say that the bombing of Europe is not a blessing, because at least it will give the architects there a chance to start all over again." Similarly, in 1942 Walter MacCornack, dean of the Massachusetts Institute of Technology and vice president of the American Institute of Architects, called the bombing of London a potential source of the city's eventual beautification: "Today, however, the bombing of some of [the area around Saint Paul's] gives London once again the opportunity to replan large areas of its city on a more perfect basis and there seems to be a good chance that long-range planning in England will have its day and out of the chaos of destruction of the war will come better cities." MacCornack contended that such opportunities would extend globally, reaching the United States as well. The result, as described in a contemporary trade journal, would be "the mobilization after the war of thousands of technicians, architects and representatives of finance to bring world-wide order out of chaos." That same year, *Architectural Forum* made this prospect tangible by inviting architects to propose designs for a hypothetical future American city that had lost its central core through bombing.[51]

European ruins brought into starker relief the damaged state of midcentury American cities, even as these landscapes were exempted from the kinds of wartime destruction that befell international terrain. The modernist architect Martin Wagner, for example, drew parallels between the states of blighted American cities and bombed European ones, recommending a solution of clearance and rebuilding for both. Similarly, American politicians and planners saw the reflection of their own municipalities' crowded tenements, dirty streets, and aging infrastructure in wartime ruins. If a silver lining of the war was the chance to plan European cities anew, so too might a more proactive approach to clearance in the United States realize similar opportunities. Federal legislation even required the large-scale demolition of hundreds of thousands of temporary wartime dwelling units within two years after the conflict ended. In this way, the government extended wartime destruction into the postwar domestic landscape. As planning intellectuals and the public increasingly spoke of waging war on the nation's "sick," declining cities, they often used military and medical metaphors to make their case. Phrases like the "war on slums" and precision strikes of "slum surgery" appeared in government documents, trade publications, and popular periodicals. Such language flattened the challenges of the postwar city. Instead of multilayered social and economic issues, urban problems became largely material enemies to be destroyed, as the nation had successfully destroyed a previous wartime foe.[52]

The scars wrought by World War II on the global natural environment have received less attention than the selective depiction of European urban destruction. Even Michener's story succumbed to this trend, with popular stage (1949) and screen (1958) versions of *South Pacific* omitting the pine tree and exploded bulldozer episode. For those who looked, however, there were "natural" ruins to be found. Robert Sherrod, a journalist for *Life* and *Time* magazines, traveled to the Tarawa atoll, in the Pacific, a year after the end of the war. Sherrod had first landed there with the marines in November 1943. In the postwar article that grew out of this later trip, he reported that twenty-five of Tarawa's twenty-six islands appeared "lush and green against the blue ocean" from the air. But the "stark, glittering white" form of Betio, the twenty-sixth island, was different. Betio bore the scars of its wartime service as the site of an airfield and an epic battle. On the ground, the journalist found a rusted tank offshore, guns and "the skeletons of amphibious tractors" scattered on the beach, and an overall obliterated natural landscape. He discovered that the war had "cut down most of Betio's thousands of palm trees, and now even freshly planted trees show a reluctance to take root in the shattered soil." The result, he concluded, was "an island of desolation and unshaded heat." Yet Sherrod's postwar visit and report were the exception. Although some soldiers remained stationed in the Pacific after the war, most journalists returned home with the cessation of hostilities. Popular period histories of key island battles show similar disregard for the conflict's enduring environmental damages. Replete with details of combat, these accounts typically devote scant space to the physical aftermath.[53]

Echoes of the decimation of Betio resonated on other islands. In the Pacific the weapons of battle, chemical agents, transplanted weeds and pests, and daily confrontations between wildlife and newly arrived militaries all caused damage. Wartime construction also produced its own environmental consequences. In the Gilbert Islands, for example, the combination of airfields, fortifications, and bombing destroyed sixty thousand coconut palms and two thousand of the pits islanders had arduously dug to nourish the palms with fresh water. In the short term, these losses deprived natives of a major source of food and shelter. Even worse, although hardy vegetation gradually restored the landscapes of uninhabited atolls, many compacted corals never recovered. Clearance for roads and airfields often permanently diminished the vegetative potential of these and more conventional soils, costing the islands some of their most fertile ground. The displacement of jungle and earth for roadways also destabilized terrain. On the island of Trinidad, Seabees slashed through jungle and excavated over 1.3 million cubic yards of material to create a path for the seven-and-a-half-mile Maracas Road. The combination of torrential rains and the soldiers' landscape interventions led to repeated landslides. On one notable occasion, workers experienced nineteen slides in a single day. While islanders enjoyed continued use of some completed infrastructure, weather and lack of maintenance led to its deterioration over time. Wildlife also suffered. Animals dependent upon damaged trees and soils died off. Bulldozers destroyed the eggs of birds, and the roadbeds they created replaced avian breeding grounds. Both building and fighting for war laid waste to great swaths of island landscapes.[54]

Substantial European and Japanese woodlands also fell. In France, for example, bombing, shelling, clear-cutting, and other military acts destroyed nearly a million acres of forest, while fires claimed another quarter million. Overexploitation of forest resources for fuel and wood products—by the French and Axis alike—took an additional toll. Similarly, the Japanese pursued accelerated logging campaigns when they captured the Philippines, Java, and Burma. When the Allies disrupted transport of the resultant materials, the Japanese toppled 15 percent of their own forests, clear-cutting up to fifty square miles per week by 1945. Native farmers in the Philippines compounded wartime losses when they responded to Japanese capture of their rice and sugar fields by clear-cutting acres of parkland to plant crops there instead.[55]

These rural ruins attracted less attention because they were much less visible. Relative to urban ruins, they were less easily measured and had a less immediate impact on daily social life. Clear-cutting aside, war-related damages to the natural environment also often produced less dramatic ruins than those of the built environment. The slow death of wildlife and diminishment of soil fertility lacked obvious visual markers and only revealed themselves over time. New tropical growth could cover over changes in topography before they were even seen. But the neglect of these environmental consequences was also a matter of priorities, reflecting diminished concern for the natural landscape at that moment. Perhaps surprisingly, domestic conservationist concerns found successful advocates in Germany, even in an environment characterized by "total war." Yet the American military gave environmental issues short shrift in its work on largely international soil. It was not until 1971, for example, that the Army Corps of Engineers issued the report

"Engineering a Victory for Our Environment." In this document, the organization responded to growing criticism by attempting to convey its recent development of "an encouraging capacity to understand the environmental movement." This capacity had been much less evident some twenty-five years earlier, during and after World War II. Its absence also came to characterize broader construction efforts that would follow on the postwar home front.[56]

War proved an ideal opportunity for rapidly incubating the practices of large-scale landscape destruction. It offered expansive, disposable terrain and urgency for quick action, turning any impediment to victory into an enemy to be conquered. Such obstacles included the troops, weapons, and infrastructure of the Axis, as well as the obstructions of the natural and built environment. Neither challenging terrain nor existing occupants could thwart the mission. With enhanced skills, practices, and—as the next chapter will show—equipment, the scale of American clearance and construction capacity seemed limited only by the imagination. In fact, a 1954 *Time* magazine profile of recent developments in earthmoving practices attributed the greatest responsibility for the dramatic changes taking place on the postwar domestic landscape to the recent "revolution in construction thinking." "Today," the article concluded, "there is almost no project too big to tackle, no reasonable limit to reshaping the earth to make it more productive." That thinking was rooted in the experiences of war. As heroicized operators and machines engaged in war work, they gained skills and earned glory while modeling—in full scale—the possibilities for future such activities at home.[57]

Chapter 2

Prime Movers

Equipment Manufacturers Prepare for Postwar Prosperity

Four wartime scenes punctuate the Caterpillar Tractor Company's 1944 annual report. In a drawing appearing on the front cover, army engineers operate bulldozers on a World War II Pacific battlefront. As the air battle rages above and bombs fall all around, orderly men and machines move earth to erect a roadway across the water. A few pages in, another image depicts similar kinds of equipment battling enemy fire and powerful surf. These bulldozers demonstrate heroic might by shoving landing craft loaded with wounded Allied soldiers out into deep water. Four pages later, Seabees deploy tractors, bulldozers, scrapers, and motor graders to clear jungle and level land for the creation of an airfield. The final image, included just before the financial statements, transports the reader from embattled Pacific islands to war-torn European cities. There, construction equipment restores the urban environment by "bulldozing debris out of the streets, tearing down tottering walls, bringing order out of chaos." Amid devastation, American men operating Caterpillar machines clear the way for a brighter future.[1]

As these boosterish images suggest, the wartime marriage of the military and construction equipment manufacturers proved vital to Allied victory. In the service of national defense, companies like Caterpillar, International

On the cover of its 1944 annual report, Caterpillar depicts World War II army engineers deploying the company's tractors to move earth for the erection of a roadway across Pacific Ocean waters. (Reprinted courtesy of Caterpillar Inc.)

Harvester, Allis-Chalmers, and LeTourneau produced massive quantities of construction machinery and earthmoving implements. This equipment provided the mechanical muscle behind the vaunted efforts of the Seabees and Engineers. Yet wartime accounts of American industrial and technological might largely omit their story. The "arsenal of democracy" has primarily been associated with the accomplishments of shipbuilders and auto manufacturers (who retooled their factories to produce tanks, bombers, and other armaments). Meanwhile, the development of the atomic bomb has overshadowed most other technological advancements from the war. Alongside all these, the

Another image in Caterpillar's 1944 annual report shows Allied soldiers operating tractors and bulldozers on the European front to clear rubble-strewn landscapes of destroyed cities. (Reprinted courtesy of Caterpillar Inc.)

construction equipment industry contributed vitally to what the historian David Kennedy has called the "war of machines," with important military, technological, and cultural consequences.[2]

This war work simultaneously benefited the companies undertaking it. While supplying crucial equipment to the military, manufacturers advanced product development, expanded production capacity, and increased skilled labor. These activities primed them for postwar commercial operations. From a public relations perspective, the companies used wartime advertising, media coverage, and shareholder communications to promote their role in the bulldozer's battlefield triumphs. In these ways, they enhanced their brands' reputations and patriotic associations. Not least, wartime partnerships forged between private equipment manufacturers and the state yielded important political benefits, paying off in the form of postwar policies that fueled future domestic demand. The gains were substantial. The construction equipment industry grew from about $200 million in annual sales before Pearl Harbor to $720 million in 1943; by 1948, sales exceeded a billion dollars. Longer-term performance, even after the fulfillment of war-induced backlogs, remained impressive by prewar standards. For many construction equipment manufacturers, therefore, the war was not a distraction from business as usual but an intensification of everyday life that positioned them well for the return to peacetime operations.[3]

"For the Nation's Arms and the Nation's Farms"

A 1941 advertisement for International Harvester made visible the ways in which heavy equipment manufacturers initially expected to contribute to the war. Based on their participation in previous conflicts such as World War I, manufacturers anticipated three roles: producing and maintaining farm equipment to fuel the war effort biologically through domestic food production, repurposing factories to manufacture ammunition, and producing machinery to serve as prime movers for pulling heavy artillery out in the field. The advertisement summed up these roles in its tagline: "For the nation's farms and the nation's arms." During World War II, construction equipment makers did all these things, while also producing slight variants on their core commercial products for expansive military construction–related

International Harvester crawler tractors, equipped with bulldozer blades and other implements, await shipment to the Pacific battlefront in 1944. (Wisconsin Historical Society, WHi-63853)

use. Ultimately, this last role surpassed the others in its significance to both the military and the manufacturers.[4]

Strong agricultural roots aptly positioned International Harvester for the farm portion of this work. The company's history dates back to Cyrus McCormick's invention of the horse-drawn reaper in the mid-nineteenth century. In 1902, the McCormick Harvesting Company merged with Deering Harvester to form the International Harvester Company (IH). Over time, IH further mechanized its farm implement offerings and expanded into tractors and trucks. On the eve of World War II, the company's domestic sales were apportioned roughly evenly among farm implements, tractors, and trucks. While the War Production Board (WPB) limited the company's ability to produce commercial products, Harvester continued to serve existing customers. The company maintained some wartime commercial production—albeit at lower levels of output—since farm implement factories were less well suited to war work than those producing tractors and trucks. Replacement parts represented the majority of commercial sales. To compensate for limited supplies of raw materials, Harvester organized a system for collecting, storing, and reselling farm scrap metal. The company also initiated a program to train Tractorettes (farm women and girls) and boys in the operation of tractors while the men were away. Through such efforts, agricultural equipment manufacturers like Harvester used the war to enhance their reputations for customer service, product durability, and patriotism.[5]

Like other wartime manufacturers, International Harvester also turned its industrial capacity to military-specific products. The company's sixteen U.S. factories—including two purpose-built for the conflict—participated in war production. On average, military output made up half of IH's wartime sales. Harvester manufactured armaments like torpedoes, cannons, guns, shells, and tanks, as well as modestly tailored versions of existing product lines that they had only to camouflage with "a liberal dousing of G.I. olive drab paint." The company also conducted research to better equip its machines for military applications. It increased the maneuverability and all-drive stability of its vehicles and supplied nearly all the trucks for the Marine Corps and navy, as well as some for the army and lend-lease partners. Trucks ultimately made up more than half of Harvester's sales of its "regular" product lines to the war effort. Harvester also produced the half-track, a vehicle that combined the front-end maneuverability of a truck's steering axle with the load-bearing and multi-terrain capacity of a set of tracks at the back. It was the company's biggest-selling single war item. In addition, Harvester was among eight construction equipment and automobile manufacturers that produced diesel engines for the M-4 Medium Tank. Finally, the company manufactured tractors, including a new product, the M-5 High Speed Medium Tractor, based on an existing crawler with modified steering, brakes, clutch, and transmission. This machine became Harvester's most significant tractor contribution to the war. An army colonel once noted of the machine, in *Field Artillery Journal,* "Any unit which is equipped with the M-5 tractor for prime movers has no real marching problem. As far as physical movement of the pieces is concerned, we are agreed that the finest article of equipment in the US Army is the M-1 Howitzer and the second best is the M-5 tractor."[6]

By contrast with World War I, the World War II military utilized earthmoving equipment for more than just weapons movement. They deployed the machines to topple trees, clear fields, and level hills as construction became a more integral military activity. These were activities at which the products of Caterpillar Tractor Company, a primarily construction-oriented manufacturer, particularly excelled. From 1942 through the end of the war, 85 percent of Caterpillar's production went to the government. Its equipment dominated the World War II construction arsenals of both the army and the navy. Although Caterpillar, like Harvester, initially produced some standard armaments in its factories, the government gradually relieved the company of such obligations in order to maximize production of construction vehicles. In 1942, for example, it ordered Caterpillar to stop work on diesel tank engines so that the company could retool its plant for expanded D7 tractor production instead.[7]

Tractors derived value not only from the power of their engines but also from the versatility of their attachments. These included winches for towing, hoists for lifting, scrapers for hauling, and bulldozer blades for pushing. Through the early 1930s, nearly a dozen companies produced bulldozer blades for Caterpillar. Beginning in 1935, however, the firm formally partnered with a single manufacturer—LeTourneau—for the majority of its implement needs. Other tractor makers allied similarly: Bucyrus-Erie supplied many implements for International Harvester, and Gar Wood and Baker served Allis-Chalmers. The entrepreneur and inventor R. G. LeTourneau had founded his namesake company in 1929 after having tried his own hand at earthmoving work. Although he was not the first to develop scraper and dozer-blade technology, LeTourneau secured several patents and advanced scraper design sufficiently to stimulate widespread adoption. During the 1930s, he created bulldozer blades with multiple angles of rotation and scrapers with increased capacity, maneuverability, and speed. His company put these products to use during the war, supplying the army alone with fourteen thousand bulldozers and eight thousand scrapers over the course of the conflict. The symbiotic relationship between LeTourneau and World War II's largest tractor producer, Caterpillar, helped make the attachment manufacturer the conflict's major producer of earthmoving implements.[8]

Despite unprecedented levels of demand, equipment manufacturers reaped only modest wartime financial rewards. This was due to both government-set price levels and excess profits taxes. While the average American corporation experienced relatively flat wartime profit margins, companies engaged predominantly in war work often saw their margins decline. Between 1940 and 1943, for example, sales among the fifty largest companies engaged primarily in war production grew by nearly 150 percent, while net profit margin fell from 9 percent to 3 percent. Caterpillar's experience was consistent with these averages. Although 1944 sales reached a new historical high, net profit margin declined to its lowest point since the early years of the Depression. Had taxes remained at their prewar rate, Caterpillar's wartime profits would have increased nearly threefold. Instead of near-term profits, however, construction equipment manufacturers derived longer-term rewards in the form of technological advancement, the training of laborers, and an enhanced popular depiction of the value of their brands and the work of their machines. They would put these less tangible benefits to productive use when they returned to peacetime operations.[9]

Wartime Advancements in Equipment Technology

Even as they focused on wartime production, construction equipment manufacturers turned their attention to the postwar years. In its 1942 annual report, Caterpillar offered an optimistic prediction: "Although research and engineering are now wholly in war service, much of the work now being done will be of incalculable commercial benefit after the war. It is certain that there will be, during the war years, a vast advancement in the accumulation of knowledge which, even though much of it may have no direct applicability, will aid our Company in the building of better products and the development of new products and the opening of new markets." This forecast proved fairly accurate.[10]

Although the war would produce no radical new construction-industry technologies, it spurred steady equipment advances that held great significance for the postwar world. Perhaps equally significant, the war popularized those advancements for widespread audiences. The prewar years had been a period of stabilization in construction equipment and methods. There was little consumer demand for improvement in the relatively slow-operating, medium-size crawler tractors that relied upon favorable conditions of weather and terrain for their successful operation. But World War II, as one equipment expert put it, was "the crucible which resulted in the fielding of a wide series of truly capable and useful machines." It stimulated the intensity of research activity to move beyond this static, prewar moment with a large volume of improved products. In particular, wartime innovation increased engine power and efficiency, introduced lighter and more compact frames, enhanced mobility, improved material quality, expanded the capabilities and versatility of various construction implements, and hastened the transition from cable to hydraulic controls. As manufacturers refocused the immediate aims of their existing peacetime research activities, the necessity of war sped up the process.[11]

Not all war work benefited manufacturers commercially. When International Harvester developed twenty different truck models to meet the military's requests for lower profile, less easily targeted vehicles, the designs proved useless both militarily and commercially. The project was soon abandoned. In addition, some more successful examples of product development, such as armored bulldozer cages, had limited applications outside the military. Even worse, some companies had to retool so dramatically for war work that they were poorly positioned to return to commercial operations afterward. One of Caterpillar's parent companies, Holt, had diverted so much of its World War I production toward military-specific tractors that at the war's end, as one sales manager put it, the company "ceased to be an active factor in the tractor field, both in domestic and foreign markets." Cleveland Tractor (Cletrac) experienced a similar fate following World War II. The devotion of Cletrac's entire wartime production to a military-specific aircraft-towing tractor with a tank-type undercarriage destroyed the company's civilian business. It was effectively forced to merge with another manufacturer in 1944. Along with enthusiastic predictions about wartime advances, therefore, most manufacturers also emphasized to investors the limited nature of the war's interruption and their anticipated rapid return to commercial operations. In its 1943 annual report, for example, Caterpillar noted, "The War Department required no radical changes in 'Cat' product design or construction and no immediate changes will be necessary to fit these products for postwar commercial uses." With this assessment, the company tried to manage investor anxieties in an uncertain time, even as it benefited from war-driven technological advances.[12]

As wartime research increased the power and capabilities of diesel engines, the improvements were sometimes adapted to more compact and mobile vehicles. Although ships transported increasingly heavy machinery, some divisions of the military required lightweight equipment. Airborne Engineers, for example, needed to airlift their machines in one piece and then immediately commence construction once they hit the ground. Developments in design and alternative metals helped manufacturers turn out equipment that weighed less and fit within a relatively tight envelope. Some of the tractors proved as productive as commercial vehicles of twice their weight. Other types of "baby earthmoving equipment" included scrapers, shovels, cranes, and rollers. In the postwar commercial market, these lightweight machines would appeal especially to small contractors, local governments, and anyone desiring ease of maneuverability in the development of compact or isolated locations. Advances in the size and capacity of rubber tires, led largely by LeTourneau, further aided this cause. In subsequent postwar domestic construction, wheel-based equipment—capable of quick and easy transport on standard roadways without damaging the surface with track treads—would prove to be suited to rapidly moving excavated dirt on large-scale land-clearing projects. Compared with tracked vehicles, those with rubber tires reduced freight costs and delivery time.[13]

Developments in equipment implements, including bulldozer blades, proved particularly significant to the popular image of wartime construction machinery. In late 1941, after learning that the British were mounting bulldozers on tanks in North Africa, army engineers commissioned research on a specially designed blade of their own. Although tractor-mounted blades had proven optimal for most construction work, a tank-mounted version offered the prospect of more effectively clearing obstacles like mines. A major from the Mechanical Equipment Section of the Engineer Board worked with Caterpillar and two bulldozer manufacturers—LeTourneau and LaPlante-Choate—to experiment with possible designs. After earlier attempts with a V-shaped blade successfully excavated mines but proved ill-suited to dirt and rubble removal, the Engineers hoped to develop a straight blade that could do it all. At no cost to the government, the major convinced each of the two implement manufacturers to construct two pilot models with different blades. The army then tested these on domestic beach obstacles and other proving grounds. Large-scale production began in late 1943, and the resultant tank-dozers saw action on the European front the following year. In Italy, Engineers were conducting similar experiments to adapt tanks with blades for fighting ammunition fires. Eventually, all troops adopted standardized modification kits. General Dwight D. Eisenhower later remarked of this "godsend" piece of equipment: "Some imaginative and sensible man on the home front, hearing of this difficulty [of small-arms enemy fire on Engineers operating bulldozers], solved the problem by merely converting a number of Sherman tanks into bulldozers." The full story was more complicated than that, joining military necessity with private sector cooperation to achieve innovation in bulldozer blade design.[14]

Although home front construction men had no use for tank-mounted blades, they still valued several of the improvements to blade technology. Both military and civilian operators wanted lightweight blades to improve handling and minimize maintenance. Less weight also meant lower costs. And both groups desired quick and easy blade release, although the military measured the requisite timespan in seconds, not minutes. In fact, in its 1944 annual report International Harvester identified "the problem of simplifying the attachment and detachment of implements to the tractor" as an object of "much study." Hydraulic controls offered the technical means to help achieve this last goal. As Harvester noted, these controls would "enable the farmer or his wife, or his boy or girl, to make any desired adjustments, either lifting or downward pressure, by merely touching a lever, without getting off the tractor and without stopping the tractor."[15]

While hydraulic technology had existed before the war, the conflict spurred such widespread adoption that by its end, bulldozers with cable lifts were relatively uncommon. The simple touch of a lever replaced the heavy application of force required on older steel cables. Hydraulics also increased speed, shortened response times, and smoothed cycling. Not everyone agreed that hydraulics were a step forward. Some Seabees and Engineers preferred cables for their fewer parts, simpler connections, and easier repair in the field. LeTourneau also spurned hydraulics for their perceived slowness and inefficiency. His company continued to incorporate cable controls into equipment designs well into the postwar years. This enduring preference for an outdated technology would eventually hamper LeTourneau, however, as hydraulics went on to dominate the postwar construction market.[16]

Cranes also proved to be important tractor attachments during the war. While the military deployed heavy, crawler tractor-mounted cranes for much of their work, demand for increased maneuverability led to the introduction of wheeled units. In addition, the invention of a screw-luffing device for big crane booms (in place of some boom cables) offered smoother handling, more accurate positioning of loads, and the diminished risk of frayed cable strands dropping the load. With some wartime cranes called upon to lift dramatic weights, such as thirty-ton aircraft in need of salvage, manufacturers also innovated the crane boom. Their work ensured that lifting capacity would not be a limiting factor for the Allies. Because of their experiences, as the equipment historian Fred Crismon noted, these cranes "emerged from the war with an excellent reputation for mobility and easy maintenance." In the postwar period, contractors would attach wrecking balls and clamshells to wheeled cranes possessing rubber tires, rapid speed, and heavy load capacity for productive use on home front construction sites.[17]

Training in Equipment Manufacture and Maintenance

World War II also provided training grounds for an army of laborers. Wartime necessities demanded that large numbers of military men and civilians advance their skills in equipment manufacture and maintenance. Companies demonstrated their patriotism through the integral role they played in developing this expertise. These institutions bore the costs of increased training at home as many existing employees left their jobs for wartime duties, but they would later benefit from an expanded skilled labor pool when war was through.

To maximize military production, wartime employment at construction equipment manufacturers swelled. International Harvester, for example, averaged nearly 70,000 employees in 1945, up from 33,500 in late 1938. The trend at Caterpillar was similar, with employees more than doubling, from approximately 9,500 in 1938 to nearly 20,500 in 1944. Turnover was also higher during this period: some 27,000 IH and Caterpillar employees left their jobs for military service. Thus, the total number of new workers trained in the manufacturing trades exceeded what the net employment figures initially suggest. In 1943, for example, Caterpillar employees increased by 2,646 over the course of the year. With nearly 9,000 employees leaving during this same period, however, the company had to recruit and train nearly 12,000 workers in order to achieve this growth.[18]

Training necessarily expanded in the war years, in response to both the influx of new employees—30 percent of whom, at Caterpillar, were women—and the transfer of existing employees from other tasks to factory-based work. International Harvester offered special training in welding at one of its factories, while Caterpillar increased the number, variety, and pace of its existing training programs for both skilled and low-skilled employees. The objective of these training programs, according to Caterpillar's 1942 annual report, was "not only to make more people fitted for skilled work, but to improve skills so that there will be a decreased percentage of spoiled work, a reduced number of lost man-hours and machine hours, and a greater total of production."[19]

Employees who departed for the war supported the military effort in multiple ways. Former Caterpillar men staffed the Ordnance Department and national and regional offices of the War Production Board. In early 1942, junior executives from the company also began overseeing a newly formed subsidiary corporation, known as the Caterpillar Military Engine Company, established to plan, construct, and operate a government-financed Victory Ordnance Plant in Decatur, Illinois. International Harvester led a similar undertaking, the Harvester War Depot, when it accepted the army's request to manage an existing ordnance depot outside Toledo, Ohio, in 1943. As the military reaped the benefits of outsourcing these administrative duties to the organizational know-how of private companies, manufacturers could offer diverse on-the-job training to their workers while gaining positive associations for their brand.[20]

The war also meant the relocation of some employees out of the factory and into the field for real-world training in equipment maintenance and repair. Beginning in 1942, International Harvester created the first of six "Harvester Battalions." The 695 men, age eighteen to forty-five, who initially made up the Maintenance Battalion, 12th Armored Division, came entirely from the ranks of Harvester employees, dealers, and sons of dealers. Although three-quarters of the applicants brought mechanical capabilities to their service, the remainder were white-collar workers eager to learn new skills. According to one Harvester promotional brochure, these men were, "in the truest sense of the term, 'fighting machines.'" Caterpillar offered similar assistance, answering the Army Corps of Engineers' call to arms with the formation of the 497th Heavy Shop Company. Of the company's 199 members, 158 were volunteers from Caterpillar's existing employees. They focused largely on building the Ledo Road in Burma. The skilled mechanical labor of these men helped reduce equipment downtimes and lengthen their lifespans. In tandem, manufacturers ensured that existing employees kept their skills sharp while helping their products achieve peak performance on the fighting front.[21]

When these soldiers returned home, they brought new, war-tested skills with them. A correspondent for *Engineering News-Record* waxed enthusiastic about the commercial benefits of this homecoming: "When they return to the places whence they came (state highway departments, contracting firms, machinery service companies, etc.), they will bring with them a background of experience that has been varied and broad. Petty trade jealousies have been ironed out, and new and better methods have been developed; what these men have gone through has strengthened them in character. The construction industry contributed to the war the very best it had in men and equipment; the return to the industry will be a wealth of experience of the kind that seasons and balances men's skills and judgment." These soldiers' former employers were typically the beneficiaries of this wartime experience. By early 1945, for example, nearly 95 percent of International Harvester's discharged employee veterans had applied for reemployment. The company found positions for nearly all of them. Owing to "training or experience acquired in service," a number of these veterans returned to positions of increased rank and responsibility.[22]

Selling the Bulldozer

Boosterish media coverage of the wartime bulldozer effectively offered construction equipment manufacturers free advertising. Alongside stories of battlefield heroism by the likes of Aurelio Tassone, many wartime accounts focused on the machines' construction-related capabilities. As an author in *Southwest Builder and Contractor* wrote of the machinery, "Often called upon to do seemingly impossible chores, [it] throbs its way through jobs far beyond its rated power." The results impressed even those familiar with commercial construction work. As this same author continued, "It is amazing, even to experienced contractors, accustomed to tough road construction jobs, what modern American machinery and equipment can do." As just one example of such coverage of the machines' many exploits, *Engineering News-Record* reported on an incident in Alaska and the Aleutians in which Engineers perched at the top of a hill needed to move their tractors to the valley floor. When a steep and swampy pass threatened to slow their advance, they opted to push the six tractors over the brink and hope for the survival of a few. The gamble paid off. As the magazine correspondent concluded, "American construction equipment is built to take punishment, however, and *every one* of the six tractors was able to run when turned right side up at the bottom."[23]

Critiques of enemy construction equipment and practices frequently included praise for American machinery. The Japanese were a particular target. *Life* reported that Seabees on Guadalcanal could repair a five-hundred-pound bomb-hit to an airfield in just forty minutes, whereas the Japanese, lacking compressors and pneumatic hammers, needed three or four hours to complete a similar task. Moreover, while the Japanese simply left the crater filled with dirt, the Americans finished their work with sturdy Marston mat. Caterpillar proudly noted in an annual report that the Japanese had spent eleven months attempting to "scratch out a landing field" on Attu Island, in the Aleutians, while on the equally rocky nearby Adak Island, army engineers operating Caterpillar tractors had progressed from beachfront landing to completed airstrip in only ninety-six hours. When American troops took the New Georgia Islands in 1944, they discovered similar imbalances between their own capabilities and those of the previously occupant Japanese. As one Pacific correspondent noted, with unveiled racism, "The Nips had not a single dump truck; instead, great quantities of picks and shovels and innumerable baskets were used by coolies for carrying material from borrow to fill. When our combat troops came ashore, on the other hand, they were closely followed by our largest size heavy construction equipment that immediately began clearing the way for an air strip more than twice the length of the one the Nips had." In fact, one of the few pieces of modern equipment the Americans found abandoned by the Japanese in New Georgia—a roller—was powered by an American-made engine.[24]

While the substantial limitations on wartime commercial production might have made such public accolades sufficient to meet companies' marketing needs, manufacturers chose to do more. They were stimulated, in part, by the Internal Revenue Service's introduction in 1942 of a business tax deduction on wartime image advertising, particularly in magazines. Another motivation was the companies' focus on postwar growth. Since the full financial benefits of wartime participation would have to come during the postwar period, manufacturers invested in product and brand marketing during the war itself to help ensure those future rewards. Marketing campaigns amplified the public's awareness of manufacturers' contributions to the conflict and preparedness for its aftermath by portraying American construction equipment as durable, powerful, and patriotic technology.[25]

As a foundation for much of this work, the advertising industry urged companies to write their wartime histories. The trade journal *Industrial Marketing* argued that the requisite information gathering would acquaint advertising departments more closely with their companies' technical aspects—features that postwar buyers would appreciate after wartime fighting had increased their familiarity with many products' nuts and bolts. The industry further suggested that the written records might prove useful as references of past practice, guidebooks for future wars, and public relations pieces generally. Probably also motivating this proposal was many advertising men's fear for job security—in the context of declining wartime commercial sales and the wary eyes of the war-focused administration, ad men served their own interests by arguing for the continued relevance of their output. While companies like Ford led in the implementation of this advice by establishing an entire history department, construction equipment manufacturers also made strides. International Harvester, for example, hired a historian, who summarized the company's major wartime projects, including the war depot, maintenance battalion, and key product lines. Companies put such information to immediate promotional use in annual reports, journalistic articles they wrote about their exploits, and scores of war-inflected ad campaigns.[26]

Equipment manufacturers produced five major categories of wartime advertising copy. First, particularly at the war's outset, companies developed institutional campaigns designed to link their brands with patriotism, while veiling commercial interests. By foregrounding the brand, manufacturers hoped to remain in the positive public eye during the long period when new product sales remained relatively rare. These ads often included generic imagery and text advocating patience, commitment, and investment in war bonds. They underscored their message with a modest attribution line, such as "Contributed to the War Effort by Caterpillar Tractor Co." Such statements were the sole recognition that a particular company had helped make the advertisement possible. Related ad campaigns introduced greater specificity to their declarations of wartime support by announcing the conversion of factories to wartime production or the enlistment of employees in the military.[27]

A second class of advertisements was aimed at the strained wartime relationship with existing customers. In early 1942, the Construction Machinery Division of the WPB restricted the sale of crawler tractors to the highest-priority customers; a few months later, the Board added all tractors, shovels, and cranes to the list. By 1944, nearly three-quarters of all new domestic earthmoving equipment production went to U.S. military services, with lend-lease garnering an additional 5 percent. Unable to purchase new machines and hampered by an insufficient supply of replacement parts, commercial customers had reason to grow frustrated with their current equipment and the companies that manufactured them. Caterpillar responded with ads that advised on the "fighting four" steps to good wartime maintenance. Since such tips were hardly trade secrets, the ads' real purpose was to celebrate the company's wartime service and promote its postwar products, albeit under the guise of customer care, rather than profits. The company also reassured customers that product lines would transition quickly from military to commercial production after the war's end. In short, as one ad declared, "'Caterpillar' Diesels Are Worth Waiting For!"[28]

A third group of advertisements showcased the wartime exploits of particular products, suggesting corollary postwar applications and patriotic associations. These ads often offered romantic depictions of military construction work, and their messages grew more explicitly commercial as the end of the war approached. As an ad for Thew-Lorain published in *Construction Methods* noted of its machines, "the whole world is their proving ground." The copy elaborated, "Extra strength, power and speed are qualities that have always been built into Lorains. They are qualities that are being tested on a world-wide basis today and are being proved under conditions that are tougher than you'll ever face. They are qualities to remember when cranes and shovels are available for peacetime projects." A Caterpillar ad published in *Excavating Engineer* in 1945 tried to substantiate future purchases not only through references to past performance but also by suggesting the military's implicit endorsement. The text read, "The thousands of 'Caterpillar' Diesels serving our Armed Forces in this war offer added evidence that these sturdy machines rate foremost preference when you put in your orders for new contracting equipment." An ad for Gar Wood Industries, a tractor-accessory maker for Allis-Chalmers machines, used similar imagery to make this case. Dual scenes depict dirt moving on a Pacific battlefront and a domestic construction site. Only the background setting and operators' attire seem to have changed between the two images, while the machines diligently go about their comparable tasks, oblivious to environment. International Harvester conveyed this same trope in another *Construction Methods* ad. At the start of the war, the copy announced, "Nothing changed but the paint." By midway through it, however, engineering innovations and new product development were resulting in a very different product line. The ad thus concluded that by 1943, "EVERYTHING changed but the paint."[29]

The fourth class of ads attempted to amplify the patriotic associations of construction equipment brands by portraying the whippings their products were dealing a much-mocked enemy. An ad for Baker hydraulic bulldozers included a photograph of a bulldozer pushing a pile of dirt containing the cartoon bodies of German, Italian, and Japanese leaders. The headline announced that Baker equipment provides "direct down-pressure on the Axis." The artist Boris Artzybasheff created a whole portfolio of satirical drawings

---

(*Opposite*) A 1944 ad for International Harvester, published in *Construction Methods,* draws parallels between wartime and peacetime crawler tractors, suggesting plentiful postwar opportunities for war-honed machinery that had advanced in more ways than paint color alone. (Wisconsin Historical Society, WHi-101244)

EVERYTHING changed but the paint

LONG before Pearl Harbor your government called International Diesels to the nation's defense. Regulation olive-drab replaced the familiar red, and the big tractors of industry went to war. *Almost nothing was changed but the paint.*

These peacetime crawlers are writing war history in stirring action. They're pulling big guns, handling bombs for the Air Forces, smoothing shell-torn landing fields, clearing jungles, building roads. Tens of thousands of such International Tractors are valiantly supporting the Marines, the Navy, and the Army. Night and day we're building more. They're tops on every fighting front.

But that's not enough for American resourcefulness under the spur of war.

In 1943 a new "prime mover" rolled from the International assembly lines—a tractor in which *EVERYTHING was changed but the paint!*

Here's a revelation in crawler power, maneuverability and fighting quality . . . a high-speed performer under heavy load . . . a go-getter whose rugged construction and ease of handling will carry far beyond the Victory.

Victory is its one job, but there'll be a world to rebuild later. Then we'll build the International Diesel Crawlers you need. And you'll know why Harvester men are saying today: *"We've got a lot of things packed into this big baby that we'll use when the war is over!"*

INTERNATIONAL HARVESTER COMPANY
180 North Michigan Avenue Chicago 1, Illinois

BUY BONDS···BUY MORE BONDS

INTERNATIONAL
HARVESTER

Direct Down-Pressure on the Axis!

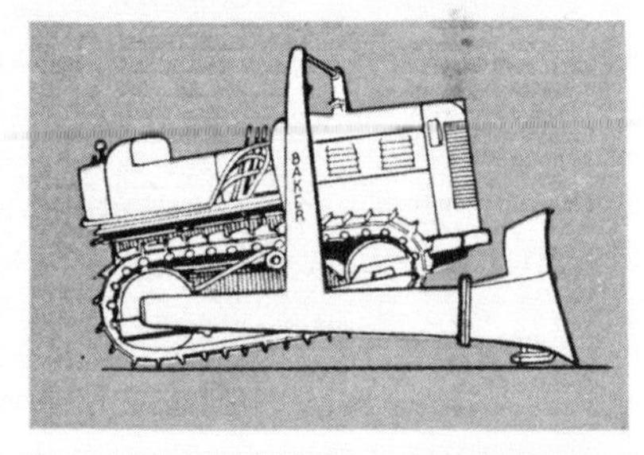

Baker Hydraulic Bulldozers *really* bear down!

The exclusive direct lift and down-pressure of the blade on Baker Hydraulic Bulldozers and Gradebuilders is a pain in the neck to Axis chest thumpers.

This simple, positive method of control permits the entire weight of the tractor front end to be exerted on the blade—the blade does not depend on its weight alone to force it into the ground. That's why Bakers get out bigger loads faster, every trip, all through each shift.

Their ruggedness and simplicity cuts maintenance costs to the bone. Their design makes the tractor engine more accessible. And they have all of the other features—moldboard level on rough ground, quick interchangeability of moldboards, tractor back end available for winch, etc. that you want in a bulldozer.

Beside rushing camps, landing fields and war plant sites to completion, Bakers are making landing strips in dense jungles and on desert sands, clearing debris from bombed cities, extending military roads in the frozen North and bringing direct down-pressure to bear on foes of democracy in other ways.

THE BAKER MANUFACTURING CO.
567 Stanford Avenue Springfield, Illinois

No combination of mechanical principles can equal the effectiveness of direct hydraulic lift and down-pressure.

The Modern Tractor Equipment Line for
EARTH MOVING
LEVELING AND GRADE BUILDING
SNOW REMOVAL
ROAD MAINTENANCE

for Wickwire Spencer Steel advertisements that followed similar lines. The machinery in his depictions, however, comes to life as it attacks its Axis enemies; in one ad, for example, big, hungry shovels tower over their terror-stricken human targets. Another ad, for Caterpillar, recounted the story of the Japanese army's trapping of British and Indian troops between the ocean and the Himalayan Hills in Rangoon. "The little Jap general must have licked his lips over the anticipated slaughter," the ad read. "But he had overlooked one fact. The British-Indian command had 'Caterpillar' Diesels." Another Caterpillar ad focused on the Germans instead: "Maybe you overlooked this machine when you planned your world conquest, Adolf. But soon you'll be seeing it in your nightmares—seventeen tons of 'Caterpillar' Diesel D8 Tractor burying you under earth and rock!" The ad continued, echoing International Harvester's proclamation of a "Dirt Moving War": "But there was one important part of war we didn't have to learn, and that was *moving dirt.* Did you ever stop to figure out how many million tons of it would have to be moved to win this war, Adolf? You'd be surprised."[30]

Rather than trumpeting specific products and brands, a final genre of advertisements depicted postwar building demolition and land clearance, more generally, as progress. A wartime pamphlet series produced by Bucyrus-Erie, a tractor attachment manufacturer, declared, "In War and Peace Progress Starts with Excavation." Another manufacturer—albeit not of earthmoving equipment—made the comparable case for the mass demolitions of slum clearance. Its 1944 ad depicted two giant hands at work in a nameless American neighborhood. The left hand employed what the architectural historian Andrew Shanken has described as a "bulldozer-like fist"—although the wrecker's clamshell might be a better comparison—to violently scoop up a swath of row houses. Meanwhile, the right hand delicately replaced the empty slate with a more ordered array of buildings and trees. Another ad produced that same year by Rodgers Hydraulic, a tractor attachment manufacturer, offered a more ambitious, though slightly grotesque, portrayal of the postwar clearance opportunity with its depiction of a bulldozer and operator

---

(*Opposite*) A 1943 ad for Baker earthmoving equipment, published in *Engineering News-Record,* equates the powerful pressure of wartime bulldozing with a direct assault on Axis powers. (Reprinted courtesy of AGCO Corporation)

Satirical cartoon portrayals of Allied equipment amplified the patriotism of American construction equipment. In this image, which appeared in a 1943 Wickwire Spencer Steel advertisement, the artist Boris Artzybasheff depicts anthropomorphized power shovels attacking Axis leaders. Wickwire also included this image in an assemblage of the artist's war work, "Axis in Agony!," published in 1944.

clearing the entire surface of the globe. Anticipating the medical metaphors later applied to the bulldozing of decayed postwar cities for urban renewal, the ad equated earthmoving with plastic surgery, announcing, "Mother Earth is going to have her face lifted!" Mother Earth's despondent facial expression aside, the ad's tone was enthusiastic. Its text elaborated, "Sounds like a rather ambitious undertaking, doesn't it, but that is more or less what is going to happen after this World Struggle is over. The Earth is in for a tremendous resurfacing operation. The construction, road building and grading jobs for Crawler type tractors in the not too distant period are colossal."[31]

The Specter of Surplus Equipment

As the war's end became imminent, manufacturers worked to limit the supply of surplus war equipment that threatened to swamp domestic commercial demand. After World War I, the U.S. government had donated to

Let YOURS be these helping hands

Neighborhoods are living things, not just inert lumber, metal and stone.

And, like all living things, they must grow or perish.

Take *your* neighborhood, for instance. It must improve, or it will deteriorate. Often, in a few years a neighborhood becomes unrecognizable, its values completely destroyed. An idle loft building or factory becomes a festering sore, spreading the infection of obsolescence. People who dwell in unimproved houses soon breed diseases and delinquency. Shrinking playgrounds drive the children into side streets or empty lots.

City planners, no matter how much they want to, cannot prevent this. They cannot protect individuals or groups unless they know what is needed. Once they do know, they can usually help.

Why don't *you* let them know about *your* neighborhood, since someone must? Let yours be the hands to set in motion the forces needed for its reshaping . . .

To help you, we have prepared a book that tells in simple language how *you* and a few friends can help your neighborhood help itself. You'll need no technical background to understand this book. You'll need no technical experience to follow its instructions. Every other one of its ninety-six pages is an illustration, a diagram or a plan. Color makes these pictures unmistakably clear. And page by page it sets forth each succeeding step you must take to achieve a revitalized neighborhood.

A better school, a larger playground, modern housing, a shopping center, restricted traffic, park areas . . . every neighborhood can rightly claim these. This book, "You and Your Neighborhood — A Primer" can help you get them. It is an expression of faith in the American man or woman's willingness to assume responsibility, and capacity to get things done.

It is an expression of faith in *you*.

Oscar Stonorov • Louis Kahn

★ ★ ★

Revere asked Messrs. Stonorov and Kahn to write "You and Your Neighborhood — A Primer" because of the many letters we received asking for such help. We had already presented a project of theirs on local rehabilitation which had met with tremendous response. And we felt that such a primer fitted in with the contribution we are trying to make to better living.

At the same time we knew that once citizens became interested in lasting life for their neighborhoods, they must also become interested in copper. For copper, enemy of rust and weather, is the metal of lasting life. Consider! If a house has been built for permanence, if its roof, flashing, plumbing and heating system are in good shape, how simple it is to renovate it. A coat of paint outside, some paint and new paper inside, a partition removed, and the job is done at comparatively small price. That is why, in building or modernizing, the wide use of copper and copper alloys makes any house better to live in, to own, or to sell.

The primer turned out to be a far greater undertaking than we had imagined. It had to be simple, it had to be specific, it had to be complete. And so it grew and grew until there were ninety-six pages! But when we brought it to the printer he became very stern on the subject of paper, and forced us to publish only a limited edition. For that reason, though we invite inquiries from any individual or any group interested in housing and planning, we shall only be able to send one copy to any one who writes. Just send your request to the address below.

BUY MORE WAR BONDS TODAY

COPPER AND BRASS INCORPORATED

Founded by Paul Revere in 1801

Executive Offices: 230 Park Ave., New York 17, N. Y.

A 1944 ad for Revere Copper and Brass, published in the *Saturday Evening Post,* promotes the postwar opportunities for large-scale slum clearance projects to be obtained at the hands of bulldozers and wrecking cranes. (Louis I. Kahn Collection, The University of Pennsylvania and the Pennsylvania Historical and Museum Collection)

MOTHER EARTH IS GOING TO HAVE HER FACE LIFTED!

Manufacturers of:

UNIVERSAL HYDRAULIC PRESSES
TRACK PRESS EQUIPMENT
HYDRAULIC KEEL BENDERS
HYDROSTATIC TEST UNITS
POWER TRACK WRENCHES
HYDRAULIC PLASTIC PRESSES
PORTABLE STRAIGHTENER FOR PIPE AND KELLYS

SOUNDS like a rather ambitious undertaking, doesn't it, but that is more or less what is going to happen after this World Struggle is over. The Earth is in for a tremendous resurfacing operation.

The construction, road building and grading jobs for Crawler type tractors in that not too distant period are colossal. While our Plant is now engaged in essential work for the Armed Forces, we are not forgetting for a moment our duty and obligation to those who depend on Rodgers presses for quick repair of crawler tracks and other heavy machinery. Right *now,* all engaged in essential work are eligible for Rodgers Hydraulic Track and Universal Presses. Wire or write for full information and prices. *If it's a Rodgers, it's the best in Hydraulics.* Rodgers Hydraulic Inc., St. Louis Park, Minneapolis 16, Minnesota.

Rodgers HYDRAULIC Inc.

state highway departments thousands of tractors that had been purchased for the war. Some of these units had never left France, while others had yet to see active service. In the long term, manufacturers had benefited from what essentially functioned as a product-sampling program. Nonetheless they had suffered in the short term from a diminished postwar market for road-building equipment. In the words of a 1943 *Engineering News-Record* editorial, manufacturers were determined to ensure that "that mistake should not be repeated following this war." The article offered two rationales for why "there should be no round trip tickets for construction machines." First, such a policy would give manufacturers the opportunity to produce and sell new, improved equipment, while stimulating employment. Second, disposal of this equipment abroad would familiarize international contractors with American machinery, spurring future export demands. A reader wrote in to add a third justification: leaving the equipment where it was would be "a far more substantial, solid and permanent way of raising standards of living in other countries than by giving them free food and free clothing (at the expense of the American tax payer)."[32]

The government began the equipment-disposal process while the conflict was still ongoing. Domestic redistribution centers received used, but still usable, machines from around the world. After cataloguing the items, the centers stored or shipped them wherever they were needed in order to maximize the machines' wartime potential. The government also directed to the underserved domestic consumer market items that were no longer useful on the battlefield. Unlike during World War I, these redistributed machines were neither new nor lightly used, but rather, according to *Engineering News-Record,* the "battered dregs" of "scorched and sorry-looking" equipment that the military no longer deemed worthy of redeployment. Domestic demand was such that over a thousand items of farm and construction equipment offered through a sale held in Kearney, Nebraska, in February 1945, sold out to the fifteen hundred eager attendees at a rate of a unit every thirty-seven seconds. "Rugged" tractors, graders, and scrapers were some of the most

(*Opposite*) Despite the forlorn facial expression of Mother Earth, this 1944 ad for hydraulic tractor equipment, published in the *Military Engineer,* depicts a world of opportunity for postwar earthmoving work. (Courtesy of Granite Fluid Power, Granite Falls, Minn.)

popular items, receiving up to 150 bids each. Sales like this helped diminish the size of the postwar equipment glut, while also meeting critical wartime equipment demand at home.[33]

Despite active management of equipment stores during the war, a substantial arsenal of machinery still required disposal at the end of hostilities. In total, the United States spent roughly $300 billion on all wartime supplies, construction and otherwise; $260 billion of this had been consumed by the end of 1945 through a combination of lend-lease, exhaustion by U.S. forces, or simply outliving its useful life. On the construction equipment side, in particular, roughly a third of all machines in use by 1944 were over fourteen years old—well above the ten-year age limit recommended by manufacturers. The government hoped to unload the remaining new and used items worthy of resale. Anxious manufacturers and industry organizations recommended the distribution of these items in a way that would encourage fair prices and protect the welfare of small business.[34]

Some of these fears proved unfounded since half the remaining materials resided abroad and were not worth bringing back home. While wartime exigencies had justified the transoceanic transportation of such items, peace changed the equation. Federal law prohibited the importation of any materials that remained in Europe. Instead, with the United Nations Relief and Rehabilitation Administration (UNRRA) as buyer, many of these machines found ready use in rebuilding war-torn cities. The scale of postwar foreign reconstruction was such that international buyers required new equipment as well. Between 10 and 15 percent of immediate postwar production went to these foreign customers. Meanwhile, on the Pacific front, the Foreign Liquidation Commission collected the best of the equipment stranded there at war's end. The commission sold some of these items within the region, including to the UNRRA and the Philippines, at a return of twenty-five cents on the dollar. Given the high cost of repatriating the remainder, its minimal value as scrap, and the desire to protect domestic manufacturers, hundreds of thousands of bulldozers, airplanes, and their abandoned parts stayed in the Pacific. Soldiers never even retrieved the most decrepit machines from their battlefield locations. They buried others on-site. As one Seabee recalled, "They took bulldozers. The big bombers, the airplanes, they stripped out—the B-29s—they stripped out the instrumentation, shoved 'em over into the ravines, and buried 'em. They didn't come back." The military sank other equipment in order to create breakwaters. Construction equipment both facilitated this work and suffered this fate. Heat, humidity, and rain took their toll, and some of the machines still rust and rot on jungle floors today.[35]

Just four years after the war's end, when the military reactivated Seabee battalions in anticipation of the Korean conflict, the decision to abandon this equipment no longer seemed so wise. Operating under budget pressures, the navy looked to forsaken Pacific stockpiles as cheap alternatives to increasingly expensive and in-demand equipment at home. It sent crews to recoup whatever machinery local work crews had not yet poached and then to assemble it for inspection on bases in Guam and Hawaii. If a machine could be made "like new" for up to 30 percent of its current replacement cost, assessors marked it for transport to Port Hueneme. Otherwise, they scrapped it. Roughly three hundred thousand tons of heavy equipment made the belated trip eastward. After navy employees and private contractors conducted repairs, Seabees "exercised" the equipment on the grounds of the base until soldiers needed it on the battlefield again.[36]

The other remaining surplus equipment awaiting postwar redistribution resided on domestic soil. In accordance with industry requests, branded items returned to the original manufacturers and their distributors for resale through normal trade channels. The government sold non–brand name items directly to consumers at prices that reflected the machines' age and use but did not undercut the new equipment market. Two groups received special priority in purchasing surplus: small businesses and World War II veterans. With construction equipment, however, the typically poor condition of the remaining items meant few real benefits for potential buyers. As one veteran recalled of his experiences traveling around to inspect surplus equipment with the CEO of his equipment distribution company, "We didn't buy all that much, however, maybe a couple of motor graders, that was about it. The equipment was beat up and beginning to rust out." Thus, in the aftermath of war, worn-out construction equipment littered Pacific shores and domestic soil. As manufacturers started replacing these machines with shiny new versions, they simultaneously transformed the entire industry.[37]

Postwar Reconversion and Growth

Construction equipment manufacturers pursued several strategies to prepare for postwar growth. Not only did they reconvert factory floors through modernization, retooling, and overall physical expansion, but executives also formulated even bigger plans. These included increasing investment in their construction-products divisions, growing through acquisition, vertically integrating product lines, and expanding internationally. As the positive outlook for the postwar market attracted new entrants, it spurred overall industry consolidation into a few major firms. Consequently, ten of approximately three hundred construction equipment companies accounted for more than half the immediate postwar volume.[38]

The termination of war contracts initiated massive reconversion efforts. The unwinding of International Harvester's M-7 tank contract, the largest the company held during the war, demonstrates the many steps that such terminations triggered. When the War Department canceled the M-7 contract in 1943, 12 Harvester plants and 437 subcontractors were engaged in the work. The company set up a special organization—composed of manufacturing, engineering, storekeeping, legal, and accounting personnel—to dispose of the materials and machine tools and to settle claims. It also made mechanical changes on factory floors, including realigning facilities for different machines, modernizing production equipment, and adjusting product models from armed service to civilian requirements. On V-J Day, in August 1945, Harvester still held roughly 565 government contracts and purchase orders representing approximately $206 million in future revenue. The government promptly canceled all of them. The unraveling of these arrangements required numerous repetitions of the M-7 process. Reconversion was somewhat simpler at companies like Caterpillar and LeTourneau, where most wartime products more closely resembled peacetime models. Nevertheless, all manufacturers endured some transition in moving from war to peace.[39]

With their eyes fixed firmly on the future, virtually all major equipment manufacturers enlarged their physical footprints following the war. In 1945, International Harvester purchased a former Buick aircraft engine plant in Melrose Park, Illinois, for crawler tractor and diesel engine production. That same year, LeTourneau began construction of a fabricating plant in Longview, Texas. In 1946, Caterpillar announced its biggest plant expansion yet. The company enlarged its facilities in East Peoria, Illinois, by more than 50 percent during the three years following the war, increasing its factory footprint from 79 to 128 acres. It also grew outside Peoria, with expansions in Joliet and Decatur, Illinois, and in York, Pennsylvania, during the first postwar decade. Although International Harvester already had foreign subsidiaries before the war, and Caterpillar had sold products internationally since the company's early days, many heavy-equipment manufacturers established their first foreign production sites during this period. The United Kingdom was the most popular destination, with LeTourneau, Allis-Chalmers, and Caterpillar all setting up factories there. Caterpillar even went so far as to create a British subsidiary.[40]

Postwar equipment manufacturers also expanded their participation in the construction industry. For companies like International Harvester, which focused primarily on farm applications, this meant intensifying investment in their construction equipment line. Although Harvester had been building crawler tractors since 1928, they aimed them predominantly at agricultural customers. It was not until 1935 that the company specifically adapted some of these farm tractors for industrial use by increasing their power and improving the provisions for a wider range of mounted implements, from dozer blades to cranes and front-end loaders. While still engaged in the production of heavy-duty crawler tractors for the war, the company formally established its long-term participation in construction by introducing a new division in 1944. The acquisition of the Melrose Park plant provided the division with a headquarters, while in 1947 came the introduction of its first major new construction product, the TD-24 tracked tractor. International Harvester continued to expand its product line over time; by 1956, it had retired the new division's initial name (Industrial Products) in favor of the more descriptive "Construction Equipment Division." Overall, however, the division suffered from under-investment relative to the farm equipment and truck divisions and never posed a real threat to rival Caterpillar.[41]

Nearly all major construction equipment firms expanded their postwar product lines through vertical integration. Before the war, dealers had sold most of their tractor units with tailored, third-party attachments. In February 1944, however, LeTourneau announced its expansion into prime mover production and the cessation of implement sales through Caterpillar dealers. Three months later, with LeTourneau's earliest patents due to expire, Caterpillar reciprocated. The company announced that it would begin building its own bulldozers, scrapers, rippers, and cable controls. Other manufacturers followed suit, albeit through acquisition rather than in-house product development. In the 1950s, International Harvester acquired the Heil Company (scrapers), the rights to Bucyrus-Erie attachments, and the Frank G. Hough Company, with its distinctive self-propelled, rubber-tired shovel loader known as a payloader. At the same time, Allis-Chalmers acquired the LaPlante-Choate Manufacturing Company (including its scrapers), Baker Manufacturing Company (bulldozer blades), and Buda Company (diesel engines). Thus, whereas LeTourneau had dominated scraper technology before the war, by less than a decade afterward nearly every major manufacturer was offering its own models.[42]

As existing companies consolidated, new ones entered the field. This growth testified to the attractiveness of the postwar construction equipment industry. In 1953, General Motors ventured into the business by announcing plans to buy Euclid Road Machinery Company. That same year, Westinghouse Air Brake purchased LeTourneau's entire earthmoving line. Similarly, Clark Equipment Company introduced a new Construction Machinery Division into an organization previously dominated by forklifts and other generic materials-handling machines. The fairly regular circulation of key industry personnel among companies helped facilitate the growth of these new entries.[43]

Not all construction industry expansion encompassed big machinery; new entrants also developed the compact, small-scale market. In 1947, the North Dakota–based Melroe Company was formed to produce agricultural equipment. A decade later, two Minnesota brothers, Louis and Cyril Keller, invented a farm-focused compact loader; Melroe soon acquired their enterprise. In 1962, the company branded a four-wheel version of the skid-steer loader with the Bobcat name in the hope of evoking toughness, quickness, and agility. Although the idea for the machine had originated with turkey farmers needing to remove manure from barns, its relatively high price tag better suited it to industrial than agricultural use. The powerful yet maneuverable machine performed many tasks, including construction, landscaping,

The Melroe BOBCAT . . . Lovely little homewrecker

The Melroe BOBCAT Loader is a lot of fun to play with . . . but really serious about her work. Demolition contractors can hardly keep their hands off her. She can make a shambles out of a ground floor kitchen in seconds or go up 20 stories to wreck an entire bachelor pad in minutes. Barely 5 feet high and 54 inches across the hips, the M-600 can really maneuver in those hard-to-get-at places.

You wouldn't carry her across the threshold but the M-600 weighs a mere 3,269 lbs. . . . so hoist her to the roof or sneak her up on the elevator. Give her a big bucket or Ho-Ram attachment and the Melroe BOBCAT will take down walls and dump debris over the side in a hurry. And you could get attached to 19 other easy-on attachments. She's no pushover either . . . positive four-wheel drive gives sure traction on any surface.

Cut your hand labor costs in half — finish every demolition job ahead of schedule . . . and below your bid projections. Just for fun, call your local Melroe BOBCAT dealer for a FREE on-the-job demonstration. Or send us a love-letter direct.

MELROE
BOBCAT
Loader

WSJ

MELROE COMPANY
GWINNER, NORTH DAKOTA 58040

Please send free illustrated literature and price list on the Melroe BOBCAT Loader . . . and details on Free demonstration.

Firm Name ____________

Address ____________

City ____________

State ____________ Zip ____________

Sales and Service Centers throughout the United States and Canada

Melroe Company marketed the demolition applications of its versatile compact skid-steer loader, the Bobcat, in this provocative advertisement published in *Wrecking and Salvage Journal* in 1969. (Photo courtesy of Bobcat Company archives)

and demolition—one ad described it as a "lovely little homewrecker." Depicting the machine next to a provocative-looking woman, the text extolled the equipment's wrecking prowess: "The Melroe BOBCAT Loader is a lot of fun to play with . . . but really serious about her work. Demolition contractors can hardly keep their hands off her. She can make a shambles out of a ground floor kitchen in seconds or go up 20 stories to wreck an entire bachelor pad in minutes." After Melroe grew to dominate the compact loader market, Clark Equipment Company eventually acquired the company in 1969; it later became known as the Bobcat Company.[44]

After a slightly bumpy start, then, substantial growth characterized the early postwar years. Civilian demand provided a strong domestic market. Increasing numbers of experienced workers and growing volumes of available materials all helped reduce costs. The long-term elimination of wartime excise taxes took less off the bottom line. Technologically improved products and gradually increasing prices further enhanced the opportunity. Of course, there were also obstacles. Labor unions that had stayed silent during the urgency of war eventually agitated over flat wage rates, and 1946 was a record year for work hours lost to strikes. At the same time, military spending was curtailed after the war, and the onset of the Korean War in 1950 never came close to mobilizing the equipment industry the way World War II had done.[45]

Still, the net impacts on manufacturers were positive. At Caterpillar, for example, sales grew by an average rate of 10 percent per year in the decade following the end of the war. Average annual growth in profitability was twice that. By the end of this period, the number of employees at the company exceeded the wartime peak by 50 percent. The company's immediate postwar bump was impressive, and the ensuing decades would continue on a positive trend: on average, Caterpillar's profits grew by 19 percent annually in the first postwar decade, and then dropped slightly to increases averaging 16 percent per year in the decade after that.[46]

New large-scale domestic construction projects fueled this continued demand. Across cities and countryside, the postwar growth of suburban housing, industry, and commercial sites; urban renewal construction; and interstate highways ensured a profitable future for construction equipment manufacturers and contractors. These developments owed much to the ideology, technology, policy, and practice of large-scale clearance that had grown out of World War II. Wartime-induced advances in technology, growth of skilled labor, and positive depictions of individual equipment brands and the clearance work those machines performed helped facilitate the redeployment of construction equipment from the battlefront to the home front. During the first postwar decade, the federal government established several new policies that provided powerful political and economic incentives for large-scale domestic clearance. The Housing Acts of 1949 and 1954, the G.I. Bill of 1944, the Federal-Aid Highway Act of 1956, and changes to the 1954 Internal Revenue Code all bolstered postwar projects. Many of these policies represented the culmination of a lengthy process of advocacy. In some cases, the public-private partnerships cemented during the war helped turn that long history into legislation. In this way, these policies provided the ultimate reward for construction equipment manufacturers' many years of wartime service.

But corporate manufacturers and federal policy makers did not achieve these landscape changes on their own. Local governments, developers, state highway engineers, and building and land-clearance contractors ultimately put the instruments of clearance into action on the ground in locations spanning the country. All these actors combined to implement the culture of clearance in U.S. cities, suburbs, and rural hinterlands. Collectively, they set the wartime bulldozer to work on the postwar domestic landscape.

Part Two

Bulldozers at Work

(*Overleaf*): An operator from the Bronx-based Palma Contracting Company puts an International Harvester TD14A bulldozer to work on New York's Cross Bronx Expressway in 1952. (Courtesy of Edgar Browning Collection)

Chapter 3

Grading Groves and Moving Mountains

Suburban Land Clearance in Orange County, California

Typifying the connection between wartime activities and the postwar culture of clearance, it was a Seabee who perhaps most influentially shaped the postwar domestic landscape. Although his residential aesthetic of peaceful nuclear family living was a far cry from the destructive battlefields on which he had furthered his technical training, William Levitt created the Long Island town that came to epitomize large-scale postwar residential suburbs. Completed in 1951, Levittown, New York, consisted of more than seventeen thousand ranch and Cape Cod–style houses constructed en masse on more than a thousand acres of former potato fields.

In the early days of the conflict, Levitt supported the war effort by mass-producing war workers' housing in Norfolk, Virginia. Soon afterward, he joined the Construction Battalions in the Pacific and helped clear land for island airfields. Levitt spent many nights talking with fellow Seabees on the battlefront. Drawing upon their diverse backgrounds in construction, they brainstormed about how to complete these vast projects better and faster. Levitt found that this wartime experience offered him an unparalleled laboratory for experimentation and analysis with his construction industry peers. When he returned home, he put these experiences to practical use. The Long

Island project, begun in 1948, was soon followed by similar projects in New Jersey and Pennsylvania.[1]

While Levitt's contribution to the postwar built environment is well known, many lesser-known American homebuilders followed his lead. Hugh Codding, for example, spent three years in the Seabees. Working as a carpenter's assistant, he built a hospital, underground storage tanks, and hundreds of Quonset huts in Maui, the Marshall Islands, and Guam. In his first year back home in northern California after the war, Codding built fifty houses. The following year, he added another two hundred. With the help of veterans he hired through the G.I. Bill, Codding quickly moved to building thousand-home subdivisions and to creating some of the first covered shopping centers in the region—including a mall called Coddington. In the process, acres of plum and peach orchards fell. His primary skill, he explained in one interview, was "to take raw ground and try to make it productive." He also drew upon the "speed building experience" he had gained in the Seabees to give his building activities a promotional flare. For a 1950s publicity stunt, his crew assembled an entire house in just over three hours. Codding's projects helped transform the landscape of postwar Sonoma County from orchards into suburbs.[2]

Servicemen's World War II experiences also shaped postwar construction in areas beyond homebuilding. When Ray Pitman returned to Kansas City, Missouri, and founded a contracting company, he drew upon the time he had spent as an army bulldozer operator. The entrepreneurial twenty-one-year-old began by purchasing a bulldozer and an International Harvester truck. Having served part of his wartime service digging ditches on the island of New Caledonia, he applied these skills to excavating for gas and water lines that served domestic developments. Another former Seabee, Gerald Halpin, followed his World War II service with a long career in real estate development. He built large office complexes, manufacturing facilities, and an early regional mall in northern Virginia. Later, in the early 1960s, Halpin and his partners purchased hundreds of acres of dairy farmland, on which they developed some of the first office and apartment buildings that would become the edge city of Tysons Corner.[3]

For all these individuals, wartime construction work variously enhanced their practical skills or demonstrated efficient approaches to large-scale building. Moreover, to the civilian population, the war showcased the mighty potential of American construction. When these men and methods came home, the domestic landscape became their new enemy, and profit—rather than victory—was the new goal.

The homecoming of large-scale land clearance practices was particularly pronounced in southern California, where the war had helped catalyze a regional transformation. Many soldiers passed through California on their way to the Pacific—Seabees shipped out from Port Hueneme, in Ventura County—while other servicemen moved there temporarily to staff a growing number of military outposts. Defense-related factories spurred a similar relocation of civilians, 1.6 million of whom migrated to the state. After the cessation of hostilities, more than half these civilians—and around half a million veterans—stayed on. The Cold War made their continued residence financially viable, both through the provision of military-industrial jobs and through the local market for goods and services created by government employees.[4]

Federal programs supported the necessary increase in housing stock, with the G.I. Bill (Title III of the Servicemen's Readjustment Act of 1944) and Federal Housing Authority funding 50 percent of all new home mortgages across the state. The G.I. Bill built upon existing federal mortgage subsidies to provide returning servicemen with zero-down-payment loans for home purchases. It offered a government guarantee of up to 50 percent of qualifying home loans (capped at $2,000), limited interest rates to 4 percent, and set a maximum loan amortization period of twenty years. It also required application for the loan within two years after military discharge. While several of these terms limited utilization of the policy, key adjustments in the 1950 Housing Act addressed the shortcomings. These changes extended the maximum loan period to thirty years (thereby reducing monthly payments), raised the loan guarantee to 60 percent, and capped the maximum amount at $7,500. Although not all postwar home purchasers were veterans, and even former members of the military were able to obtain mortgages through other means, the impact of the G.I. Bill was significant. In the first postwar decade, the bill helped nearly four million World War II veterans secure government-backed mortgage loans totaling in excess of $30 billion. These were predominantly first-time home-buyers purchasing primarily new construction in new suburban communities. The bill thereby stimulated the large-scale land clearance necessary to ready the landscape for the new homes these purchasers were demanding.[5]

Changes in the postwar tax code encouraged corollary growth of suburban commercial development. The 1954 Internal Revenue Code introduced accelerated depreciation, reducing the length of time necessary to write down a building from forty years to seven. This accounting change stimulated further land clearance by making it more profitable to build new construction on greenfield sites than to reuse existing locations. It proved particularly influential in spurring new suburban shopping centers—by Hugh Codding and others.[6]

Orange County epitomized the postwar suburbanization phenomenon. Popular journalists unequivocally celebrated its growth in daily print pages and glossy magazine spreads. In a 1946 article, "California Has a New Boom," *Life* profiled various veterans, war workers, and businesses moving to California, especially its southern region. *Look* devoted an entire 1962 issue to growth across the state, calling California "A Promised Land for Millions of Migrating Americans." The headline of a 1964 *Los Angeles Times* article put it most simply: "Orange County Spells Progress." From a population of just over 216,000 in 1950, the county had more than tripled in size by 1960, and doubled again by 1970, making it the fastest-growing large county in the United States. As throughout the nation, the construction work that accompanied this growth began with the large-scale destruction of the natural environment. Beginning before the war, but accelerating from 1945 on, orchard properties were dramatically transformed. Rows of fruit and nut trees surrounding islands of houses rapidly gave way to rows of houses surrounding islands of trees.[7]

These individual subdivisions collectively transformed the pattern of the entire region into one of low-density sprawl, and by the 1960s, the destruction had spurred protests that led to the passage of agricultural-preservation legislation. As clearing the shrinking amount of orchard flatlands became more difficult and more expensive, developers utilized powerful earthmoving equipment and techniques to target new hillside construction. In the hills, postwar developers cleared trees from the land, and also took land from the land—"moving mountains" as many observers put it—in order to create flat, blank slates for new subdivisions.[8]

This close-up view of Orange, California, in 1946 (*top*) and 1972 shows its postwar transformation from rows of trees surrounding islands of houses to rows of houses surrounding islands of trees. (HistoricAerials.com)

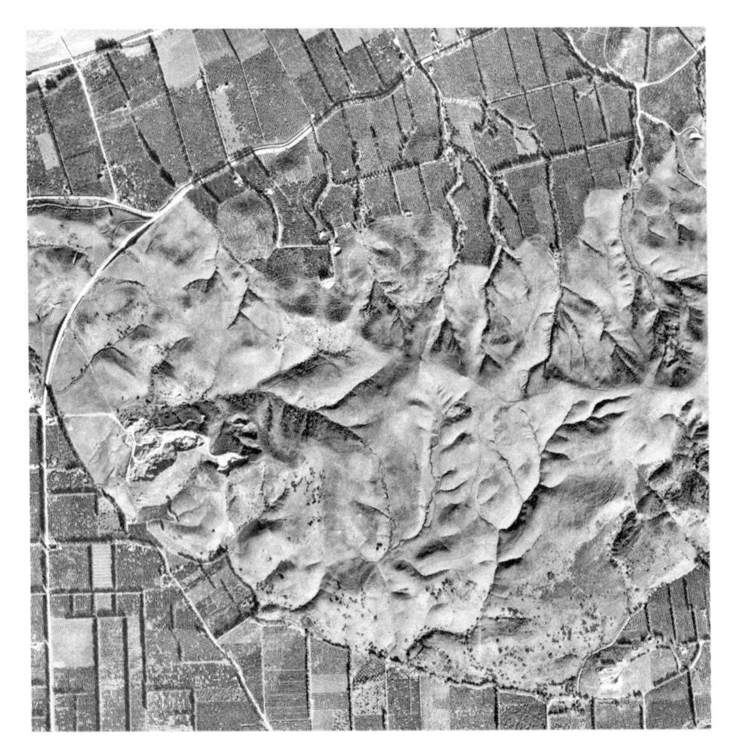

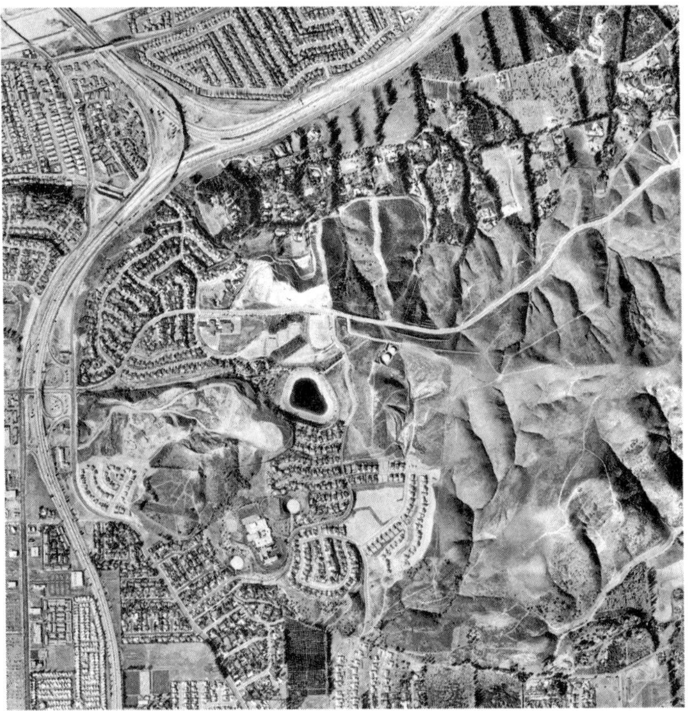

In Orange County's Santa Ana Canyon, suburban clearance also leveled previously undeveloped hillsides between 1946 (*top*) and 1972. (HistoricAerials.com)

The development of Orange County illuminates the many drivers of postwar suburbanization. Well documented elsewhere are the impacts of federal mortgage subsidies, rising automobile use, mass housing construction, and the growth of political conservatism. The Orange County story exposes the corollary significance of land-use economics, however, and the rise of a military-inflected culture of clearance—from advances in earthmoving technology and practice to popular and policy support for clearance implementation. Collectively, these factors spurred the destruction of large swaths of the natural landscape to make way for the postwar suburban built environment. What the environmental historian William Cronon has called the "first nature" of undeveloped hillsides and the "second nature" of human-made orchard growth were both sacrificed in the process. Boosters initially heralded this destruction as natural and progressive. But as large-scale land clearance gained traction, it exposed social, economic, and environmental consequences that increasingly called the sense of progress into question.[9]

Why the Orchards Fell

In the April 10, 1960, issue of the *Los Angeles Times,* columnist Walter Muhonen described the frequent sight of newly opened housing tracts. "I have seen many an orange grove in Orange County go through the process of transformation from a good source of Vitamin C into streets and lots and, loftiest of all, man's castle," he wrote. "Often I would pass an orange grove in the morning and at nightfall I would see the bulldozers had uprooted the trees and piled them in wilted disarray in the middle of the grove." The changes Muhonen observed were part of a pattern of farmland clearance that was transforming not just Orange County but the entire state and country. Between 1942 and 1955, California lost approximately sixty thousand acres of farmland per year; soon after, that rate intensified to over a hundred thousand acres per year. By the mid-1960s, bulldozers were toppling groves across California at the rate of one orange tree every fifty-five seconds. Nationally, cropland declined at a rate of more than one million acres per year from 1949 to 1969. New suburban residential, commercial, and industrial development often followed.[10]

Although a rising chorus of critics would lament this transformation, noting its negative consequences for the environment, food production, local community, and perhaps even the economy, many early observers accepted the lost groves as the necessary first step in the move toward modernization, progress, and profits. Muhonen's perspective was somewhat ambiguous, but many other journalists were blatant in their praise. Headlines like "Sylmar Olive Groves Give Way to Progress" and "Subdivision Blooms in Orange County" typify the positive sentiments. Government documents also reflected this optimism. In 1954, for example, the county formally began surveying the growth of housing, population, and industry within its borders. This annual report's title, *Orange County Progress Report,* suggested not-so-subtle political support for the process.[11]

These developments continued a long tradition of technological cultivation of the southern California landscape. The region's citrus history dates back to early-nineteenth-century Spanish missions. By the end of that century, economical irrigation and rail transportation had helped the industry take off. Boosters adopted the imagery of citrus crate labels to promote southern California as an idyllic paradise, and new arrivals soon followed. The typical ten-acre orchards on which they settled were small enough for landholders to pursue the self-sufficient agrarian ideal. The erection of their homes upon these same properties helped the new residents establish a "middle landscape" that was at once rural, pastoral, and linked to urban services and culture.

Over time, however, the reliance of commercial citrus cultivation on technology, scientific research, and migrant labor belied the widely circulated images of the land's natural fecundity. Agriculture required large-scale land clearance of its own, particularly in connection with the development of irrigation systems and land-reclamation projects. When postwar developers set their sights on these same properties, they were accelerating a longer pattern of landscape development. Through the use of more modern machines and methods, they cleared away the old growth of agriculture, woodlands, and other natural environments to impose new, built landscapes on the southern California garden.[12]

Changes to Orange County's orange tree population demonstrate the scale of the region's transformation. At the start of World War II, the county had almost five million orange trees—more than an eighth of the nation's total. The Valencia orange, a summer-ripening fruit that had originated in China, was the primary driver of Orange County's dominance. Valencia trees occupied more than 80 percent of orchard real estate. These fruit trees reached their peak acreage of 66,585 in 1945 before the rapid decline began that, by 1954, would dethrone Orange County from its position as the nation's top orange-producing county. (It dropped from first to fourth.) Between 1950 and 1960, Valencia acreage halved, from 59,393 to 30,398 acres. It nearly halved again during the following decade, when acreage dipped to 16,473. By 1980, Valencia orange trees occupied a mere 5,500 acres, and today, at roughly 85 acres and with virtually no new trees planted in decades, the fruit is almost entirely absent from the county.[13]

Land clearance removed not only orange acreage but also vast quantities of lemons, avocados, and grapefruits, among other crops. Commercial crops like walnuts, persimmons, and olives had disappeared completely from the county by the mid-1960s. Only strawberry and tomato acreage increased during this period. Newly introduced crops were equally rare, limited to mushrooms, turnips, and parsnips. While improved growing techniques—from mechanization to pesticides—simultaneously increased the productive capacity of the remaining land, these advances did not compensate for the massive loss of fruit acreage.[14]

Residential suburbs accounted for the majority of the lost orchard land, while associated urban and suburban uses—such as businesses, industries, roads, freeways, schools, and churches—occupied much of the remainder. For prospective developers, southern California's orchards and ranches had many appealing attributes. Their close proximity to Los Angeles, with its growing metropolitan network of freeways, drew would-be homebuyers looking to commute from the suburbs to the city. The good soil, water, and climate that favored agriculture offered psychological allure. In addition, the location satisfied developers' wallets. By connecting to existing streets, sewers, and utilities, builders could save substantially on associated infrastructure costs. The large scale of these parcels provided yet another incentive by easing the logistics of land acquisition. At a time when individual holdouts could increase the cost and time it took to build, many developers pursued the single-owner properties that offered the most acreage for their efforts. Once acquired, the vast areas of relatively flat or gently rolling terrain minimized land-preparation costs. This land also allowed for easily choreographed construction of similar homes without having to make modifications for the eccentricities of more individualized tracts.[15]

As developers began acting on this real estate potential, they drove up land valuations. In the county's Garden Grove area, for example, prices soared from two thousand dollars per acre in 1952 to eight thousand in 1958. By the end of the decade, prices in some areas had doubled on top of that. The value of potential commercial property grew even more rapidly. For some orchard owners, easy land sales seemed like a great opportunity. But for those who wanted to stay in business, escalating land assessments meant higher property taxes. In aggregate, taxes were growers' largest single expense. Between 1950 and 1965, taxes on California's agricultural land grew from 6.7 percent to over 15 percent of net farm income. As Florida growers avoided these escalating expenses while also enjoying cheaper land and lower maintenance costs, they further challenged the economics of the citrus market in which Orange County growers struggled to survive.[16]

Other challenges befell postwar orchard owners. Growers complained that suburban land clearance caused dust storms. They criticized new suburban residents for trespassing on their property, damaging and pilfering crops, and lobbying to restrict farming practices like the use of pesticides and the heating of orchards. Suburban development also brought damaging pollution from factories and automobiles into close proximity to crops. This pollution reduced overall yields, diminished individual fruit size, and spurred defoliation. Age-old natural enemies took a further toll. The freeze of 1948–1949 struck as developers were setting their sights on the county, nudging some growers toward land sales. One former orchard resident recalled the consequences of the freeze: "That morning after the snow, my Dad took movies of all of us gleefully playing in the snow, not realizing at the time that we would be losing our way of life pretty soon. There was another big freeze in 1949 as well, but our's [*sic*] were ruined. Oranges were stripped from the trees quickly to hopefully salvage some of them to send to the packing house. After that the trees were all pulled out, and the part of the property was sold off and subdivided." Finally, by 1960, almost all the remaining citrus trees in southern California had been infected with the quick-decline virus, the treatment for which typically required uprooting an entire orchard.[17]

Groves often were left dormant during the land-acquisition phase. Growers sometimes ceased orchard operations or neglected their groves while waiting for a buyer. Similarly, some developers left newly acquired lands untended until they were ready to begin construction. By 1961, decaying orchards had led one Tustin resident to write a letter to the *Los Angeles Times* urging local officials to take action lest the dead trees create a fire hazard for surrounding properties. These unmanaged properties also became host to undesirable activities, including vandalism, off-road vehicle use, and underage drinking. In their physical embodiment of abandonment and neglect, the rural landscapes resembled contemporaneous urban neighborhoods, which were then awaiting their own postwar redevelopment.[18]

From Grove to Grove Street

Once developers were ready to work, they cleared the orchards quickly and inexpensively. As Muhonen observed, an entire grove could fall in a day. According to the trade magazine *California Citrograph,* during the mid-1950s the complete land-clearance process—including tree uprooting, refuse removal, and land grading—typically cost developers about a dollar per tree, or roughly a hundred dollars per acre. This was a tiny part of the total development costs, making the destruction of orchards an economical—and consequently, almost forgettable—step in the home-building process. In fact, William Garnett's famous aerial photographs of the construction process at Lakewood, California, omit this stage entirely. Crawler tractor tracks appear in the first frame of the series, suggesting that the machine had passed through. But the images do not depict the clearance of the site's bean fields. While such a scene would have added to the survey's completeness, its absence is unsurprising. Starting with destruction would disrupt this developer-funded documentary project's intended portrayal of suburban construction as unambiguous progress.[19]

Multiple pieces of tractor-based equipment assisted with tree removal. This task, also typically undertaken in small areas to remove diseased or unproductive trees, was a regular part of orchard maintenance. A grove worker might affix a bulldozer blade to his versatile orchard tractor, or he might saw off trees at ground level using a brush cutter and chain saw. A tree cutter

The first in William Garnett's series of aerial photographs of Lakewood, California, from 1950 reveals construction vehicle tracks on newly cleared bean fields. (William A. Garnett, "Grading, Lakewood, California," 1950, gelatin silver print, 18.9 × 24 cm, The J. Paul Getty Museum, Los Angeles. William A. Garnett © Estate and Heirs of William A. Garnett.)

mounted on a bulldozer was even better suited to the task of clearing trees and brush, although it would leave stumps and roots in place. Other implements, like Caterpillar's treedozer, could remove entire trees, from branches to roots. This attachment consisted of a toothed horizontal beam placed in front of a bulldozer or V-blade. After the knockdown beam bent over the tree's trunk, the plow would dig under the roots and lift the tree out of the ground.[20]

Weldon Field, an Orange County, California, mechanic, tinkered with his own Caterpillar crawler tractor to construct a stump puller tailored to uprooting southern California fruit trees. The stump puller consisted of a tractor engine, a V-shaped pulling implement, ten cables, and connecting gears. Once the tree was caught within the steel V, the machine's approximately

Weldon Field uses his tractor-mounted stump puller to uproot an Orange County orange tree. (Photo courtesy Orange County Archives)

sixty-ton force lifted it straight up from the ground. Then the operator would use a shovel and chain to pull up any roots that had broken off during the process. Beginning in the late 1930s, Field applied his stump puller to the removal of trees damaged by disease, gophers, or frost in Orange, Los Angeles, and Riverside Counties. The postwar period offered even greater opportunities for his stump puller. As Field recalled, "I pulled the old seedling walnuts when they were taking out the walnut trees around here and putting in citrus. I pulled a lot of those with it. Then eventually I pulled the citrus trees out for houses."[21]

Tree removal created a need for refuse disposal. Since labor costs made salvaging the wood uneconomical, some builders burned the pulled trees on-site. In an interview, Field recalled one landowner's attempt to burn his trees on a clearance project in Capistrano. When spraying the trees with orchard heater oil succeeded only in burning off the leaves, Field offered the landowner some advice. "I said, 'Go get yourselves some old tires and stuff and stick them in underneath those and get them on a bed of coals, and then you can get going.' Well, they finally did. Of course, the law now, you can't use old tires; it makes too much smog, but at that time, they didn't bother with that. They finally got them burned up." By the mid-1960s, the burning of pulled trees was so widespread that Orange County was issuing an average of five hundred burning permits per month.[22]

As Field's reflections suggest, the environmental consequences of all this burning soon raised concerns. Orange County began establishing "no burn days," and by the end of 1967 county officials had banned open burning entirely. Still, this regulation did not eliminate the practice. One builder who needed to clear twenty acres of orange groves for construction of a drive-in theater in nearby La Verne opted to pay a fine rather than incur the excessive costs of debris removal. An associate of his recalled, "The developer called us, and he says, uh, 'We're gonna wait til a foggy morning, and I'm gonna take a tanker truck in there with diesel fuel. We're gonna spray all those piles with diesel fuel, and set 'em on fire.' They did it and paid the bill. $500."[23]

Sometimes builders practiced selective clearance, preserving small numbers of mature trees as landscaping for future homes. In San Mateo, for example, the developer Bill Holden subdivided an orchard to maintain six to eight fruit trees on each lot. Such design decisions catered to homeowners' visions of buying into California's citrus landscape, even as their arrival ultimately destroyed it. Orchards with widely spaced walnut trees were especially well suited to selective salvage—so much so that one observer cynically noted that they "virtually demanded invasion." Even across the country, on Long Island, Bill Levitt imported up to four fruit trees for each of his suburban properties. California developers enjoyed practical and economic advantages by preserving similar landscaping already present on their land.[24]

House and Home magazine encouraged the preservation of trees, whether on orchard land or elsewhere, as a best practice in postwar homebuilding. Trees could shade, beautify, block noise and objectionable views, maintain property values, and project a sense of stability, history, and community, from a development's earliest days. On less stable soil, trees and root systems could also help the earth cohere. The writers feared that too few builders were preserving trees on their construction sites, instead choosing "to bulldoze [the land] down to the raw earth," and lamented that builders would often replant with "invariably buggy-whip saplings that will take years to achieve effective size." Although Lakewood, California, was notable for its inclusion of street trees, these imports covered over the destructive land-clearance process that had spared few of the older trees. William Garnett witnessed these practices through his photography work. "I was hired commercially to illustrate the growth of that housing project," he recounted. "I didn't approve of what they were doing. Seventeen thousand houses with five floor plans, and they all looked alike, and there was not a tree in sight when they got through."[25]

Long-term landscape conservation required better stewardship than selective preservation and new plantings. Builders who scraped and stock-piled topsoil for redistribution on completed lots, for example, would often deplete the soil of nutrients through inappropriate interim storage. In addition, fruit trees sometimes struggled to survive outside the nurturing orchard environment. One *Los Angeles Times* columnist observed in 1948: "Three years ago you bought a lot in a new subdivision of an old walnut or citrus orchard. . . . But now your trees are dying. At best you face the prospect of withered and misshapen trees. At worst your property will be denuded. Your dream trees are gone. So, you ask, why?" As the author went on to explain, subdivision development had destroyed the orchard's irrigation pipelines and canals, and now houses and pavement obstructed the absorption of groundwater. Homeowners were also likely to irrigate inappropriately, creating a lack of water in some cases and the onset of fungal disease in others. In addition, if homeowners did not protect their ornamental trees against pests, then they could pose a threat to remaining nearby commercial orchards. Thus, even when some suburban clearance appeared more selective than absolute, reconstituting the natural landscape proved a larger challenge.[26]

The Backlash

Cultivation of new orchard land in the state's Central Valley, combined with improved growing techniques, helped California maintain its total orange production over time. Yet these statistics tell only part of the story. In the two decades following World War II, the state lost roughly one-seventh of its highest-grade soils to urbanization. Creating replacement farmland from more marginal lands meant the sacrifice of wild lands and substantially increased agricultural costs. In Ventura County, for example, farmers established replacement orchards on steep hillsides with shallow, rocky soils, resulting in higher irrigation and erosion-control costs, more time-intensive pesticide application, and more dangerous labor practices. Terrace agriculture had prospered in California since at least the 1920s, as farmers looked to the hillsides for frost protection. Postwar advances in earthmoving made these terraces both more prevalent and more expansive.[27]

Orchard migration led to a decline in overall Valencia orange production and raised fears about other specialty crops for which California was the country's sole domestic supplier. More locally, the departure of the citrus industry from Orange County transformed the community. As growers' associations closed down, jobs departed, leaving abandoned packinghouses in their wake. Long part of local culture, annual citrus festivals and growers' conventions also disappeared. The loss of orchards terminated not only a regional industry but also a local way of life.[28]

Given these rapid transformations, it should not be surprising that many rural residents failed to celebrate this change as progress. Some pulled up surveyors' stakes in futile attempts to derail plans for new tract houses. As one resident wistfully recalled, "We nearby kids were not happy to lose our orange grove, and our tree house, high up in an avocado tree." As sites awaited construction, children often coopted the interim landscapes as temporary play spaces—which they were no less ready to forsake for the brevity of their existence. Even worse, injuries sometimes resulted from the children's occupation of the construction zone. In one tragic instance, a bulldozer that was clearing a Los Angeles County orange grove for a new housing project accidentally scooped up and critically injured a three-and-a-half year-old boy who had been playing there.[29]

To these physical and emotional victims of the orchards' demise, the rapid suburbanization of farmland was akin to the plant diseases they were accustomed to combating—a veritable "expansionitis" or "subdivision fever," as one farmer put it in *California Farmer.* But the new condition was not easily treated. "This one can't be sprayed, vaccinated for, or eliminated in the ordinary ways that physical diseases are controlled," he observed. "Yet, in many respects, it is as deadly to agriculture as a true disease." Relatively elite groups, from homeowners to equestrian clubs, offered more cultural and economic rationales for resisting development. In the interest of advancing their property values, they advocated for rural preservation grounded in carefully crafted—but racially exclusionist—myths of an enduring western heritage.[30]

Concerns over lost farmland and open space eventually spurred reforms. Individual localities proposed—and sometimes enacted—limits on municipal annexations, public purchase of susceptible properties, increases in local land-use controls and zoning, right-to-farm laws, and changes in the methods of assessing farmland. When these local- and county-level reforms achieved limited success, attempts at regional planning ensued. A coalition between planners and fruit growers, based largely in northern California's industrializing Santa Clara County, lobbied for the passage of the 1955 Greenbelt Act, the first successful state-level regulation. The Greenbelt Act restricted cities from annexing lands with exclusive agricultural-use zoning. The act's influence was limited, however, by an amendment restricting its application to places where the master land-use plan at the end of 1954 included an exclusive agricultural zone classification.

The act's lack of influence on property taxes and assessments presented another challenge, as these were two key factors behind the land-development pressure. Several amendments to the Revenue and Tax Code sought to address this. These amendments required that certain categories of land—including agricultural lands, airports, and recreational areas—be assessed based on their legally acceptable uses rather than future potential uses. This effort also proved ineffective, as assessors failed to treat the restricted uses as permanent. A third attempt at open-space preservation came through the Scenic Easement Deed Act of 1959, under which cities or counties could acquire real property on which to enforce restricted use. But the absence of public funds (for purchases) and tax incentives for landowners also limited the impact of this act.[31]

After ten years of active debate, the state passed the more influential California Land Conservation, or Williamson, Act of 1965. This act enabled property owners to partner with city and county governments to restrict parcels of land to agricultural use for a renewable ten-year period. In return, landowners would pay reduced taxes based on the income produced by these rural uses, rather than on the land's maximum development potential. In

1966 the act's protection expanded to include nonagricultural open space as well. Changes over the succeeding years led to legislation for creating revised assessments (that is, forbidding sales data–based valuations and replacing assessor discretion with procedural guidelines) and expanded the types of land eligible for protection (wildlife habitats, highway corridors, and recreation areas). By 1980, the Williamson Act was protecting over 16 million acres from the bulldozer.[32]

Even though other states implemented legislation similar to the Williamson Act, none of these policies provided a long-term solution to the challenges of suburban growth. The act established agricultural preserves for ten-year time spans. After that development could continue as before. In addition, by protecting only selected tracts, the act placed no limits on total rural development. Instead, it shifted development from one rural area to another. In addition, little of the protected property included prime agricultural land on the urban fringe, where landowners often opted to forgo short-term tax benefits to maintain control over the immediate use and potential resale of their land for urban purposes. Thus, few of the lands conserved by the Williamson Act may have actually been threatened at all during their short period of protection.[33]

One byproduct of these reforms was to push suburban development farther into the hills. In defense of their orchards and groves, farmers had been encouraging builders to refocus their sights on other natural landscapes. The headline of a letter published in *California Farmer* in 1953 summarized this position: "Put Houses, Not Farms, on Hillsides." In the wake of the Williamson Act, as the availability of central, relatively flat land dwindled, and as the land that remained escalated in price, such prospects proved more palatable. An unintended consequence of land "conservation" was to make large swaths of undeveloped, less expensive hillside land more attractive to developers. In the absence of affordable large-scale hillside land-clearance practices, it had not made physical and economic sense to develop such sites before. Now, thanks to advances in earthmoving technology, the increased use of engineers and geologists as homebuilding consultants, and the spread of best practices in manuals published by trade organizations like the National Association of Home Builders (NAHB), this was no longer the case. As the association prophetically declared in 1969, "There is hardly a site that cannot be developed to serve man."[34]

Bringing "New Life to the Lands of Anaheim Hills"

A full-page advertisement appearing in a January 1973 issue of the *Los Angeles Times* promoted the first of several developments that would make up the new southern California community of Anaheim Hills. The text proclaimed, "For centuries, this acreage has lain untrod and unenjoyed. It may be part of Orange County's only Spanish land grant, but it's never been part of the experience of any but a few people. It's been locked-up and shut-off for years and years. Now, all of that is changing for the better. . . . Anaheim Hills, Inc., has launched what may be the boldest plan ever for the development of this land. It is a plan that from its inception, foresaw the opportunity to transform the fallow hills by adding to them the critical elements of water—and people. Water to bring the earth to greenness. People to give it life and meaning." Located thirty miles south of Los Angeles on the eastern fringe of Anaheim and in the unincorporated area outside Orange, the Anaheim Hills project spanned 4,200 acres. The developer envisioned it as eventually containing commercial and civic infrastructure, thirteen thousand new homes, and a population of over forty-five thousand residents. A 236-acre municipal golf course, equestrian center, lakes, parks, and hiking trails would fill the rest. As this ad went to press, the first subdivision of three hundred houses, known as Westridge, was opening for viewing.[35]

Although the scale of Anaheim Hills was typical of the developments built in the initial postwar decades, its topography was not. True to its name, Anaheim Hills was a foothill development, where rocky, wild terrain contrasted with the well-ordered and flat rows of orchards-turned-houses located below. Developers sited this suburban housing upon land that logistics and economics had previously deemed unbuildable. But the project was not the first hillside development to arise. During the 1950s and 1960s, pioneering builders carved out more than sixty thousand hillside home sites in Los Angeles County alone. These projects, however, had primarily involved the construction of hill-specific homes atop inclined terrain. These often incorporated multilevel living and staircases for accessing spaces both inside and out. Relatively low densities permitted builders to fit buildings to existing terrain without having to grade extensively. Yet by the late 1960s and early 1970s, builders were leveling the hilltops for denser, more uniform development in a process dubbed "mountain cropping." Their environmental modifications were more extreme: instead of adapting homes for the topography, they scraped away the mountains to create vast flat pads to accommodate the repetitive building tracts characteristic of low-lying, level lands.[36]

The Anaheim Hills development exemplified these later postwar practices, reflecting not only technical advances but also recognition that the developers had to take into account the emerging environmental movement. It was not enough for Anaheim Hills, Inc., to suggest that houses were better than undeveloped land and that growth was good for its own sake. That was a story builders had been telling for decades. With the Williamson Act in effect for five years at this point, and environmentalism more prominent, developers and their marketers felt obligated to engage more fully with their site's natural landscape. Only in this way could they hope to prove that development and nature were complementary, rather than adversarial.

With a sleight-of-hand that obscured the extensive environmental transformations inherent in the project, the Anaheim Hills advertisement focused more on the growth of trees and enhancement of the landscape than on the erection of buildings. It posited that untouched land—"locked-up and shut-off for years and years"—was wasteful, undemocratic, and incomplete. In the absence of people, the "fallow hills" were falling short of their potential. Without larger supplies of water, they were also being deprived of their latent natural aesthetic. Anaheim Hills, Inc., privately estimated that subdividing the hillsides would require substantial land clearance. On roughly half of the terrain, the company expected to topple scores of trees, move tons of dirt and rock, and grade thousands of construction pads—all at a cost of more than $100 million. Once in progress, earthmoving for the project proceeded at rates of up to fifty-five thousand cubic yards of earth per day. But the ad made no mention of this forceful destruction of nature, or of the visual blight—in the opinion of some—that the construction process and its resultant housing imposed upon the natural landscape.[37]

Instead, the advertisement suggested that development improved upon nature. "Each new home means more new trees, more shrubs, more green lawns, more beauty," the ad read. "Beauty is becoming a reality where it once barely existed." While the greenways and outdoor recreations included in the Anaheim Hills master plan were innovative, describing residential grounds as more beautiful than unbuilt hillsides was a subjective judgment at best. It prioritized orderly built development over the disorderly—and, therefore, unbeautiful—natural environment. The ad concluded, "At Anaheim Hills it is our intention to enhance the land in ways that bring forth the land's best." Or as the architect Hugh Hardy put it several years earlier, regarding earth-moving in general, "You can very easily make the point that we're messing up the landscape—ping, ping, ping—but if you drew back and saw it in a large enough scale, we're providing more landscape, as man and nature have always done." From these perspectives, postwar bulldozers were not razing unspoiled landscapes but discovering "new land" and perfecting it in a way that Mother Nature had not.[38]

The landscape transformation at Anaheim Hills demonstrated the pitfalls of converting hillsides into habitation. The clearing and grading of soil on level orchards paled in comparison with this mammoth endeavor. Given the relatively new nature of hillside development, local regulations and oversight often lagged behind practice. They placed few constraints upon the builders' pursuit of profit. In the hills, the machines were more powerful, the labor more intensive, and the changes more dramatic. And—as one might expect—the collateral damage was more substantial as well, particularly for the environment. The opponents of agricultural clearance continued to achieve only partial success in winning over the public, as detractors argued that farms could be relocated and that the food supply had not suffered much from the decline of prime acreage. The case for more tightly controlled hillside clearance, however, was harder to ignore. The physical destruction wrought by its excesses was more visible, even if the extent of the loss revealed itself only over time.[39]

The Mechanics of Moving Mountains

A 1958 advertisement in *Southwest Builder and Contractor* magazine features a southern California contractor praising the superior capabilities of

LeTourneau scrapers for grading hillside development. He notes, "Our work is mainly subdivisions with terraced lots—LeTourneau C Fullpacks have a great advantage on account of their maneuverability. Our operators also like their good visibility, easy loading, and ease of operation. They are great earthmovers." The rest of the copy elaborates upon the stable machines' skill at "moving big yardages fast," and at high speed. LeTourneau had designed the Model C Tournapull pictured in the ad on the eve of Pearl Harbor, and the machine had proved to be an important earthmoving tool for the Allies during World War II. Owing to strong and enduring demand for large-scale earthmoving, sales continued to grow after the war. By the mid-1960s, scrapers could travel over forty miles per hour, carry up to seventy-five leveled cubic yards, and slice layers of earth more than a foot deep. Three main operating parts made this possible: a bowl with a cutting edge for digging into and carrying the earth; a rotating apron to form the front of the bowl; and an ejector to form the bowl's rear edge and to push materials horizontally over the cutting edge and out of the bowl.[40]

The bulldozer complemented the scraper by providing the engine to pull nonmotorized scrapers across the land. Even on its own, the dozer's powerful engine and varied attachments could clear and grub the land, remove and later respread topsoil, and rip and transport earth and rock short distances. On hillsides, in particular, the bulldozer was well suited to digging out benches in bedrock. When equipped with a slope board at its side, it could also cut at a slope. Alternative attachments facilitated other maneuvers: angle dozer blades pivoted to the side, and U-blades assisted in moving large volumes of material with minimal side spillage. Together, the bulldozer and scraper were the two pieces of equipment most critical to postwar earthmoving work.[41]

Earthmoving machinery advanced materially after the war. While many of the underlying technologies had existed prior to the conflict, it was only afterward that manufacturers regularly integrated them into their commercial products. Hydraulic controls became increasingly common, providing

---

(*Opposite*) Earthmoving machines appear "slicing away at hills" in Los Angeles in 1959. As *Life* reported, "In the week ahead, if the weather happens to be good, another 10,000 acres will go under scrapers." (Leonard McCombe/The LIFE Picture Collection/Getty Images)

A tractor with a blade attached at front pulls a scraper on a 1965 construction training site. (Cloyd Teter/Denver Post/Getty Images)

more power, faster responses, and smoother cycling between motions than had previous, cable-based operations. By the 1970s, new cable dozers had become rare. New metallurgy made the machines more durable and stronger while remaining light. Engines increased in power and speed. The postwar introduction of torque converters helped match that power to the load in order to maintain speed. Pairing torque converters with power shifts in implements like front-end loaders helped prevent clutch slipping while also improving flexibility. Meanwhile, the concurrent development of smaller machines for selected applications offered more versatility and the opportunity to apply mechanization in relatively confined spaces. Up-and-coming manufacturers like Bobcat supplied the market with some of these smaller machines. Finally, as the performance of any piece of equipment ultimately relies upon the performance of its operator, postwar machines improved the human experience as well. Cabs increasingly offered cool air, cushioned seats, noise blocking, and more responsive controls. The incorporation of foot-operated steering freed the operator's hands for the many other functions of these increasingly complex machines.[42]

Even the deceptively simple bulldozer blade received close research and design attention during this period. A 1951 Caterpillar marketing publication titled "Let's Talk Bulldozers" detailed 13 separate blade attributes that contributed to its effectiveness. These ranged from braces, which allowed for quick and easy tilt, high-tensile steel construction, and ball-and-socket joints for precision. The company justified its close attention to detail: "No other tool affects the earthmover's cost records so much as does the bulldozer." In addition to surveying the range of the company's blade line, the pamphlet introduced a new model, the 8U, designed for versatile and strong long-haul dozing. The 8U joined 26 existing Caterpillar blades and 277 competitor blades also on the market, whose list prices ranged from a thousand dollars to over four thousand. With stiff competition among at least a dozen different manufacturers, the postwar marketplace demanded constant innovation if a company wanted to distinguish its products from those of its competitors.[43]

The advances across a range of products made the 1960s a period of "unprecedented growth," in the words of the construction equipment expert Eric Orlemann, in both the size and power of this machinery. Stimulated by the demands of the era's many large-scale projects—including highways, dams, mining, and other construction—manufacturers focused their design and production on several much larger machines. Caterpillar, Euclid, Allis-Chalmers, and LeTourneau (with its aptly named "Goliath") produced their largest "monster scrapers" during this decade. As maximum crawler tractor weights increased from about twenty tons to more than fifty between 1955 and 1976, these vehicles expanded what was physically and economically possible. Although many equipment companies participated in the trend, Caterpillar led the way. In 1959 it began rolling out its brand-new 600-series motor scrapers, several of which would form the core of the fleet of equipment at Anaheim Hills.[44]

Les McCoy & Sons, one of the excavating companies on that project, put these machines to work. The company owned only Caterpillar products. Like many contractors, the firm felt that a relationship with a single dealer would minimize work stoppages while awaiting replacement parts. It was also easier for mechanics to service equipment from just one manufacturer. The firm owned most of its equipment outright, since it could write off roughly two-thirds of purchase costs through depreciation allowances in the first two years. McCoy's mechanical fleet included seven of Caterpillar's top-line two-axle scrapers (with 900 horsepower), three of the largest standard scrapers the company had ever produced (with forty to fifty-four cubic yards of bowl capacity), two smaller scrapers, four crawler tractors, two wheeled tractors, and a single D7 track-mounted medium bulldozer. The D7 line had been the workhorse of the U.S. military for World War II earthmoving operations (during which it was called the M1 heavy tractor), but postwar earthmoving companies like McCoy's gave the dozer civilian use as well.[45]

As equipment steadily increased in power and capacity, the cost of moving dirt effectively held flat. Between 1950 and 1960, Caterpillar's tractor-scraper combination increased its horsepower by 50 percent, from 225 to 345, and its capacity by 30 percent, from 15 to 19.5 level cubic yards. International Harvester experienced similar advances. In 1965, its largest machines generated 50 to 100 percent more productivity than they had in 1956. The company also noted that these improvements were obtained "with greater comfort and less operating skill and effort" than in machines of the past. These gains more than made up for the growth in the hourly ownership costs of this equipment. As a result, the work of earthmoving—grading, filling, and compacting—was no more expensive in 1960 than it had been three decades earlier. As *House and Home* proclaimed, "New machines are so efficient that grading, filling, and compacting are the only homebuilding costs that are lower today than in 1929."[46]

But economy was not synonymous with cheapness. The relative inexpensiveness of earthmoving derived, in part, from the machines' productivity, which offset the dramatic growth of labor rates since the war. Moreover, these new machines added power without comparable increases in heft, which would have limited operators' ability to move them around home sites. In the first dozen years after the war, the largest crawlers tripled in power, while only doubling in weight. The architect Philip Johnson articulated the psychic appeal of this work and the machines behind it. Specifically, he called the bulldozer "the tool of our day" and elaborated, "The bulldozer is the only thing that hasn't gone up in price—the way crafts have. You see, crafts have disappeared; good materials are too expensive to use; and you think right away of how to cheapen them. One of the luxurious ways and one of the American ways, due to our great road-building programs, is the bulldozer. And that is a perfectly natural way to think about it. That is one element. This is a romantic idea. I am not sure that it is cheap at all." While acknowledging the significance of earthmoving's low price to its postwar ascendance, Johnson recognized the bulldozer's less tangible appeal as well.[47]

For other observers, the allure of earthmoving work derived less from romance than from excitement and awe. One journalist writing in *Nation's Business,* for example, likened excavating machines to the dinosaurs: "Spread out before us we saw a panorama of devastation, where a vast herd of modern excavation machines roared, gouged and roamed like prehistoric monsters grazing." An operator commented on the "fun" of operating a shovel. "The big machine gave me a sense of power," he told the reporter. "I, a man of average size, could walk up to a mountain—and move it!" A writer for *Landscape Architecture* magazine captured similar euphoria in her depiction of an earthmoving contractor at work in the hills of San Francisco. "'You ought to see my "cats" move that dirt!' declared the proud engineer, his eyes shining across the table at the slightly less enthusiastic planners."[48]

This characterization of proud earthmoving men fit with widely held stereotypes of the profession. As one wartime editorial in *Southwest Builder and Contractor* noted, "Americans who operate a construction machine get their greatest thrill out of power and speed which responds to their will and slightest touch. One of the most intriguing sights on a big construction job is the sangfroid of the operator of a powerful tractor which pushes aside huge rocks, levels trees and climbs precipitous hills." In 1960 a writer for *Architectural Forum* further observed, "Where contractors in other trades are usually cautious conservatives, the earthmoving entrepreneur, often up-from-the-ranks himself, is generally an open, energetic individual with something of the air of a river-boat gambler about him." Postwar earthmoving work derived its appeal not just from its affordable price tag and completed projects but also from the innate thrill felt by those involved in the process.[49]

Despite such simplistic adoration, rigorous technical skill underpinned much large-scale postwar earthmoving work. In reference to the bulldozer, a 1957 article in *Architectural Forum* observed, "In the years ahead the real measure of the big blade will be not how much it can cut, but how well."

The quality of that performance was a product of both man and machine. A large number of builders were able to look to the hillsides for postwar building, owing to the application of these tools by growing ranks of skilled earthmoving operators following the techniques of increasingly engaged geologists and soil engineers. This period offered opportunities for real-world experimentation that facilitated the rapid advancement of a previously underdeveloped field.[50]

The creation of equipment-operating schools during the postwar years ensured the availability of skilled machine operators. One such facility was the National School of Heavy Earthmoving Operation, established outside Charlotte, North Carolina, in 1955. On seventy-one acres of hilly soil, classes of approximately thirty students each spent four to eight weeks in a boot camp environment that focused on the art of cut-and-fill. Their ten-hour days consisted of two-thirds fieldwork and one-third lecture and observation. When not in class, students lived in bunkhouses and bought their meals in the mess hall. Owing to the Mormon background of the school's founder, strict rules rounded out the military-style experience. The school prohibited gambling, hard drinking, and smoking during class. Instructors also wrote up students who used "excessive profanity." Despite these potential deterrents to future enrollments—as well as the mechanical aptitude tests and character references required for admission—"Bulldozer U" (as one journalist called it) thrived. Because of interest from inspired applicants and students sent by contractors, most classes were filled to capacity, sometimes months in advance.[51]

Over time, similar programs emerged elsewhere. On 640 acres near Braidwood, Illinois, students at the Greer Earth Moving School were soon receiving their own education in the "ABCs of earthmoving." Individual equipment makers also ran their own schools for training operators and mechanics. In keeping with the demographics of most equipment operators at that time, the students trained in these programs were largely white. In fact,

---

(*Opposite*) In this scene from a 1957 issue of *Life,* students at Greer Earth Moving School in Braidwood, Illinois, learn the ABCs of heavy equipment operation. (Al Fenn/The LIFE Picture Collection/Getty Images)

when an earthmoving training program started in Alabama in 1973 targeted specifically at rural blacks, it was unusual enough to merit a profile in *Ebony* magazine.[52]

Plentiful postwar earthmoving work stimulated the growth of the industry. According to the Census of Construction Industries, between 1967 and 1972 national receipts for excavating and foundation work nearly doubled, from $1.65 billion to $3.05 billion. Single-family housing accounted for approximately one-sixth of the total, and almost all these receipts—housing related and otherwise—came from privately funded new construction projects. During this same period, the number of employees in the industry grew by a third, reaching nearly 115,000 workers during the peak summer season. Most companies had fewer than 5 employees and performed the bulk of their work in state. The small number of firms with 10 or more employees, however, accounted for more than half of total revenues. In 1972, nearly 6 percent of the companies were based in California, the state that generated the greatest volume of annual receipts (over $300 million).[53]

Contractors like Les McCoy & Sons, working on Anaheim Hills, were part of this trend. The company was a family operation, with Les at the helm, his brother as chief oiler, and three sons working out in the field. Mechanics, operators, foremen, and supervisors rounded out their crew. McCoy had been in business for himself since the 1960s, although his roots in the industry dated back two decades before that. Before founding his own company, he and his family briefly worked the land by farming. While some of the skills and equipment overlapped, McCoy ultimately chose to pursue contract engineering. It was the more reliable and profitable enterprise. By the end of 1973, McCoy & Sons was well into its third earthmoving project at Anaheim Hills. Collectively, on these three projects, the company had moved eight million cubic yards of earth.[54]

Around this same time, the National Association of Home Builders of the United States published its first extensive guidelines for hillside development. The NAHB had spun off from the National Association of Realtors in 1942, and it served as the main professional organization of the growing residential building industry. In 1950 the organization published *Home Builders Manual for Land Development,* which made only brief reference to project siting and grading. The manual advised readers to avoid excessive clearing, lot grading, and other practices that would "raise development costs, lower final selling prices, or both." When the organization expanded its manual into a more comprehensive volume in 1953, it devoted the first significant space to hillside development. The treatment was still relatively brief: three pages on the subject, lumped under the heading "Miscellaneous Problems in Land Planning." This limited attention probably reflected the less complicated terrain most often tackled then by builders. At that time, knowledge about residential earthmoving was minimal, relative, for example, to grading for highway work. By 1969, however, NAHB's subsequent *Land Development Manual* devoted multiple sections to the increasingly complex process of land preparation. The manual advised hillside developers to maximize economy while minimizing liability by efficiently executing five major stages in the land-preparation process: grading-plan development, clearing and grubbing, cut-and-fill, soil compaction, and landscaping.[55]

After a land survey and the creation of a topographic map helped establish a feasible grading plan, site clearance could begin. The leveling of trees, shrubs, and other growth was similar to the major clearance work required on former orchard land. For the wholesale removal of existing groundcover, a combination of two bulldozers and a ball-and-drag chain was an expeditious way to clear large landscapes quickly. This setup could topple small trees and pull out others with diameters of up to thirty inches. When necessary, controlled dynamiting or a tractor equipped with a stump splitter could dislodge even larger trees. Then a bulldozer would topple the tree before a tractor-mounted ripper pulled out the stump. The ripper also removed all organic and compressible material, including felled trees, peat, and soils with high void ratios. The topsoil would be stockpiled for reapplication after grading was complete.[56]

The real earthmoving occurred in the cut-and-fill phase. Bulldozers and scrapers dug into steep hillsides and relocated the excavated material to flat or below-grade areas. Whereas in 1955 the widely read excavating workbook *Moving the Earth* advised that dozers be used on hills of up to 20-degree slopes, the third edition of the workbook, published two decades later, had upped that range to 30 degrees. Crawler tracks helped the bulldozer grip the angled hillside, while varied attachments equipped it for the specific tasks involved. An angled blade or slope board assisted with terracing, while a

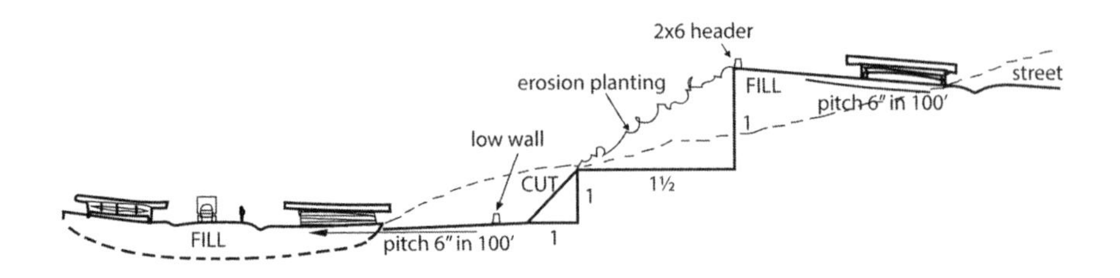

This cross-section view of the cut-and-fill technique, based on an instructional sketch published in *House and Home* magazine in 1953, shows how postwar builders turned sloped hillsides into stepped terraces suitable for new construction. (Drawing by Madeleine Helmer)

U-shaped blade could push large quantities of dirt. Rock removal required either drilling and blasting or the application of a tractor-mounted ripper. Over short distances of up to 150 feet, the bulldozer could transport dirt and rocks on its own. For transport of up to a mile, the wheeled scraper was more effective. Ideally, volumes of cut-and-fill from hills and valleys balanced out; when they did not, excavators had to import or export earth from the site by truck, sometimes at significant expense. As an example of the earthmoving volumes achieved during this period, on its second Anaheim Hills project, McCoy completed cuts of up to 100 feet in depth and fills of up to 150 feet in height. On an average day, their machines moved 35,000–40,000 cubic yards of earth—enough dirt to cover an entire football field, more than two stories high. Had complaints from valley residents about noise and lights not forced the discontinuation of their nighttime work, the contractor's daily volumes would have been even greater.[57]

Earthmovers could reshape the hillsides into several possible profiles. They typically produced a step-terrace form, with patterns of flat and sloped land alternating along the height. This configuration maximized the number of flat, buildable pads without the need to level the hillside entirely. It also helped slow water flow through a stepped descent. Streets generally followed existing contours; when they did not, retaining walls were necessary. The most common profile was a conventionally graded slope, where rectilinear shapes and unvarying gradients dominated. As an alternative, contour grading, or sculpting, disrupted some of the linearity of conventional grading by bringing some of the practices of golf course design to residential development. Curvilinear slopes produced a wavelike effect between flat top and bot-

Hillside grading has created a series of flat building pads upon which new suburban homes are being constructed in the Trousdale Estates neighborhood of Beverly Hills, California, ca. 1960. (UCLA Department of Geography, Benjamin and Gladys Thomas Air Photo Archives, Spence Air Photo Collection)

tom steps, upon which crews applied slope drainage devices and vegetation in a regular pattern.[58]

Beginning in 1973, Anaheim Hills became a key site for experimentation with a new kind of land shaping known as landform grading. In a technique developed by the Anaheim-based consulting engineer Horst Schor, excavators attempted to mimic the irregularities of the natural landscape by engineering asymmetrical sequences of undulations, peaks, and gullies. The goal was to achieve the inherent stability and aesthetic appeal of natural forms while still maintaining the hillside's capacity for buildable land. Positive byproducts of this approach included greater tolerances in grading work, wider variety in house lot shapes, and overall increased visual diversity of the ensuing built community. While landform grading was more environmentally sensitive, the entrenchment of more traditional design attitudes limited its widespread application. A developer might have to demonstrate this relatively new practice on experimental slopes in order to gain county engineers' approval. Then, in the field, he sometimes had to bypass grading contractors in order to educate equipment operators directly with drawings and photographs. As a bulldozer operator on a later California landforming project observed, this work was both challenging and rewarding. He told a *Los Angeles Times* reporter, "It's a lot more challenging, because you're not just going in straight lines. You just have to get set in your mind how it's supposed to look. If you've got a good grading plan and a good set of stakes to follow, then it's not a lot harder. But [it] sure is interesting work, trying to make it look natural."[59]

Following grading of this new profile, compaction helped stabilize the reshaped terrain for future building. Ninety percent compaction densities were common in highway design, and that standard carried over to residential work. Various pieces of equipment helped achieve this result. As heavy scrapers moved dirt from the cut to the fill site, their weight simultaneously began the compaction process. Water trucks added the liquid necessary for denser packing. For the completion of large-scale plateaus or building pads, excavators typically also ran heavy rubber or sheepsfoot rollers over the land in repeated sweeps.[60]

Drainage systems and landscaping finished the land-preparation work. Following a watershed study at the project's outset, excavators at Anaheim Hills, for example, had routed storm water through temporary drains for the duration of earthmoving work. At the conclusion of the project, they incorporated long-term drainage systems into the completed slopes, factoring in both upstream flows and downstream effects. In general, since the rate of erosion was at least two thousand times greater on construction sites than on healthy, vegetated sites, phased construction minimized the length of time when land was exposed to the elements. Temporary ground cover—from fast-growing rye and grass seed or manufactured coverings—provided short-term soil cohesion. Where it was possible to salvage trees, their root structures offered additional assistance. Or, owing to the rise of the landscape con-

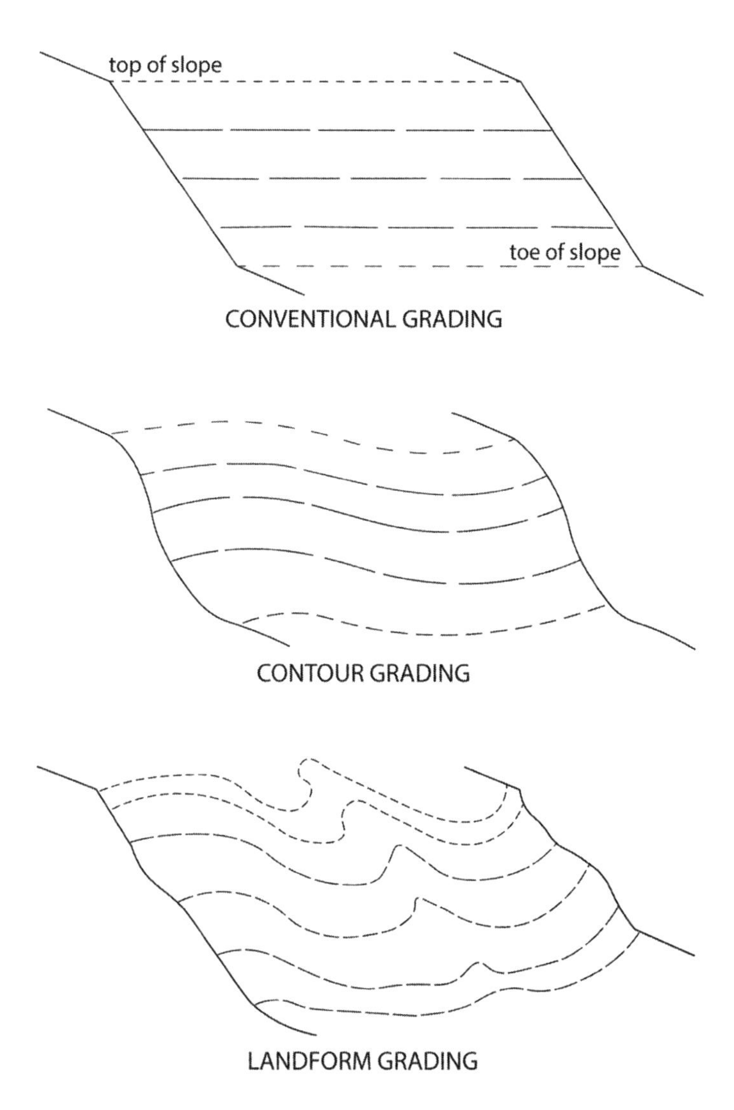

By the 1970s, in place of angular shapes and planar faces with unvarying slope gradients, some engineers were proposing compound shapes and variable slope gradients for hillside excavation. These slope profiles—ranging from rectilinear to more organic—illustrate evolving thinking about how to reshape hillsides for profitable, but also increasingly stable, development. (Drawing by Madeleine Helmer; adapted from Horst J. Schor and Donald H. Gray, *Landforming: An Environmental Approach to Hillside Development, Mine Reclamation and Watershed Restoration* [Hoboken, N.J.: Wiley, 2007], 187. Courtesy of Horst J. Schor. Copyright © 2007 by John Wiley & Sons, Inc. All rights reserved.)

Tractors pulling heavy sheepsfoot rollers, as depicted in this 1958 scene, helped flatten and compact soil in advance of future construction. (Nat Farbman/The LIFE Picture Collection/Getty Images)

tracting business with its own heavy equipment, site-specific trees might be removed, stored, and replanted by developers who had the foresight to do so. Planting more substantial vegetation after the final topsoil had been applied further stabilized the soil. Sometimes developers took care of this work themselves. On other occasions, landscaping fell to the homeowner, a newly formed homeowners' association, or even the benevolence of the local Boy Scout chapter. As time would show, however, replanting the denuded hillsides was not a luxury task to be tackled as an afterthought. Rather, it was a necessary step in combating some of the hazards introduced by repurposing the earth from the service of nature to the service of the builder.[61]

The Consequences of Hillside Grading

Just because earthmovers *could* move the landscape to build seemingly anywhere did not mean that they should. Unstable ground was the most catastrophic among a long list of concerns—physical, economic, environmental, and aesthetic—prompted by large-scale earthmoving and subsequent increases in development and population. Anaheim Hills was not immune to these problems. As communities began to experience the various unintended consequences, they worked to limit development, obtain compensation when damages occurred, and sometimes even change regulations that controlled hillside development in the long term.

Hillside development was difficult to fit into existing infrastructure, especially in cases where no future growth had been intended. Residents living just below the new properties had not anticipated having higher-altitude neighbors. Streets planned by earlier builders for more limited use now had to serve increased traffic—from earthmoving equipment to the cars of uphill neighbors. Nor were the requisite utilities in place for the new developments. The costly extension of these services to higher elevations increased taxes and other expenses on their downhill neighbors. It could also require additional land clearance. For all these reasons, groups like the Santa Ana Canyon Improvement Association tried—unsuccessfully—to block development at Anaheim Hills. Another oppositional group, the Orange Unified School District, filed a lawsuit claiming that the city of Anaheim had violated California's 1970 Environmental Quality Act by failing to consider and disclose the impact of new development on school facilities. Although this group was also unsuccessful, its unusual legal action demonstrated the expanding scope of environmentalists' concerns.[62]

Surrounding communities were also troubled by the Anaheim Hills project's more overt environmental impacts. A 1972 Environmental Impact Report estimated that Anaheim Hills would emit 14.5 tons of pollution daily, primarily from automobiles. These forecasts concerned residents of Riverside County, located downwind of the planned community, as their own officials estimated substantially higher levels of pollution (up to 94 tons per day). As one Riverside councilwoman lamented to the *Los Angeles Times,* "We can't seem to get your local governments to understand the consequences to us—as remote as we may seem—when you allow these big developments to go in." Air quality concerns prompted a state congressman to ask the Environmental Protection Agency to explore classifying the development as a "stationary" source of pollutants—a first for a housing project—which would put residential air conditioners on par with industrial smokestacks. Although he did not succeed, the EPA responded by more seriously considering the classification of Anaheim Hills as a "complex source," in recognition of its role as a direct generator of large traffic volumes, rather than pollution. In these ways, large-scale housing development at Anaheim Hills highlighted the absence of regional governance, particularly with regard to land use and the environment.[63]

Land clearance created mixed cultural consequences. At Anaheim Hills, a bulldozer flattened a quarter-mile-high peak formerly known as Robber's Roost (in recognition of its having served as a hideout for stagecoach robbers in the 1850s). The developer responded to environmentalists' complaints by paying a fine and attempting, unsuccessfully, to "restore" the peak. Other landscapes suffered when bulldozers destroyed artifacts buried under the dirt. As a silver lining, some workers safely uncovered Native American relics and the foundations of early California missions. Even the threat of land clearance could create an urgent need to fund advance exploratory digs. Anaheim Hills, Inc., for example, retained Archaeological Research, Inc., to examine the project's 4,200 acres in advance of development. Their survey uncovered one archaeological "site" and two artifacts: a hand-stone for seed grinding and a milling stone. Once the firm concluded that the property contained no archaeological deposits that would interfere with development, clearance continued as planned.[64]

Hillside clearance also had more immediate aesthetic and environmental consequences. As an article in *Landscape Architecture* noted, "The miles of ugly, bare, steep slopes are an undeniable eyesore, and the new home owners are baffled at the enormity of the job of landscaping them." One city official expressed similar concerns at Anaheim Hills, noting, "We all feel rather bad about some of the things they are doing to the hills." The dull vision of the completed housing did not always improve these perceptions. Anaheim Hills's first residential neighborhood, Westridge, populated the landscape with repetitive housing forms sitting atop stark, planar slopes. As the conservationist architect Malcolm Wells observed about this type of postwar development, "The simple fact remains there just isn't any building as beautiful, or as appropriate, or as important, as the bit of forest it replaces." A host of architectural and environmental critics supported this position, from William Bronson's critiques of "mountain cropping" in *How to Kill a Golden State* (1968) to Peter Blake's extended visual essay on the blight of postwar development, *God's Own Junkyard* (1964).[65]

These texts joined a panoply of environmental literature that emerged during this era criticizing the increasingly maligned suburban bulldozer. In 1959, William Whyte advocated for the "vanishing countryside" in *Life* magazine. Robert Cubbedge decried the rape of the land in his 1964 *Destroyers of America.* In *No Place to Play,* from 1966, Margo Tupper objected to the continued advance of suburban development into the vicinity of her own suburban home. Raymond Dasmann's *The Destruction of California* (1965) and Richard Lillard's *Eden in Jeopardy* (1966) added further California-specific treatises to this list. Notably, these critics included both outsiders who disdained the advance of suburban sprawl and suburbanites themselves—who viewed the nature of suburbia as something in need of saving as well.[66]

Suburban clearance and building practices sometimes fueled the brush- and wildfires endemic to southern California's hills and canyons. Some homeowners permitted hillside brush to grow wild in an attempt to create an aesthetically appealing environment. The combination of wood roof shingles, flammable growth, and dry, warm, high-velocity canyon winds could prove dangerous. Limited local supplies of water made battling these fires more challenging than on lowlands. A fire in the vicinity of Anaheim Hills in 1976, for example, spanned 2,280 acres, heavily damaged 22 homes, and displaced hundreds of residents. It ultimately took 400 firemen and 100 marines to contain the conflagration.[67]

Erosion and landslides were the most directly damaging environmental consequences of large-scale hillside development. Earthmoving for roads and buildings replaced the irregular forms of nature with the standard geometries of the builder, disrupting slope stability. In addition, lawn water, septic tank seepage, leaks from underground piping, and the stresses imposed by development could turn a normally firm combination of earth, rock, and shale into a slippery surface. This occurred in sections of Los Angeles shortly after hillside development had begun there in earnest. One particularly damaging slide, during a 1952 downpour in the county's Hollywood Hills, dumped rocks, mud, and assorted debris onto the streets and houses of newly built subdivisions. Later that year, residents formed the Federation of Hillside and Canyon Association, which passed the city's Hillside Grading Regulations. Theirs were the first such regulations in the country. They established maximum angles of slope, instituted a grading permit process, and required the review and approval of grading plans and inspection of completed work. Code enforcement and increased availability of geological and soil information led to dramatic improvements. While more than 10 percent of sites developed before the regulations' institution had experienced some landslide-related damage—on the order of three hundred dollars' worth per damaged site—the incidence was down to 0.15 percent (and seven dollars' worth of damages per affected site) just ten years later. Still, during the winter of 1961–1962, there were seventeen hundred reports of slide damage in the region.[68]

Although California became a leader in grading regulation—requiring the adoption of a minimum set of grading standards in localities across the state by 1974—landslides continued. A 1978 journalistic investigation uncovered shortcomings in both regulations and their implementation. Local authorities maintained significant discretion in incorporating scientific knowledge into their codes, and developers sometimes tried to thwart proposed changes. Even when stringent regulations were in place, inadequate code enforcement sometimes negated their potential benefits. Moreover, most cities had no geologists or grading inspectors on staff, so they had to rely on the

Following a 1958 rock slide in Glendale, California, the *Los Angeles Examiner* reported that young Rodney Wald unsuccessfully deployed his toy bulldozer in an "attempt to move [the] mountain of earth" that surrounded his home. (Courtesy of University of Southern California, on behalf of the USC Libraries Special Collections)

ones hired by developers. The consequences of the use of the builders' staff and associates included the certification of inappropriate land as buildable, insufficient compaction of fill, and the creation of slopes that were too close to construction sites. Since developers' liability for geological weaknesses lasted only ten years, by the time these physical problems emerged, the responsible parties were often in the clear. A 1979 grand jury investigation exposed these practices as part of a larger collusion between builders and government in the development of southern California's open space. The result, the investigation concluded, was corruption of the entire regional planning system.[69]

Anaheim Hills was not immune to the challenges of landslides. In January 1993, heavy rains caused the slumping of a twenty-five-acre hill slope. This slide affected over two hundred homes, whose owners sued the city. Anaheim blamed the movement of earth on residents' overwatered lawns and leaky swimming pools. Homeowners, in turn, faulted leakage from plastic water conduits while also challenging the city for knowingly building on ancient landslides without adequately informing prospective residents. Although the two parties ultimately reached a settlement, the homeowners' compensation did not cover their damages and lost property value. The incident exposed a loophole in California real estate disclosure laws, which had allowed developers and agents to sell homes to unwitting residents without fully disclosing the risks the homeowners were assuming. The slide also demonstrated that large-scale hillside earthmoving was not only a technical challenge but a political and social one as well.[70]

An irony of the environmental damages wrought by development-fueled fires and landslides was that they required the return of the bulldozer. These were the very same kinds of vehicles that had helped create the dangerous landscapes they now sought to save. Four bulldozers joined the firemen and marines who squelched the 1976 fire. Bulldozers also helped clear fifty-foot-wide firebreaks surrounding Anaheim Hills to prevent future fires. In this case, large-scale land clearance begot more of the same. Similarly, bulldozers helped clean up homes destroyed by landslides. On those occasions, their purpose was not solely to clear away nature from the land but to remove its built features as well. Thus, as the machine facilitated development that called into question the making of bulldozer-based landscapes, it created further applications for itself.[71]

When the bulldozer arrived to demolish damaged built fabric following a landslide, it signaled the end of development on a site where geology and construction practice had proven its folly. But in cities across the nation during this same period, the bulldozer was participating in demolition on a much larger scale as part of a new beginning, rather than an end, to postwar construction. Just as the clearance of nature was making way for suburban development, the triumphant demolition of existing city buildings was readying the land for urban renewal.

The machine's significance to the urban renewal project, however, was different. Advances in postwar bulldozer and scraper technology facilitated both practical and economical clearing of suburban landscapes, yet the machines often operated in the background of public consciousness. In cities the bulldozer's role was less technological than visibly symbolic. Certainly it knocked down buildings and pushed away rubble as a vital part of urban renewal's large-scale demolition practice. But torches, cranes, and wrecking balls were equally important. The program's physical accomplishments owed more to new policies and levels of popular support than it did to mechanical innovations. In the case of urban renewal, bulldozers made good metaphors, but they did not radically influence the evolution of urban clearance practices. Yet for all their metaphoric power as instruments of erasure—recall the giant hands in the 1944 Revere advertisement—they could not keep demolition from being a financially, socially, and logistically messy business.

Chapter 4

"Armies of Bulldozers Smashing Down Acres of Slums"

Urban Renewal Demolition in New Haven, Connecticut

On a January afternoon in 1958, Mayor Dick Lee of New Haven, Connecticut, climbed into the operator's cab of a wrecking crane. With a smile on his face and a supportive crowd assembled around him, Lee swung the crane's skull cracker against the brick walls of an aging tenement building. In almost no time, the demolition was complete. Subsequent rebuilding has erased any remnants of this scene from the city's physical landscape. Yet its memory endures owing to the *Life* reporting team that was on-site that day. One month later, the magazine published a three-page photo-essay on the mayor under the enthusiastic headline: "City Clean-up Champion."[1]

The most striking image in this photo spread depicts Lee seated comfortably at the controls of the crane. While the click of the camera's shutter suspends the wrecking ball in midair, the pile of displaced rubble beside it and the largely cleared landscape reflected in the crane's back window testify to both this machine's—and the mayor's—latent power. Lee can hardly contain his pleasure as he unleashes that potential, turning toward onlookers located outside the photographic frame, a satisfied smile sweeping across his face. Outtakes from the photo shoot reveal the careful selection of this image. One of the photographs shows the wrecking ball even closer to its target and the

In 1958, Mayor Dick Lee, of New Haven, Connecticut, swings the skull cracker of a wrecking crane in a building demolition performed for *Life* magazine reporters. (Robert W. Kelley/The LIFE Picture Collection/Getty Images)

mayor more focused on his task but obscures his facial expression. Another captures his beaming smile but leaves out the wrecking ball. Yet a third discard depicts the mayor standing triumphantly in front of the wreckage after the demolition is through. The chosen image, however, foregrounds the key players, both human and mechanical, demonstrates the mayor's personal participation in this work, and offers a hint of the pleasure he takes in the destruction. Among the various options, it is the most triumphal portrait of state-sponsored building demolition. This visual embodiment of the culture of clearance is the image that *Life* chose to put forth for its readers.[2]

It was no accident that the journalists were on hand to witness this scene. Lee had orchestrated it specifically for them. His staff proposed their itinerary and scheduled the building wrecking as one of the stops. Preparations taken in advance of the city's opening urban renewal demolition ceremony suggest the additional steps that probably preceded the official, public wrecking. The demolition contractor had cleared out the building's interior of all pipes, radiators, sinks, and plumbing. Overall, the crew had "prepared it to

Outtakes from the *Life* photo shoot foreground varied aspects and emotions of the carefully orchestrated event. By contrast with the published image, none so actively, explicitly, and triumphally portrays state-sponsored building demolition. (Robert W. Kelley, Richard Charles Lee Papers [MS 318]. Manuscripts and Archives, Yale University Library. Courtesy Estate of Robert W. Kelley.)

the extent it will take only a few minutes to make its walls cave in." Meanwhile, the electric and telephone companies had removed the wires located at the front of the building. This practical step also helped eliminate clutter from the planned photographic depictions. As a result, according to the New Haven Redevelopment Agency's daily demolition progress report, "the building was collapsed in a flourishing manner."[3]

Such public spectacles of the seemingly fast and effortless pace of the city's urban renewal demolition campaign were a common practice. Critics even accused the mayor of purposely leaving scattered, vacant eyesores standing "for propaganda purposes and no other." The sight of all these decayed buildings would bolster the case for his large-scale program of destruction. They would also assist the mayor in more practical ways by giving him a ready supply of structures to showcase in demolition displays. In fact, Lee regularly arranged for the destruction of New Haven's most imposing tenement buildings on occasions when visiting dignitaries and journalists joined him on his daily site visits. His guests included Adlai Stevenson, foreign ambassadors, and journalists from *Life, Harper's,* and the *Saturday Evening Post.* Lee once made light of this practice, noting, "Some mayors give out keys to the city. We knock down buildings for our guests."[4]

More than three thousand copies of the magazine issue—nearly a 60 percent increase over normal sales—sold across the region. But Lee did not rest after the publication was in print. He congratulated the author on a "job well done" and then orchestrated a letter-writing campaign to *Life*'s editors. Lee told his assistant, "I want to write 12 or 15 letters of varying lengths, saying different things, but all complimentary, about the *LIFE* article. I want half of them in the mail over the weekend, and the other half the first part of the week. If they get 15 or 20 letters from us praising the article, they may run one or two." Sixteen influential New Haveners responded to Lee's call and then forwarded carbons of their letters to the mayor. Lee's office supplied another handful of responses for others to sign and mail in as their own. Two letters ultimately appeared in print, the second of which lauded the feature as "a well-deserved honor for the man who has given this old New England city a decidedly new and hopeful look." The writer—identified to *Life* readers only as "L. S. Rowe, of New Haven, Conn."—was an unsurprising cheerleader for the article and its subject. Lucius Rowe was head of Southern New England

Telephone Company, the future occupant of the first new building erected in the city's inaugural urban renewal project area. He was also chairman of the Citizens Action Committee charged by Lee with guiding the urban renewal process, and his was the second name appearing on the assistant's list of solicited letter writers. This letter was Lee's final attempt to maximize the *Life* coverage in endorsement of further demolition work in the city. While historians have tended to gloss over the demolition phase of the urban renewal process, *Life* and Lee celebrated this critical work for all to see.[5]

Across the nation, scores of cities turned to urban renewal as a way to clear away the old and make way for the new. Title I of the Housing Act of 1949 initiated the federal program, providing two-thirds of the cost of clearing land and making possible the assemblage and write-down of large urban parcels to attract privately financed new development. By 1965, nearly eight hundred municipalities—located in almost every state in the country—had participated or planned to participate in urban renewal. Only half these projects were in the execution phase, yet municipal renewers had already acquired nearly fifty-seven square miles of land, upon which wrecking crews had demolished over 150,000 structures. These were typically not monumental edifices of major significance but myriad anonymous, small-scale, vernacular buildings. In contrast to the dramatic implosions that have long attracted public attention, these buildings fell largely at the hands of relatively undistinguished crews operating wrecking balls, bulldozers, and various hand tools.[6]

Each city had its story, often with a prominent local figure taking the lead. Best known is New York's experience, where the ambitious Robert Moses oversaw the clearance of vast tracts for new public and private developments, beginning well before the passage of the national legislation. Pittsburgh was another early mover, with local business leaders, organized into the Allegheny Conference, providing critical initiative and backing to remake the pivotal Golden Triangle downtown district. Once the federal government began helping to fund these projects, the reach of renewal spread even farther. Washington, D.C., offered a local example for Congress in the city's southwest quadrant, one of the earliest, largest, and most complete federally funded redevelopment projects. The leadership of Ed Logue proved crucial to the ambitious realization of many well-known projects in Boston, from downtown to ethnic residential districts. Hundreds of smaller cities also gave renewal a try, with prominent politicians sometimes following Lee's example and getting their hands dirty on the clearance site. In 1958, for example, the Alabama senator John Sparkman put a torch to slum housing in the city of Huntsville to begin the destruction of the infamous "Honey Hole" neighborhood.[7]

New Haven's experience of this national story was typical in process, if somewhat more vigorous in application. It was also widely followed on the national stage. Despite its relatively moderate size (population 164,443, in 1950), New Haven ranked sixth nationwide in levels of urban renewal grants received. In fact, New Haven received more urban renewal grants per capita than any other U.S. city. Between 1957 and 1980, officials used this funding to demolish over 3,000 buildings, located across more than three hundred acres of land. This was in addition to the 438 structures demolished by the state for Interstate 91 and a local connector, and more than 2,000 of the city's dwelling units destroyed by the Connecticut Highway Department during the 1960s. To put it differently, in its most active decade of destruction, the 1960s, New Haven eliminated roughly one out of every six dwelling units. The urban renewal program that Lee was promoting—and which most of his constituents initially supported—accounted for the majority of these losses. Although many historians and political scientists have written about New Haven, I offer here the first examination specifically of demolition in the city. Drawing upon previously overlooked urban renewal demolition records, I uncover the social, material, and economic history of the process as it played out in a vanguard postwar landscape.[8]

While the sadness of neighborhoods flattened and homes lost held the potential to lend building demolition a depressing air, *Life* turned the often-overlooked activity into a cause for celebration. Neither the city nor the magazine waited for the ceremonial ribbon cuttings or laying of cornerstones that initiated the construction phase of the projects; rather, both promoted the destruction. The boosterish record that Lee orchestrated, and which *Life* published, offers a powerful example of one of the ways a visual record of demolition was marshaled to increase its appeal. By dramatically performing his demolitions in a positive public spotlight, Lee was harnessing urban clearance as a vital political tool for physically, socially, and psychologically reshaping the American city.

Although *Life*'s photographs depict the mayor at the helm of the wrecking machine, the government did not act alone in the mass demolitions of the postwar era. Business was its critical partner. The Teamsters, Operating Engineers, manual laborers, and project supervisors responsible for this work on a daily basis remain curiously absent—not only from this particular magazine spread but also from our understanding of the history of urban renewal. We know little about these actors, their processes, or the roles they played in making large-scale building demolition physically and economically possible. Stereotypical photographic depictions of urban renewal jump quickly from "Before" (slums) to "After" (new high-rises and highways). Ground-level images have a similar effect, contrasting cramped, dirty living quarters with sleek new housing. If we focus on the time in between these scenes, we can see what "progress" looked like when it was "in progress."[9]

Photographic archives help lay bare a demolition process that has left relatively little historical evidence. The present-day built environments are often quite altered from their demolition phase. Written archives can be scarce, particularly given that most demolition companies were small firms that were unlikely to preserve and deposit their records. Documentation of the urban renewal process allowed for the unintended creation of miniature company archives at the municipal level. These extant records are invaluable. Where photographs also exist, they can leave an equally descriptive trail. Photojournalism depicts the character and ethos of postwar demolition, while images produced for more administrative ends illustrate what it was like to execute and be impacted by this work. Especially when paired with other available sources—including journalistic and trade literature accounts, vivid oral histories, and seemingly mundane municipal logs and paperwork—these photographs help recapture the ephemeral demolition landscape.

If there is a common perception of what urban renewal clearance looked like, it is an image of "armies of bulldozers smashing down acres of slums." That is how Albert M. Cole, eventual chief administrator of the Housing and Home Finance Administration, which oversaw urban renewal, once described the "early picture" of declining postwar cities' road to renaissance. But bulldozers were typically more relevant to the removal of the demolition's detritus than to building wrecking as such. Cranes, wrecking balls, and acetylene torches proved more valuable there. As in the wartime Revere ad of hands clearing the city, the bulldozer's greater utility to urban demolition was as a metaphor rather than as a unique physical actor in the clearance.[10]

Postwar demolition was a messy business, for reasons of finance and material impact, as much as it was precise engineering. Relative to Cole's simple description, the everyday experience of postwar mass building demolition was quite different for the wrecking crews, municipal administrators, and local citizens whose lives it affected. On the ground, demolition proved slow, dirty, difficult, and environmentally and socially damaging. The actual felling of the building—that instant which momentarily transfixed *Life*'s photographers and New Haven's viewing audience—was one small tale in a longer, more complicated saga. Both administrators and practitioners learned as they went, often undertaking large-scale demolition for the first time thanks to the boon of redevelopment funding. The rules changed over time, altering the economics of an already competitive industry and bankrupting some wrecking companies in the process. The work itself also proved more physically complicated than the image suggested by the swift and powerful sweep of a wrecking ball or bulldozer blade. Further, physical destruction wrought traumatic social loss. As clearance became more targeted than wholesale, even residents who had been spared increasingly found themselves living and working alongside structures undergoing the demolition process and suffering collateral damage as a result. Examining the interplay among the residents, bureaucrats, and hard hats in this one city allows us to uncover the mechanics of urban renewal demolition. The contrast between celebratory promotion and harsh practice, revealed within these, helps explain the enthusiastic ascent of mass building clearance, as well as its relatively swift decline, in the postwar period.

The Postwar Demolition Company

Building wrecking was not an activity unique to the postwar period. Baron von Haussmann's dramatic mid-nineteenth-century remaking of Paris was one of the most notable large-scale examples of this work. Across the

Atlantic, throughout the nineteenth century and increasingly in the early twentieth American cities incorporated demolition into their housing codes. Places like New York, Chicago, Detroit, and Milwaukee tackled dilapidated or abandoned building stock with ambitious slum-clearance programs. During the Depression, federal workers provided a substantial supply of unskilled labor to fuel these efforts. But private enterprise also proved critical. Many small, private, family-owned companies got their start in the 1920s and 1930s. The Housing Act of 1937 increased the scale of their efforts by providing federal funding for slum clearance in advance of public-housing development. Owing in part to the expense of moving their equipment and the local nature of their markets for salvaged materials, most of these businesses sought work in nearby or regional areas. The local, small-scale character of demolition companies persisted into the postwar period. In New Haven, all the successful urban renewal bidders were based relatively nearby. The majority came from Connecticut, and all were based within a hundred-mile radius of the city.[11]

The postwar demolition industry expanded off this base. Existing companies increased in size, and many new firms formed. The majority of the forty-plus firms that bid on New Haven's urban renewal contracts, for example, got their start after the war. The industry had also reached such a critical mass by this time that it began to organize. The National Association of Wrecking Contractors was founded in 1946. Although the fledgling trade organization eventually died out, it was the forerunner of today's more enduring National Demolition Association. By 1967 there were over 1,000 wrecking and demolition companies nationwide, employing a total of about 7,500 construction workers. Three-quarters of these businesses had fewer than 10 employees (although the larger companies accounted for more than half the receipts). The firms were located throughout the country, but were concentrated in the Northeast.[12]

This growth had practical and economic roots. State and federal redevelopment laws, including the long-awaited Housing Act of 1949, enabled an increase in large-scale urban clearance activities by making demolition work more affordable to cities: they had to pay only one-third of urban renewal clearance costs. As a complement, legions of soldiers, trained for work in World War II and skilled in facets of wrecking work ranging from heavy equipment operation to bridge and building demolition, were now available. The principals of one typical wrecking company hired in New Haven, for example, had worked in the Army Corps of Engineers, supervising building, runway, dock, and road construction during the war, as well as the maintenance, rental, and purchase of heavy machinery. After returning home, they established their own wrecking and construction company in 1946.[13]

Not only the number of wrecking companies but also the nature of their work changed in the postwar period. Projects grew larger in scale. Companies whose biggest projects to date had been on the order of tens of thousands of dollars in fees now found themselves bidding on urban renewal contracts measured in the hundreds of thousands. This increased scale lent itself to more mechanized operations. The typical postwar demolition company owned or could access a loader, a crane, a bulldozer, several types of trucks, and assorted hand tools. When necessary, it could supplement these with rental equipment. Larger companies sometimes also owned generators, compressors, and welders, in addition to greater quantities of the other equipment. General construction equipment manufacturers produced many of these items. Caterpillar even featured one of its bulldozers engaged in New Haven's urban renewal work in an advertisement published in 1959. The text accompanying the image explains that "progress is being made" on the demolition site. Equipment utilization was critical to the contractors' bottom lines. In 1960 one wrecker estimated the daily costs of idleness at $96 for a truck-mounted crane, $120 for a bulldozer, and $136 for a crawler crane. These were substantial costs when demolition fees at that time averaged around $700 per building.[14]

The practice of paying contractors to tear down buildings, rather than being paid *by them* for the opportunity to do the work, was another difference between pre- and postwar wrecking. Early-twentieth-century wrecking companies earned their income through the resale of salvaged materials. By the post–World War II period, however, increasing salvage market supply and innovations in building materials had made recycling less valuable. Used bricks, which had sold for three dollars per thousand at the start of the century, had become so ubiquitous as to be worthless. Other lightly used items—including electrical fixtures ($7), plumbing fixtures ($165), and boilers ($375)—maintained modest values. But most postwar cities prioritized speed

He's renewing a city's pride

Take a look at the city where you live, work or go for shopping and entertainment. Do you see overcrowded housing, run-down commercial areas, dreary factory lofts . . . filth, misery, danger?

If so, something can be done about it. And in many cases progress is being made. Today there are 385 cities with urban redevelopment programs underway or in the planning stage.

But it's slow going. By 1975 our population will increase to 235 million. Unless something is done, 30 million of these people will live in slums.

Now is the time to make your community a better place to live. There are men like the one in the picture. There are planners, architects, city officials and citizens groups with foresight and enthusiasm. There is a Federal Urban Redevelopment Program. And there are fleets of mighty Caterpillar machines to tackle the job.

But most of all . . . urban redevelopment starts with you . . . and your pride in your city. By rebuilding our cities today, you will avoid having to rehabilitate *people* tomorrow.

Preparing for the future needs of our cities—and of our entire nation—is a job for all of us. After all, if we don't do it . . . who will?

Caterpillar Tractor Co., General Offices, Peoria, Illinois, U.S.A.

CATERPILLAR
REG. U. S. PAT. OFF.

Diesel Engines · Tractors · Motor Graders
Earthmoving Equipment

Progress Report. New Haven, Connecticut, is "building for the future" with a redevelopment plan encompassing 725 acres. Redevelopment of cities is only one of the needs we must meet by 1975.

Robert is two months old. When he is 16 years our nation must have . . . tens of thousands of miles of new roads · double our present water supply · double our school facilities · 20 million new homes · 20% of our present housing rebuilt · 2½ times more oil · 60% more lumber and twice as much pulpwood · 55% more metal ores · soil conservation on 1,159,000,000 farm acres · 123,300 new dams, 1,200 miles of levees · double our present hospital facilities · triple our electric power.

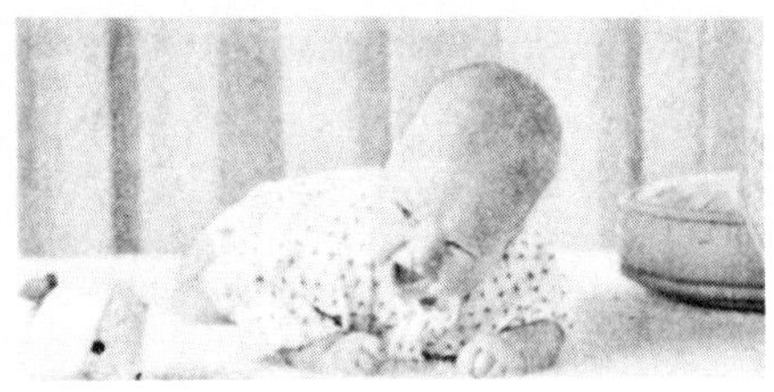

This Caterpillar advertisement, published in *U.S. News and World Report* in 1959, proclaims that "progress is being made" by two of the major machines of urban renewal—the bulldozer and the crane—at work on a New Haven redevelopment site. (Reprinted courtesy of Caterpillar Inc.)

over the laborious and time-intensive salvage process, and the latest machinery helped make that faster pace of work possible. Thus, for an average building torn down in New Haven's early urban renewal days, total salvaged materials represented only about 10 percent of wreckers' compensation. In short, postwar demolition was more complete in its destruction.[15]

With salvage making up a diminishing component of the income contractors could expect to receive from demolition jobs, they began increasing the fees charged to building owners. A complete bid submission included not only the bid price but also a bid bond, proof of insurance, and descriptions of current workload, available equipment, and past experience. Yet the bid price alone became the primary metric used by cities in awarding wrecking contracts. In fact, even when a proven wrecking company, experienced on New Haven projects, overbid an otherwise unknown company by less than 1 percent ($880) on one project, the city's Redevelopment Agency chose to award the contract to the lower bidder. When the runner-up company appealed to the city to reconsider, the agency responded with sympathy, but did not change its selection.[16]

The calculation of those all-important fees was more of an art than a science. The experience of Commercial Contractors, the wrecking company hired on four of New Haven's first six urban renewal demolition contracts, illustrates this well. Commercial was just a year old when it successfully bid on the city's first demolition contract, awarded in January 1957. The company generally charged a flat fee of $718 per building to demolish what were predominantly wood and brick two-story structures. On the second contract, Commercial raised its rates by 25 percent. It then charged per-floor rates of $330 and $400 on its third and fourth projects, respectively. A decade later, in 1967, the agency attempted to identify a common multiplier to use when completing in-house estimates of demolition costs. But it was unable to do so. According to internal records, experience had shown that the calculation needed to incorporate a variety of uneasily quantified factors, including "the locations of the buildings (whether in heavy traffic area or not), the amount of clear working area, if machinery or hand laborers must be used, the proximity to adjacent buildings, the availability of a dump site (its cost at the time and distance from the work site), the type of material, the height relationship to the adjacent buildings, etc."[17]

As Commercial's experience suggests, demolition costs escalated significantly over time. This was a function of contractors' greater experience in cost estimation, the diminishing value of salvage, and rapidly increasing labor costs. Both skilled and unskilled workers populated wrecking companies. A typical firm had a core group of five to fifteen skilled employees, including supervisors, foremen, and perhaps a few machinery operators. It generally hired unskilled crew members as needed for particular projects. The federal government set minimum wage rates for all urban renewal workers by employment category and monitored compliance through employee interviews. The highest-paid employees included supervisors and heavy-equipment managers, while drivers and laborers earned slightly less. All these workers, however, earned at least two to three times the minimum wage. According to agency payroll records, on the city's first contract, the company spent about forty dollars per day on skilled laborers and twenty-five dollars per day on unskilled workers, after factoring in insurance, taxes, and benefits. Between 1957 and 1975, demolition workers' wages nearly tripled, while inflation less than doubled, contributing to the dramatic increase in demolition fees over the course of New Haven's urban renewal experience.[18]

The majority of New Haven's wreckers were white union men. The Operating Engineers, which represented handlers of heavy equipment, required that their members be paid for a full forty-hour week, even when they worked fewer hours. They also shut down most demolition work in the city for a brief period in 1967 with a strike. Although members of the Teamsters, who drove the trucks on demolition sites, could be hired by the day, in practice contractors found they too required a full week of pay. The wages of unskilled laborers were the most easily managed. Contractors paid these men only for the hours they worked and could tell them not to come in the next day if necessary. This flexibility helped employers manage uncertain schedules, but it undoubtedly created challenges for workers seeking steady employment. The few African American workers hired typically occupied these lowest paying positions. Only after the passage of federal civil rights and affirmative action legislation in the mid-1960s did the city begin to actively ensure that its contractors were both employing a substantial number of minorities and offering them access to varied ranks of employment opportunity. By the time of its last urban renewal demolition contract, in 1977, New Haven required that

Map of New Haven showing the boundaries of the twelve redevelopment and renewal project areas. Over nearly a quarter century, urban renewal demolition reshaped parts of all these districts. (Drawing by Madeleine Helmer based on NHRA records)

minorities make up between a quarter and a third of total manpower. Yet city officials lamented the difficulty of finding skilled minority equipment operators, even as they increasingly sought them out.[19]

Beginning in early 1957, and lasting for over a quarter century, fourteen different wrecking companies helped realize Lee's urban renewal demolition plans in New Haven. Their work touched a range of neighborhoods throughout the city. They began with Oak Street, a centrally located, multi-ethnic community consisting largely of Eastern European residents and small-scale businesses. Wreckers conducted their most complete clearance job there. Next they expanded to the downtown, primarily commercial, Church Street project. From there, they moved cranes and wrecking balls eastward to the Italian enclave of Wooster Square, the site of the largest total number of building demolitions in the city, but also New Haven's inaugural location for federally funded rehabilitation. While the largely African American communities of Dixwell and the Hill were the last of the city's major demolition areas, wrecking crews eventually spread to all other project areas, located even farther away from the downtown core. Demolition was most common at the start of the city's redevelopment work, peaking in the first half of the 1960s; yet clearance remained a consistent practice throughout the urban renewal era. The dramatic imprint these companies left upon the city has yet to be fully erased.[20]

The Difficult and Dirty Work of Demolition

The everyday execution of the demolition revealed the complicated nature of this work. Major challenges included refuse disposal, contractor oversight, and the sequencing of individual wreckings. As ground-level experience would make clear, the simplified image created by Lee at his public wrecking performances was a far cry from the headaches, hassles, and heartaches that characterized the daily reality of clearing the city. That said, New Haven's first urban renewal demolition contract went uncharacteristically smoothly. Each day, Commercial Contractors operated a crew of four to eight men, a crane, and a bulldozer (known as "Monster") on the site. After stripping for salvage, they worked from top to bottom as they eliminated stories from individual buildings. Hand tools and crane-mounted wrecking balls accomplished building destruction, while bulldozers removed the remains. Wreckers sometimes used newly cleared spaces for burning debris or temporarily storing salvage. But their ultimate job was to deliver to the city cleared, level pieces of land that were ready for new construction. Rain and cold temperatures slowed initial progress somewhat, but by two weeks into the contract, nine buildings were down. The full contract took an entire year to complete and cost the city $62,000 to demolish eighty-six buildings. Although it had taken twice the amount of time and included twenty more buildings than originally estimated, this first contract launched New Haven's urban renewal program with a promising start.[21]

This ease of operation did not last long. By the city's second demolition contract, the disposal of massive volumes of refuse was already raising public complaints. On the first contract, Commercial had burned accumulated wood and trash on-site. Each afternoon, the crew dug a pit, filled it with debris, set fire to the pile, and by four o'clock had bulldozed it over with dirt so that the smoke would die out before the end of the day. Although the original bid documents had forbidden burning, the New Haven Redevelopment Agency made a last-minute change to permit the practice—provided that it occurred on-site and in the presence of a representative from the fire department. The agency estimated that this procedural change cut approximately 50 percent off of originally anticipated demolition expenses. Opposition increased, however, as work moved from relatively isolated areas of the city to more central commercial sites. In particular, the Chamber of Commerce argued that burning in the downtown area created an inhospitable environment and risked damaging places of business. The agency responded by restricting burning that would cause a "nuisance," although it eventually required contractors to burn all combustible rubble on a special site set up south of downtown. Ultimately, by the mid-1960s, the agency banned the practice entirely.[22]

Finding a new repository for the debris proved a challenge. The agency initially gave contractors essentially free use of the city dump, but had terminated access by 1965 as the facility filled. Wreckers then disposed of noncombustible materials in private dumps outside New Haven, but these sites offered no long-term solution. Refuse transport was time-consuming and expensive, and the facilities rarely maintained the full-time equipment support necessary for convenient use. Sometimes the owners of private dumps also just changed their minds. The town of Hamden, located north of New Haven, permitted contractors to use the town dump until local residents began complaining of rat sightings. Although the agency originally renounced responsibility for finding a new dump site, it changed course and set up a dedicated debris-disposal area in New Haven after the lack of a workable dump location brought all demolition activity to a halt. Soon, however, the dust and dirt created at this new site provoked the displeasure of nearby residents as well.[23]

Some contractors dealt with the refuse-disposal challenge in unscrupulous ways. In 1968, agency officials learned that someone was surreptitiously disposing of rubble on vacant lots in the city. Even more disruptive were contractors who buried rubble on their job sites in place of suitable fill. The agency received allegations that Commercial employees had been observed "digging up good land, putting [their] rubble in the hole, and then covering the rubble with some of the dirt taken out of the hole." Through such practices, the contractor had purportedly "left virtually every site with defective fill—wood, tanks, etc." Confirmation of these allegations arrived when property developers noticed sinkholes and settling on completed sites.[24]

Another contractor frequently found guilty of similar offenses, as well as of verbal abuse of agency inspectors, was Barnum Lumber Company. On an August morning in 1967, Nate Laden, Barnum's owner-supervisor, refused a demolition inspector's request to cap a sewer properly on one of his work sites. He told his men to "throw [the inspector] in the cellar pit and bury him if he says anything else." When a more senior agency official, Armand Aloi, confronted Laden later in the day, he received similar treatment and was ordered off the site. The situation rose to such a furor that one of Laden's associates, a Mr. Lombardi, confronted Aloi, "cursing and hollering obscenities," and placed his hand on his face as if to strike him. Aloi retreated from the scene, but afterward he wrote up the incident as the latest in a pattern of aggression from Laden toward the agency. (Six years earlier, Laden had tried to run down another inspector with one of his trucks.) The cumulative impact of these episodes, Aloi wrote, was that "the situation has now reached the point where my inspectors are frightened and are refusing to go on a Barnum Lumber site. I might add that I am not anxious to have another duel with Laden, Lombardi and company, and I recommend that Barnum's contracts be cancelled." In the end, however, the company's low bid prices kept it steadily employed by the city. In total, Barnum won ten contracts, worth nearly one million dollars. This was the second-highest volume of work (as measured in both dollars and number of contracts) of all New Haven's urban renewal wrecking companies.[25]

A more persistent challenge was the logistical coordination of the large

Although the full block of buildings at the center of this image eventually came down, they were razed not in one fell swoop but in the patchwork pattern suggested by the scattered structures still standing in 1960. (Charles B. Gunn, Richard Charles Lee Papers [MS 318]. Manuscripts and Archives, Yale University Library. Courtesy Estate of Charles B. Gunn.)

number of buildings to be wrecked. The Redevelopment Agency's requests for proposals had invited bids for sizable, largely contiguous "sections" or "groups" of buildings, averaging about sixty structures per contract. This led many bidders, in the words of one observer, to plan to complete contract demolitions in "one sweep movement. . . . There were not to be any selected buildings taken apart one at a time but it was just to be the whole mess." Yet in practice the city often released buildings on a piecemeal basis, resulting in a patchwork pattern of destruction instead. This made it difficult for demolition contractors to manage the allocation of expensive wrecking equipment efficiently between their out-of-town bases, other ongoing projects, and New Haven work. It also necessitated the substitution of smaller, slower equipment for the larger, more efficient cranes they had originally planned on using—and had previously factored into their bid prices. One wrecker sued the city for financial reparations and ultimately received forty thousand dollars on top of its original eighty-seven-thousand-dollar contract.[26]

Delays in the release of buildings derived from hold-ups in both the building acquisition and the demolition stages of work. Acquisition could stall due to challenging family relocations, occupant resistance, the escalation of eminent domain disputes into litigation, or long waits for federal approval of changes to original demolition contracts. The agency itself even sometimes prolonged acquisition proceedings in the hope of negotiating building releases rather than having to resort to condemnation. After acquisition was complete, contractors sometimes delayed demolition of available properties until enough permits had been assembled to merit moving their equipment back into the city. At times they would put off demolishing smaller buildings in order to take down larger, more profitable structures first. Although the agency discouraged such practices with the threat of fines and contract termination, contractors often held out as long as they could in order to preserve their bottom lines in the face of logistical uncertainties. By the time a building was released to the contractor, surrounding site conditions had sometimes changed sufficiently to require additional costly and time-consuming preparatory steps. When a new housing development went up next to one demolition target, for example, the wrecking company had to alter its site access and debris-disposal plans and erect scaffolding around the new structure. These changes more than doubled the time required to complete the demolition and cost the city over twenty-eight thousand dollars in extra fees.[27]

Given all these delays, it was not unusual for demolition contracts in New Haven to drag on for three or more years. This was an economic hardship for contractors. With capital tied up in a bid bond and insurance until the contract's termination, one wrecker lamented to the agency in 1965, "The cost of these two items is more than we can expect to make when we do the actual demolition." Rising demolition costs lowered or eliminated the profits on bid prices established several years earlier. As another contractor wrote to the agency in 1967, when nearly 20 percent of a contract remained outstanding after three years of work, "The time for release of buildings has stretched beyond a reasonable limit that anyone could have expected. We, therefore, request that the balance of the buildings be withdrawn, our retainers released, and the contract closed out." When possible, the agency accommodated such requests by removing the delayed buildings from the contract and transferring them to another open contract with the same company. It also adjusted demolition prices up if they were clearly outdated. By the mid-1960s, it had begun mandating contract termination after four (and later three) years, whether or not all the buildings had come down.[28]

Persistent challenges like these, combined with the competitive nature of the industry, kept most demolition contractors from profiting too greatly from their efforts. In 1965, for example, the president of one of the city's most reputable wrecking companies pleaded with the agency for quick payment so he could keep his business afloat. Three years earlier, Commercial Contractors had found itself with $125,000 in obligations to the bank and the Internal Revenue Service. To cover part of this cost, the company tried to trick the agency into reimbursing moving expenses that it had both inflated and planned to pocket rather than incur. Even after this episode, however, the agency continued to award Commercial contracts. Its bid prices, which were sometimes only half as high as those of the next lowest bidder, proved too appealing to pass up. They were also too low for the company's continued viability. On one demolition contract, for example, Commercial's payroll amounted to over $55,000, versus only $62,000 in demolition fees. Equipment costs probably added an average of about $400 per week for the year-long duration of the project. If salvage brought in another 10 percent in income, that left no additional funds to cover overhead or amass profit. By 1965, after receiving a total of nearly half a million dollars in demolition fees from the city, Commercial had disappeared from city directories. Perhaps unsurprisingly, larger, more established demolition firms such as the Cleveland Wrecking Company regularly bid 25 to 50 percent more than Commercial did. The few regional and national wreckers like Cleveland were the ones making real money in the demolition business, while the many bottom feeders, like Commercial, sucked much of the profitability out of the enterprise. Urban renewal budgets may have benefited from their low prices, but the savings came at these wrecking companies' own expense.[29]

Fire as Hazard and Opportunity

Willie Jeffries reported to work with Commercial Contractors at 8:00 A.M. on Monday, August 29, 1960, as he had done for the previous two years. The weather that morning was fair, and he could already feel the heat picking up on the seasonably warm late-summer day. At twenty-six, Jeffries was among the younger members of the wrecking crew that had been responsible for taking down most of the buildings in New Haven's urban renewal program to date. Of the fifteen Commercial employees involved in demolition work in the city that week, he was one of only three African Americans. The others included a fellow laborer and a driver, the two lowest-paid occupations on the site. Jeffries commuted to work each morning from Father Panik Village, a public housing complex located in Bridgeport, Connecticut, twenty miles southwest of New Haven. As a laborer and "burner" (torch operator), Jeffries was one of the least senior men on the project. But manual laborers made up the bulk of the crew and played a critical role in the demolition process, especially when the acquisition of salvage was involved.[30]

This particular morning was Commercial Contractors' eighth day of work on the demolition of a hundred-year-old, five-story former factory known as the Shoninger Building. The structure, which had a brick exterior and predominantly wooden interior, was located in the Wooster Square neighborhood. Jeffries began his work that day by using an acetylene torch to remove the metal pipes of one of the building's old sprinkler systems. Afterward, he moved to another side of the structure and applied his torch to the amputation of a metal fire escape. Just before 11:30, when in proximity of his original project, Jeffries noticed smoke wafting from waste materials in the area. He quickly slipped away for a bucket of water and tossed it on the source of the smoke. Unable to squelch the incipient fire, he dashed off again for a second supply of water. But his efforts came too late. By the time he returned, the fire had ignited the building's oil-soaked floors and was spreading up a nearby elevator shaft. With the blaze now beyond his control, Jeffries summoned the assistance of the project superintendent, Pat Potopowitz, and the on-site city inspector. At about the same time, eight-year-old Carol Longobardi was playing in front of her home across the street when she noticed smoke pouring out of the vacant building. Soon her mother, the inspector, and several other local residents were alerting the fire department to what would eventually grow into a four-alarm blaze.[31]

The strained efforts of nine engine companies, four truck companies, and two deluge units finally contained the Shoninger Building fire. Although twelve firemen were hospitalized, there was no loss of life. Damage to neighborhood property came to over three hundred thousand dollars and included the destruction of two cars parked on the street, collateral injuries to surrounding buildings, the burning of the Shoninger Building itself, and—most significant—loss of the furnishings, ceiling murals, structural elements, and prized stained glass of neighboring Saint Louis's Church. Pastor Lawrence W. Doucelie was unreachable (he was traveling in Italy) when the fire occurred, and he did not learn of the disaster until the cab driver taking him back from the New Haven train station broke the sad news.[32]

Some of Mayor Lee's political opponents seized on the fire as an opportunity to criticize demolition activity in the city. In an effort possibly aimed at boosting his public profile in his campaign for the State Senate, Alderman Salvatore Ferraoulo called for a probe into the qualifications of the workers and the levels of staffing on the city's demolition contracts. Wooster Square alderwoman Carmel Scoppetta went farther, complaining to the *Journal-Courier* that the redevelopment and relocation offices were "enacting the destruction of the city under the guise of redevelopment. They are responsible only for the destruction of the city, its housing, and its citizens."[33]

Yet fire had not always been regarded as a symptom of the failings of urban renewal. In the early days of the program, Lee had invoked it as the very justification for proactive building destruction. During a speech at a luncheon following the inaugural urban renewal demolition, the mayor turned briefly away from the morning's excitement to focus on the "shadow" that lurked over the gathering: "The shadow, of course, to which I refer is the shadow of the Franklin Street tragedy, which struck this city just six days ago and left in its wake death and disaster of a kind not dreamed of in a peaceful community such as ours." The Franklin Street fire had ignited an eighty-six-year-old factory building in the Wooster Square neighborhood, killing fifteen people. Unregulated working conditions—including inadequate or nonexistent fire escapes, alarms, sprinkler systems, and exits—contributed to the tragedy. Yet Lee blamed only the buildings themselves. In his view, the aging built environment was the enemy, and it was the government's role to save the community from that specter through victorious building demolition. Lee took the occasion of each of the era's fires to remind people of the risk. Making buildings the enemy also diverted attention away from their largely minority and working-class occupants, although demolition removed them as well.[34]

There was reason to believe, however, that the Shoninger fire was the consequence of demolition work. Site evidence suggested as much, and past precedent at demolition sites located outside the city also pointed in that direction. New Haven's fire chief was aware of one remarkably similar fire that had occurred at a Harlem warehouse less than three years earlier. In that case, sparks from a wrecker's acetylene torch had caused the building to go up in flames. The fire chief had promptly forwarded Lee the *New York Daily News*'s front-page story on the incident, appending a cover note that urged, in all caps, "WE DON'T WANT THIS TO HAPPEN IN NEW HAVEN."[35]

Yet when history did repeat itself, the city officially declared the Shoninger fire as "undetermined" in cause, and perhaps the result of mischievous acts. This finding permitted Lee to position the incident in a politically expedient light. Instead of responding with a more cautious approach to demolition and its dangers, Lee argued that the fire justified increasing the pace of work. The fire marshal's official report noted that the Shoninger fire would have had human, as well as physical, costs had it occurred just three years earlier. Back then, more than two hundred employees of nine commercial tenants still occupied the building. This situation might have created "probably one of the worst fires in the history of our City." Therefore, as Lee reiterated on the morning after the blaze, "Thank Heaven yesterday's fire occurred when the building was being torn down, not when it was still in use." As city residents weighed the benefits of a more deliberative approach to demolition, including the possibility of renewal through targeted building rehabilitation, city officials used the fire to pose the counterargument. There was no time for slower alternatives, they seemed to suggest, lest New Haven go up in flames before urban renewal could be implemented.[36]

One of the fire's most direct outcomes was a costly, short-term increase in demolition work. To minimize public hazards from the damaged structures, as well as prevent further fires in the area, the city marked several buildings for immediate destruction. The urgency was particularly acute given the two parochial schools located across the street, with the start of the academic year only a week away. The agency ordered the contractor to do whatever was necessary, including working overtime on Labor Day weekend, to eliminate the fire hazard. Local newspapers highlighted the unusual nature of this weekend work as their journalists photographed the "engines of destruction" performing "demolition even on Sunday." This rushed activity resulted in sixty-one thousand dollars in excess costs due to the requisite police surveillance; overtime wages, equipment use, and overhead for the wreckings; and compensation to Commercial for lost salvage.[37]

Going forward, the agency took a more proactive approach to fire prevention. Mayor Lee announced the formation of a "fire safety squad," which would inspect the sites, keep a lookout for negligence, and help prevent arson. Over time, firemen also conducted fire watches of buildings either awaiting demolition or otherwise considered at risk. More significant was the soon standard practice of rushed demolition preceding any potentially dangerous time period. The Fourth of July provided the major annual impetus for these measures. In the summer following the Shoninger fire, the marshal advised the mayor that dry weather conditions combined with activities anticipated over the holiday weekend posed a major fire hazard to vacant buildings then awaiting demolition. The agency authorized the immediate destruction of more than twenty of these structures, even rushing acquisition proceedings on a few of them in order to address as many potential problems as possible. This process was repeated annually. In 1964, for example, sixty buildings made the holiday fire watch list. While disputes initially emerged over responsibility for these costs, the agency began contractually charging its wreckers with the expense from mid-1961 on. By then, these procedures had become a regular part of the escalating costs of urban renewal building demolition.[38]

Living in the "Fear and Dust" of Demolition

In the wake of the Shoninger fire, the Board of Aldermen charged the Redevelopment Agency with improving the policing of its demolition contractors. The board's initial draft resolution bluntly asserted that the agency had allowed these companies to disregard relevant policies, resulting in "severe damage . . . to one of the finest churches in the city" and causing "people in the demolition area [to] live in fear and dust." The resolution did not pass in this exact form. Nevertheless, its early language captured something of the physical and emotional reality of the landscapes of building demolition for residents, businesspeople, and passersby. Demolition brought disruption, damage, dirt, fear, and, for some, the uprooting of businesses and lives.[39]

Urban renewal demolition meant relocation for roughly 20 percent of New Haven's population, or about 30,000 individuals. This came on top of the already disruptive effects of the highway clearance—for both the interstates and the Oak Street Connector—also occurring during this period. The urban renewal of Oak Street, the city's first and most fully demolished project area, displaced 886 families, or about 3,000 residents. The agency offered them modest moving stipends and help finding new, physically improved homes at comparable costs. Nearly half of Oak Street's families were eligible for public housing, typically located in Elm Haven, a new high-rise development that was itself the outcome of a pre–urban renewal slum-clearance project in the city. In the end, however, two-thirds of the families chose to move to private rentals located elsewhere within New Haven, including the Hill and Dixwell neighborhoods. Although the Relocation Office—by law—offered them assistance, most found new residences on their own. With sufficient quantities of housing available, the pace of relocation proceeded as quickly as building acquisition and demolition schedules required, with about 50 families relocating a month. This rapid pace was a function of the early timing of the work. Over time, decreasing availability of good, affordable housing, coupled with real estate market segregation, made it increasingly difficult for those displaced by demolition to find new residences.[40]

A disproportionately large number of those relocated were minorities. Nationally, nearly two-thirds of residents displaced for urban renewal were nonwhite, leading some to nickname the program "Negro removal." New Haven's postwar demographics shift that story statistically, but not substantively. As 85 percent of the city's residents at that time were white, they made up the bulk—56 percent—of those displaced. Roughly 42 percent of displacees were black. Yet despite the smaller representation of African Americans within the population, they were 4.5 times more likely to be displaced than white residents. Among those relocated, blacks were also more than three times more likely than whites to end up in public housing, and more than three times less likely to purchase a home in town. Further, many of the whites who were displaced, including Italians and Eastern Europeans, were minorities themselves. Finally, in New Haven's downtown area a substantial number of displaced residents had lived in single-room-occupancy households. There were 350 such roomers in the Oak Street area alone. These were mainly single men, some of whom were probably homosexual. Not only did renewal destroy housing types well suited to this population's economy and lifestyle; it also offered them no official assistance with relocation. The group did not fit traditional perceptions of family-centered domesticity and received no aid as urban renewal destroyed their homes. In New Haven, as across the country, the burdens of displacement fell unequally along lines of race, ethnicity, and sexual orientation.[41]

For all those displaced, the seemingly rushed relocation often induced shock and fear. One former resident, Teresa Argento, described the reactions of her Wooster Square neighbors: "They had nowhere to move. It was like they expected a disaster and 'We have to move! We have to move!' They gave us a deadline. Where are we going to go?" The difficulty of dealing with this news was exacerbated by the seeming abruptness of removal orders and the perceived inequity of the minimal compensation—what some described as "a pittance"—that residents received for their loss. Argento elaborated, "I still say it's one thing when you're ready to go, but when someone comes in and says, 'Lady, you have to leave and I'm giving you a year's time or six months' time. You have to leave.' That isn't fair."[42]

From a purely material perspective, relocation often improved the circumstances of the families it touched. Even Argento admitted as much, noting that her new home in the Annex section of the city was "lovely." Another former Wooster Square resident, Richard Abbatiello, described his new neighborhood of Fair Haven Heights as so foreign that it felt as if he would need a visa to go there. But he too acknowledged some of relocation's upsides, including the fact that it enabled his parents to participate in the "American Dream." They had gone from a small residence in a three-family building to a ranch-style house with two bathrooms, a garage, and a yard. As he concluded, "So I think mom and dad were trying to make a better life for their children and for themselves, too, and this was the way of doing it."[43]

An improved physical environment, however, could not make up for the traumatic sense of loss imposed by demolition and relocation. Urban renewal broke up dense neighborhoods of residences, churches, schools, and corner grocers with long-standing ties. It also separated families and friends. Follow-up studies of this process in other cities showed that most relocated residents never regained the sense of community they had lost. Displaced citizens sometimes responded to their relocation by making return visits to their former home sites. After moving out of the Hill neighborhood, young Harry DeBenedet either hitchhiked or caught the bus to make near-daily pilgrimages to his former street corners, despite the fact that he described his new home in nearby North Haven as part of a "beautiful community" with a "woodsy" character. In a comparable way, Argento's mother never came to terms with her new home. As her daughter described it, "We had all the meadows. It was really beautiful and my mother was so unhappy. So unhappy. She thought we took her to California." Argento's friend Norma Barbieri summed up the overall effect: "It was, very traumatic for all of us, all of us; a way of life went."[44]

Although these families experienced the residential losses associated with demolition, their physical relocation went fairly smoothly. Not everyone was so lucky. Some residents dismissed the options offered them as too expensive, geographically inconvenient, or otherwise undesirable, forcing the agency to evict them. Others wanted to relocate but had difficulty doing so. A common relocation challenge was the family that was too large or too affluent for public housing yet still earned too little to afford market-rate accommodations. Additionally, informal housing segregation practices by homeowners, landlords, and realtors made it tougher for nonwhite families to find neighborhoods in which to relocate.[45]

The case of Richard Durham and his family reveals some of the specific challenges that difficult relocation cases posed to both residents and officials. In 1957, Durham and his wife lived in a three-story brick multifamily building located just south of downtown. They had made their home there for nearly a decade. When the wrecking crew was ready to begin demolishing the structure in mid-December, the Durhams had yet to find new housing. So the agency moved them to a three-room apartment in a nearby block. It too was scheduled for demolition, but at a later date. Soon after, the Durhams' daughter and her seven children moved in with them. As an agency employee recorded in their file, "Her husband was in jail for non-support, and she had just come from the hospital, with twins. When we visited the apt. to let them know that the daughter and children must leave, we found her in bed with the flu. The bed was occupied by herself, the new-born twins, and 4 other children. (The 7th child was of school-age and was out)." While relocation officials saw the new arrivals as trespassers preventing the relatively straightforward relocation of the apartment's legitimate occupants, they also recognized that the parents were "loyal to the daughter" and "wouldn't leave [her] in the lurch." After further investigation, they found a two-family, eight-room house for the Durhams in the neighboring town of West Haven. Welfare assistance helped make the finances work. In some respects, the Durhams seem to have improved their housing arrangement, as their expanded family was now resettled together in a larger structure. Although relocation records end there, however, city directories continue the story. They record the Durhams living at yet another location, back in New Haven, only a few years later. For this family, redevelopment interrupted a period of extended residence with a pattern of repeated and disruptive relocation.[46]

Demolition also impinged upon those who had to live and work amid the destruction. Sometimes their presence was unintended. The clean-up of the Shoninger fire, for example, exposed three families still living in what was previously believed to be a deserted demolition area. Although the occupants claimed not to have been offered relocation assistance, the Relocation Office quickly attempted to correct the alleged oversight. More typical victims were the Wagners, a family awaiting relocation from the Oak Street neighborhood. The Wagners remained resident in the building where they had lived for more than a quarter century as they waited for their new home to be ready. In the meantime, the structures on either side of them came down, producing the mess, noise, and general disruption characteristic of the process. Even more invasive, in 1959, the agency permitted contractors to begin preliminary demolition activities on another building they knew to be still occupied by three tenants. One of the residents eventually complained that the water had been turned off, windows removed and left unprotected, and fires built, resulting in "extensive smoke damage." At that point, the agency halted further work until all tenants had moved out.[47]

There were also residents whose homes were spared the wrecking ball but who nevertheless had to endure the demolition of nearby buildings as their neighborhoods experienced urban renewal. Numerous homeowners contacted the agency about a fence, party wall, driveway, or garage that wreckers had damaged during a neighboring job. In one instance, a contractor inadvertently pushed a concrete wall against the adjacent building, damaging that structure's walls, ceilings, gutters, and siding, and destroying a tree. A woman who was surrounded by house demolitions on three sides of her property suffered similarly substantial property damage. Over time, cracks developed in her walls and ceilings, and she had to rebuild her chimney. She attributed these injuries to "the continual pounding of a weight used to crush concrete and flying debris." As her daughter wrote to the agency on her behalf, "We live in constant fear of fire and robberies. Our home was recently robbed and several attempts to break in have occurred. Amid this chaos, my mother is trying to maintain the property having recently spent $3,000 to paint and repair the exterior of the house. The money that she had set aside for her retirement is gone and there is nothing left to subsidize New Haven's redevelopment program. My grandmother is old and very sick; my mother is under a doctor's care and in poor health which is now aggravated by this constant source of irritation. This house has been our family home for many years and we stay here with the perhaps unrealistic expectation that conditions will improve." Although the resiliency of New Haven's neighborhoods depended upon both improved building conditions and the physical and emotional investments of its residents, at times the burdens of renewal seemed to crush that resolve where it was needed most.[48]

Proximity to demolition introduced less visible public health hazards. Wrecking unleashed toxic dust into the air. Present-day studies of the demolition and gut rehabilitation of buildings of similar age have revealed elevated lead levels on streets, alleys, and sidewalks located within a hundred yards of the structures. These persisted a month after the work was complete. Even worse, lead dust fall exceeded current E.P.A. standards sixfold on sites within ten yards of demolition work. This dust might stick to passersby, adjacent residents, or neighbors who hung clothes on nearby clotheslines. Recent studies have only begun to reveal the magnitude of these public health problems and suggest policies to address them. Without this knowledge, contractors working a half century earlier could not have adequately minimized the potential ill effects for themselves or the other occupants of the areas in which they worked.[49]

Postwar demolition work did have some safeguards. Once they received building access, contractors were responsible for all maintenance. They often had to board up building openings (particularly after they removed windows and doors for salvage operations), shovel snow, and hire night watchmen. These activities helped protect against vandalism, fire, accidents, and other nuisances that perpetually plagued unoccupied buildings. Once the demolition work itself got under way, wreckers had to erect barricades and hose down their sites daily to control dust. For both practical and political reasons, the agency tried to police its contractors' adherence to these responsibilities. Since urban renewal was a visible public activity, any unanticipated harms that it imposed on the general public generated "unwelcome publicity" and reflected poorly on the city. The agency increasingly formalized its project oversight over time. Official assessments of the city's first urban renewal demolition contract had included a mere line or two about progress every few days. By contrast, inspectors on later projects visited all active demolition sites daily, each time filling out a standard one-to-two-page form describing site conditions, contractor activity, equipment in use, and problems encountered. They also offered remedial recommendations as appropriate. Nonetheless, some contractors frequently ignored their maintenance responsibilities, particularly with regard to dust control, turning entire residential neighborhoods into dirty, inhospitable construction sites.[50]

The contractor's work did not end with the felling of the building and removal of its major remains. Cleared sites were sometimes left unleveled or strewn with debris, and the agency worked vigilantly to ensure that wreckers completed each clearance job. Otherwise, these potentially dangerous sites could attract unwelcome users. The police chief's reaction to one such unfinished lot demonstrates the tension that characterized the city while urban renewal demolition was under way. In 1965, Chief Francis McManus wrote to the agency about the "great number of housebricks lying on the sidewalk" in one demolition area. Not only were these "dangerous to pedestrians," he contended, "but in the event of any disturbance will afford ammunition for the participants. Please have the contractors cover these sites with sand as soon

Scattered debris and hulking remains surround Mayor Dick Lee and Frank O'Brien of the New Haven Redevelopment Agency as they survey a demolition site from the Oak Street Connector in 1956. (Charles T. Albertus, Richard Charles Lee Papers [MS 318]. Manuscripts and Archives, Yale University Library.)

as possible, and remove these missiles." Chief McManus was writing several years before the riots of the late 1960s turned pieces of New Haven's built environment into armaments and fire traps. The earlier riots experienced by other U.S. cities, emerging civil rights unrest, and tensions created by urban renewal itself seem to have prompted his perception of this relatively banal debris as potential weapons. McManus's premonitions later proved prescient when riots roiled through the city—and especially its urban renewal areas. In August 1967, fire-bombings destroyed fourteen buildings, and looting cost store owners millions of dollars in damages. The riot of April 1968 took down nine more buildings. The city responded to both of these urban uprisings by ordering the rushed, preventive wreckings of approximately seventy structures that they feared might fall into future rioters' hands. Thus, the riots brought both immediate and follow-on demolitions of their own. As in the case of clearance-induced landslides in the suburbs, over time the bulldozer unwittingly spawned occasions for its return to the sites of its urban destruction.[51]

Not only riots but also the everyday practice of urban renewal dramatically affected New Haven's businesses. The leveling of neighborhoods greatly depleted the local customer base. For landlords, it could also increase vacancies well before the time when building acquisition was to occur. Once redevelopment was announced, building improvements halted, and residents fled rather than renew their leases. In 1960, one landlord purchased and remodeled a four-story brick mixed-use building consisting of ground-floor stores and six upper-story apartments. In late 1972, while all his units were occupied, he was informed by the agency that his building would be demolished. Within a year following newspaper publication of these plans, all his tenants moved out. As vandalism and fires damaged the vacant property, the owner was unable to obtain new insurance. Not only did demolition cost him rental income, but the impending destruction probably decreased the building's valuation when the agency assessed it for acquisition. The result was further dilapidation of the city's physical infrastructure and economic limbo and loss for those with financial stakes in the affected properties.[52]

More than two thousand New Haven businesses had to move. Renewal may have simply hastened the closure of businesses already facing difficulties, but it overwhelmed more successful enterprises as well. Initial reimbursement caps of $2,500 could make the cost of moving prohibitive. Those planning to return to the redeveloped areas faced the challenge of finding temporary quarters in which to wait out the long construction process. Others made permanent moves to elsewhere in the city, giving up formerly prime downtown locations and their largely local customer bases. A former resident, Murray Trachten, recalled the dilemma that faced the owners of some of the businesses he had grown up with in the Oak Street neighborhood: "A couple of these businesses who were gonna be out of business, and they didn't know where they could relocate because so much—not so much, *all* of their business was just neighborhood-related. So that if they moved, if the City took their property and paid them, it's not like they could open up someplace close to that neighborhood, for them it would be starting a brand new business someplace else and in a completely different neighborhood. Which they weren't inclined to do." A study of business relocation's effects on the Oak Street and Church Street neighborhoods corroborated Trachten's story. By 1962, more than 30 percent of 124 displaced businesses had either gone out of business or just disappeared. For business owners like these, many of whom had been neighborhood fixtures for decades, demolition destroyed not only their *place* of business but their *entire* business. Further, the mixed-use nature of many of the city's neighborhoods meant that the owners of these typically small-scale enterprises were also often the victims of residential clearance, further compounding the burdens of displacement.[53]

Of Parking Lots and Burger Shops

Ultimately, not only the process but also the end products of urban renewal demolition disappointed many New Haven residents. Yet an image in Lee's laudatory *Life* photo-essay suggested promising—even "hopeful"—physical prospects for his redevelopment program's near future. A photograph appearing at the bottom of the spread's opening page depicts the mayor standing proudly in front of a white canvas of cleared land. While the snow cover was an artifact of the winter weather, it also seems to suggest demolition as a form of cleaning and rebirth. This was an association that had been forged during the large-scale clean-up of wartime ruins. These optimistic metaphorical depictions of urban clearance—strikingly reflected in imagery like the Revere ad showing hands clearing the city—were perhaps the war's most influential impact on postwar domestic building demolition. Relative to the case of earthmoving, where wartime advances in equipment and clearance practice spurred dramatic postwar developments, the war's influence on postwar urban demolition was more symbolic. The panoramic perspective of this long, narrow image proudly encompasses the vast scale of destruction the city had already achieved. In contrast to the varied terrain of rock formations, rising Yale University spires, and bumpy downtown skyline all appearing in the distance, the foreground is near-uniformly barren and flat. The mayor's small figure blocks part of the view of the one new building going up, instead directing the viewer's gaze back toward the empty foreground that dominates the scene. In the published version, a few letter labels indicate where future construction is to go, but the image's overall effect is to highlight the dramatic removal of all that had once existed only a stone's throw

In this image from a *Life* photo-essay from 1958, Mayor Lee stands proudly in front of the seemingly blank canvas of empty space created during the city's first urban renewal demolition project. (Robert W. Kelley/The LIFE Picture Collection/Getty Images)

away from the heart of the city. With Lee as our willing guide, the photograph asks *Life*'s readers to appreciate this vista as part of a new beginning, rather than an end, for the city's rapidly changing built environment.

But cleared landscapes such as these endured longer than anyone had anticipated or desired. Residents and businesspeople experienced relocation as a rushed process; and demolition, for all of its delays, came as a sudden blow. Yet the reconstruction process could be painfully slow. In a project in Buffalo, New York, for example, ten years after the displacement of 2,200 families from 161 acres of land, only six new single-family homes had been built. In Saint Louis, locals referred to a large overgrown and undeveloped tract in the center of their city as "Hiroshima Flats." On average, across the country, more than 20 percent of urban renewal land remained unsold by the end of the tenth year of execution. This occurred because of the "flattening out" of land disposition toward the end of a project, when only the less desirable parcels remained. It also resulted from a practice typified by the New Haven Redevelopment Agency's modus operandi of undertaking "the execution and activities of acquisition, relocation, and demolition before any serious attempts were made to market the land." As this land sat vacant, it cost the city tax revenues. After three valuable downtown New Haven blocks slated for commercial construction remained cleared for over eighteen months, one journalist crowned Lee "Mayor of the hole city."[54]

In 1967, a sea of temporary parking lots occupied prime New Haven downtown real estate that had been cleared for urban renewal but was still awaiting new construction. (Charles B. Gunn, Richard Charles Lee Papers [MS 318]. Manuscripts and Archives, Yale University Library. Courtesy Estate of Charles B. Gunn.)

The ensuing vacant, but littered, lots created eyesores for the public and served as insults to those who had called these same areas home. As one former New Haven resident recalled, "I always remember with a little bit of bitterness because the neighborhood was—was structurally the buildings were sound. And it was a nice neighborhood and it certainly was a safe area. And they knocked my house down and they didn't do anything in that neighborhood for ten years. They just left it a mess." Where new buildings had been built nearby, these empty landscapes also impeded repopulation as prospective residents felt justifiably disinclined to move next to what still looked like a slum. One former downtown worker best summed up the city's sorry state of renewal during an impromptu tour he gave an out-of-town visitor. As he later recounted, "Okay, so we, lunchtime we walked up Crown Street, all the way over to Church Street, and lo and behold, there was the New New Haven, and what was it? It was one great big immense parking lot."[55]

While cities across the country would have preferred to put urban renewal land to higher use, they often settled for interim solutions—like parking lots—instead. Parking lots mitigated the need for site maintenance while offering some revenue generation. Similar possibilities accompanied utilization of the land as storage sites. Community benefits, rather than revenue generation, motivated additional uses for these properties. These included serving as the locations for flea markets, portable library facilities, gardens, playgrounds, pocket parks, and temporary relocation housing for mobile homes. The city of Oakland, California, even turned the cleared site of a future shopping center into a putting green for a short time. While federal officials highlighted the economic rewards and goodwill that these uses could afford, they also warned local agencies to be clear from the outset about their temporary nature. Any psychological benefits could be quickly undone when the time came for the termination of the short-term uses.[56]

Worse still, cities never put some of their cleared sites to new uses at all. As late as 2004, Theresa Argento noted the continued presence of some underutilized lots in the Wooster Square neighborhood: "They said that New Haven was going to be a model city. It was not. We still have parking lots of property that [w]as demolished, beautiful homes." The Oak Street Connector, also still unfinished, is the most extreme example of this phenomenon. As former Legion Avenue resident Murray Trachten recalled, "What they did was years and years in advance of when they thought they would need any of this, they simply condemned all those buildings, they took them, they literally wiped out that whole section by taking every building there. In anticipation of there being a redevelopment of that whole area, and clearly as you go by there now, 50 years later or more than 50 years later, there's not much there. There's a four-lane road going in two directions and then there's that space in-between. But that was gonna be part of the future of New Haven, and it just never happened." In all these cases, the parking lots, empty lots, or other temporary uses that sprang up in the prolonged interim offered no substitute for what had been present before, or for what was supposed to come after.[57]

The public had largely supported Lee's early redevelopment plans, but their attitude changed to disappointment and anger as they experienced its harsh personal realities. As another former resident recalled, "Everybody, everybody thought it was a good idea when it was started but the minute it started to develop, . . . the houses would get torn down and nothing would happen." Theresa Argento perhaps best summed up the ensuing mixed emotions: "We talk about Dick Lee all the time. Poor man, I mention him so many times, but he thought he was doing well. He thought it was a good thing at the time, but it was not. It was not." Urban renewal, in New Haven and elsewhere, was not a sinister plot waged against the city and its minorities but an often well-intentioned program that proved far more successful in theory than in implementation.[58]

Soon, snow-draped scenes of the mayor standing amid renewal's destruction gave way to different representations of the process. Lee was more likely to be found posing at groundbreaking ceremonies, or in front of completed construction projects, than drawing further attention to the city's unfinished sites. In 1966, for example, Lee identified what he called "one of my favorite pictures." This image, which the New Haven public relations office distributed, shows the dignified mayor dressed in a business suit, and without a single construction implement in sight. Also appearing in the scene are a new parking garage and office tower. Instead of tearing down the old building spire that pops up in the distance, Lee seems to have purposely stepped aside to allow viewers to appreciate the harmonious juxtaposition of old and new. The contrast between this portrait of the mayor and those that appeared in the *Life* photo-essay suggests a change in how he implemented and represented his projects. Hands-on participation in the messy work of demolition had given way to hands-off administrative oversight of the end products of rebuilding. Lee also stopped condemning the old built environment in favor of the new. For him, urban renewal had become more complicated, focused less on the means than on the ends.[59]

The gradual departure of demolition from the city's publicity spotlight also reflected the ascendance of rehabilitation as a renewal strategy on par with, and even preferred over, redevelopment. Although rehabilitation had become a federally funded alternative to demolition with the passage of the Housing Act of 1954, it took most cities several years to begin to implement this new approach. New Haven's first substantial rehabilitation work occurred in its third urban renewal area, Wooster Square. When Lee went to Washington, D.C., in 1961, to report to Congress on his city's achievements, he brought evidence of this latest practice with him. Through before-and-after photographs of an exemplar rehabilitated private home, he offered visual evidence of the value of investing in existing properties. By that point, the city had targeted Wooster Square with more demolition than rehabilitation by a factor of nearly two to one. Yet only three years after his triumphant *Life* photo spread, Lee chose to emphasize rehabilitation, the less representative part of the neighborhood's renewal plan.[60]

Experience was already beginning to show that clearance was not always the optimal solution for cities. It was messy and difficult and came at a high price. These social and cultural consequences incited criticism, and by the mid-1960s, residents were less willing to acquiesce in the proposed future paths of the bulldozer. Organized resistance followed. Since much of the planning was already in place by then, however, the best for which many protestors could hope was to reroute the destruction, rather than put an end to it entirely. Or they could do as one local businessman did: work within the system to preserve whatever material traces of the past that he could. Ken Lassen's small-scale acts of preservation memorialized the very pieces of derelict rubble that the city had previously celebrated destroying.[61]

Lassen was the third-generation operator of a popular New Haven burger shop known as Louis' Lunch. The luncheonette was located across the street from much of the early downtown redevelopment work, and only three-and-a-half blocks from the site of the city's opening demolition ceremony. From his front-row vantage point, Lassen observed the large number of businesses being lost in the destruction. Having gone to school with the son of the head of Barnum Lumber Company, one of the demolition contractors, Lassen contacted his old friend to express interest in salvaging some of the bricks still lying on the sites. He later recalled his query: "Ask your father if I can work over these buildings. I want one brick from a building that only had one business, and if there was five businesses in the building, I want five bricks. I want to pick them out myself. And I don't want anybody to get in trouble. How do we work it?" Barnum granted his request; and after each building fell, Lassen would scour the site for the particular brick or bricks of interest. He reflected, "Well, everybody said 'you're crazy.' There's too many tears over there not to be crazy! All those poor people. What are you going to do? I took my bricks home and washed them and cleaned them and put them away, not knowing what I was going to do with them."[62]

Eventually, in the mid-1960s, the agency notified Lassen that his business was next. But he was adamant about not only staying in the neighborhood but also continuing to occupy his modest 12 × 18–foot, single-story brick building. Louis' Lunch, established in 1895, had by the postwar period become a local institution. It is popularly believed to have introduced the hamburger sandwich to the United States in 1900, and Lassen felt that it merited preservation. Although he failed to get a listing on the National Register of Historic Places, he waged a lengthy public campaign, which aroused substantial support locally and beyond. Lassen's objective was to acquire his own 30 × 30–foot piece of property, located in the general vicinity of his existing site, on which to relocate his building. The city was sensitive to his requests, but the challenge of securing such a humble plot of land amid the mammoth-scale medical and industrial projects already planned for the area proved almost impossible. At the last minute, Lassen obtained an acceptable property. Instead of bearing just the price of demolishing the structure—originally set at $1,000—the city agreed to split the $5,700 cost of moving the restaurant to its new home. The agency's director, William T. Donohue, remarked of the decision, "I think this action illustrates the good faith and commitment of a public agency toward a local businessman. It also goes a long way toward dispelling the notion that redevelopment is insensitive to historic preservation and favors the wrecking ball." His comments could not have diverged farther from Mayor Lee's dramatic celebration of that same implement less than two decades earlier.[63]

On July 29, 1975, the Yonkers-based house-moving firm of Nicholas Brothers placed the Louis' Lunch structure on a flatbed truck. They hauled the small building to its new location a few blocks away. Where before the restaurant had shared a party wall with a neighboring tannery, the relocated building now stood alone. The bricks that Lassen had been collecting over the years found a new home as the missing fourth wall of his relocated restaurant. As he put it, "All those tears were accounted for right then and there. I really enjoyed that, to say that these people were not forgotten. Not forgotten." Indeed, his efforts quickly began to pay dividends. Within a couple of months of reopening, a customer came in who had worked in a bakery represented by one of his bricks. Lassen knew the building. He concluded, "And one tear from the bakery traveled down his cheek and that paid for the whole thing."[64]

By the 1960s and 1970s, the tide was turning against urban renewal. On

As documented by the *Journal of Housing,* house movers relocate Louis' Lunch out of the way of redevelopment, July 29, 1975. (Courtesy National Association of Housing and Redevelopment Officials)

a national stage, urbanists like Jane Jacobs, in her *Death and Life of Great American Cities* (1961), used the case of New York City's West Village to argue for the vitality, rather than the obsolescence, of mixed-use neighborhoods. A year later, in *The Urban Villagers,* the sociologist Herbert Gans questioned the labeling of ethnically marginal communities, such as Boston's West End, as slums. In a third widely read critique, *The Federal Bulldozer* (1964), Martin Anderson took a conservative perspective to condemn the absence of realized economic benefits from urban renewal. These were only the most notable of a growing tide of discontent directed toward large-scale clearance and its prospects for renewing the postwar city.[65]

Complementing these written treatises were individual lived experiences that endorsed alternative urban futures through practice. In the case of Lassen's New Haven eatery, perseverance helped preservation and relocation win out over demolition. For decades, in fact, preservation had been on the rise throughout the nation. This was both a reaction against the inequities and material losses of urban renewal and a practical response to the process's economic costs. Philadelphia's Society Hill neighborhood pioneered the practice of preservation-based renewal. Reflecting an appreciation for the colonial period, preservationists there deemed relatively few buildings from later eras worth saving. Far fewer structures would have survived had planners not championed restoration for its relative economic benefits as well. In 1966, the National Historic Preservation Act institutionalized preservation by introducing widespread legal protection for historically significant structures.[66]

The Housing and Community Development Act of 1974 officially ended urban renewal. Policy makers designed the replacement Community Development Block Grants (CDBGs) to minimize the federal government's role and to provide support for a wider range of community-based mechanisms for urban revitalization. At the block grants' inception, with nearly a thousand urban renewal projects still under way, acquisition and clearance made up roughly 20 percent of fund allocations. A decade later, demolition was down to 4 percent, housing rehabilitation had grown from 11 to 35 percent, and a total of only twelve urban renewal projects remained nationwide. Where demolition continued apart from urban renewal, as in the form of Urban Development Action Grants (UDAGs), first established in 1977, the work shrank in size. Instead of focusing on "large-scale, longer-term development packages," UDAGs were shorter, single-project efforts. The locations of some of these projects upon previously cleared, still undeveloped land helped to right some of urban renewal's initial wrongs.[67]

Even Herbert Gans agreed that the collective efforts of ground-level protest—rather than the writings of individual critics—were most directly responsible for these policy changes. New Haven's Redevelopment Agency, for example, met with more organized resistance in the largely African American neighborhoods of Dixwell and the Hill, which were targeted for the next urban renewal projects. Part of the Dixwell dissent derived from concerns over a ring road proposed for the area. Across the nation, perhaps the most vocal and widespread protest rose up in response to proposed highway clearance such as this. As the bulldozer ravaged urban neighborhoods to make way for new housing, office buildings, and shopping centers, it was also clearing similar kinds of communities for highway construction. This clearance extended beyond urban boundaries to include suburban and rural landscapes as well. In both anticipation of and response to these practices, citizens across the nation rallied to protect their homes, communities, and the natural environment. While their success was mixed, they joined with anti-renewal voices to reframe the previously positive depictions of large-scale postwar clearance.[68]

Although planners and municipal bureaucrats became the objects of citizen protest that resulted from urban renewal, it was the engineer who largely directed the bulldozers of postwar highway development. State highway departments, staffed largely by engineers, controlled both technical expertise aimed at efficiency and order and legions of equipment capable of turning their knowledge into landscape reality. The engineer's experience in creating the interstate highways offers a third example of postwar landscape development predicated on large-scale clearance. As in the cases of urban renewal and suburban development, the engineer and his powerful machines were first celebrated and revered before they too were criticized and feared.

Chapter 5

"The Intricate Blending of Brains and Brawn"

Engineering the Postwar Highway Boom

A 1950 article in the *New York Times Magazine* vividly recounts an attack by a powerful group of invaders on one quiet postwar community: "Like a modern mechanized army, except that it was given no air support, the force arrived by road and rail, set up its camps, and took over. Officers, men and machines spread out over the terrain and, without firing a shot, captured the territory north and south of this village in the name of the Empire State. They will not depart until some time next summer, when they have tamed the wilderness and left behind another segment of the wide and handsome Thruway that some day will cross Governor Dewey's domain from Buffalo to the Bronx." As becomes clear by the end, this scene was from neither the Korean War, just then beginning, nor World War II, completed half a decade earlier. This was a home front invasion. As private contractors and civil servant engineers effectively imitated recent military practices, the era of superhighway construction had begun.[1]

More than mere allusion, the article's battlefield references reflect the look, feel, and execution of this work, as well as the cultural context within which readers might have placed such a scene. As the author notes, "Building a road, we learned, is like any major battle, a matter of careful and precise planning." Photographs of the Thruway—appearing in an array of media venues—furthered these military-highway connections. The conclusion of the *Times Magazine* piece looks longingly toward the future, when scenarios like this will soon play out across the country: "Somewhere else bulldozers will be locked in battle, power shovels will be making their passes at Mother Earth, and armies of slogging foot soldiers will be following the tractors to victory."[2]

Projects such as the Thruway paved the way for the vast postwar system of interstate highways to come. The Thruway was the first statewide limited-access expressway system. Along with similarly pioneering efforts in New Jersey, Pennsylvania, and California, it helped set subsequent interstate system standards. The American Association of State Highway Officials (AASHO)—later the American Association of State Highway and Transportation Officials (AASHTO)—adopted many of these highways' design criteria, and the interstate system eventually incorporated most of their routes. Even some of the leaders behind these early postwar roadways transferred from turnpike to interstate oversight. Charles M. Noble, for example, migrated from directing construction on the New Jersey and Pennsylvania Turnpikes to leadership on Ohio's interstates. Bertram Tallamy served as head of the New York State Thruway Authority before becoming the first federal highway administrator in 1956. Similarly, early postwar expressways in places like New York, Chicago, Detroit, and Los Angeles established templates for systems that would follow as part of the urban interstates.[3]

Assembling the more than forty-six-thousand-mile interstate network required large-scale building and land clearance. In many ways, the work mirrored ongoing practices for urban renewal and suburbanization. Yet highway clearance was also different. The federal government covered 90 percent of costs. This spurred states to pursue the greatest amount of new construction in order to maximize their shares of the Highway Trust Fund. Highway clearance also wrought its own physical and social devastation upon the land. It cut through neighborhoods, farms, woodlands, and mountains, without necessarily wiping them out. Furthering the social injustices of urban renewal, victims of highway development were compensated less substantially and uniformly than those of the former, a discrepancy that magnified its harms.

Highways offered the possibility for less destructive forms of clearance through house moving, although the realities of implementation led to demolition more often than might have been strictly necessary.

Two other critical distinctions of the highway program were its explicit link between military and civic approaches and the lead role played by engineers in establishing those connections. Wartime exigencies enabled many engineers to train for subsequent home front highway work. They brought to this task an "intricate blending of brains and brawn," as the *New York Times Magazine* put it. The combination of physical boldness and technocratic rigor proved integral to the creation of postwar highways. An extensive oral history project sponsored by the Public Works Historical Society affords direct access to the history of the highways' development recalled in the engineers' own voices. The interviews reveal a mentality that pragmatically, if sometimes naively, maximized opportunities for clearance. The engineers prioritized economy and efficiency over environmental considerations and individual social impacts. At least initially, both engineers and the general public expressed broad faith in this technical approach. Yet in practice, the process of clearing the landscape for interstate highways became as implicated in the contentious "design politics" of large-scale postwar development as its urban renewal and suburbanization contemporaries.[4]

Defense Connections and Federal Support

Both military and civilian demands propelled the rise of the National System of Interstate and Defense Highways, but the military connections have generally been less well understood. The Department of Defense signed off on every location and specification for the system, from roadway widths to clearance heights. Any design discrepancies resulted from changing contexts and opinions. The interstates also served as a job stimulus program for World

---

(*Opposite*) A front-end loader and operator battle the landscape as they clear the way for New York Thruway construction in 1954. As with her visual documentation of construction equipment during World War II, photographer Margaret Bourke-White imbued her technological subjects with triumphant grandeur. (Margaret Bourke-White/The LIFE Picture Collection/Getty Images)

War II veterans, providing an opportunity for direct application of military engineering expertise on the American landscape. After returning from the war, Seabee Jacques Yeager, for example, finished his civil engineering degree at Berkeley and then co-led the family construction business as it worked on almost every southern California interstate. Additionally, although not put to the test of a nuclear emergency, the interstates provided valuable Cold War infrastructure. In North Dakota, for example, they assisted in moving equipment to and from two of the country's largest air bases. As an engineer from that area recalled, "They haul an awful lot of heavy equipment over the highways and they usually use I-29 or else drop down to Highway 83. . . . There's no question in my mind that some of the heavy loads that they bring the equipment in on, why it's helped an awful lot."[5]

These connections were long-standing: the military had been closely embedded in the highway system from its earliest days. During World War I, the military organized truck convoys to supplement railroads for logistics. The difficulties of navigating these challenging routes highlighted the need for better infrastructure. The first Federal-Aid Highway Act passed in 1916, when Europe was at war and the United States was preparing to join in. It established a highway program funded fifty–fifty by federal and state governments but did not prescribe design standards. Three years later, the army attempted the first transcontinental truck convoy. The difficult two-month trip "dramatized the need for better highways," as one participant, Lieutenant Dwight Eisenhower, later put it. In 1922, at the request of the Bureau of Public Roads (BPR), the army produced the Pershing Map (named for the general behind its design), the first map of roads of prime importance in case of war. The military updated the map in 1935, and most of the routes would later become federal or interstate highways.[6]

The Federal-Aid Highway Act of 1938 officially set the interstate program in motion. It charged the BPR with studying the possibility of a six-route network, including three north-south and three east-west roadways. The resulting report, *Toll Roads and Free Roads,* released a year later, detailed the first formal concept of the interstate system. President Roosevelt transmitted the report to Congress, with endorsements by the secretary of agriculture and the secretary of war. For purposes of both "national defense and the needs of a growing peacetime traffic," the report recommended a non-toll interregional system, 14,336 miles long, that generally avoided urban areas. The design of these superhighways would be multilane and limited in access, and they would require right-of-way procurement at a minimum of thirty-six acres per mile in rural areas and nineteen in urban areas.[7]

Although World War II halted the implementation of these plans, it offered opportunities for military-led roadway development that continued to advance highway thinking and skills. Navy Seabees and army engineers built roads across the globe, and the BPR partnered with the army on the massive Alcan Highway. At home, the ebb in domestic highway-building activities afforded time for planning postwar projects. Ohio, for example, assigned sixteen engineers to complete reports on future interstate locations. By war's end, they had assembled engineering surveys for all rural routes in the state. At the same time, BPR officials and individual members of Congress continued to advocate for preparation of national postwar highway plans. In 1941, President Roosevelt appointed the National Interregional Highway Committee, led by BPR director Thomas H. MacDonald, to conduct a formal study. Three years later, the committee released *Interregional Highways,* a report that advocated the creation of a 33,920-mile system, in addition to 5,000 auxiliary urban miles. Following hearings, Congress incorporated the report's recommendations into the Federal-Aid Highway Act of 1944. This act was notable for designating a National System of Interstate Highways of up to 40,000 miles in length to connect metropolitan areas and serve the national defense. With the broad roadmap now in place, it became the project of the postwar era to tailor the plans at the state level and galvanize sufficient support to secure funding for their implementation.[8]

European roadways provided additional models. Although MacDonald had toured France, the United Kingdom, and Germany in 1936 to research a prospective national highway system, increased numbers of Americans and future highwaymen got a chance to see these same systems during the war. As the former secretary of the Illinois Department of Transportation later recalled, "Having served in World War II in Europe, and seeing how they utilized their system, and seeing (in) some other countries where they did have a very bad system, it just brought to mind that, upon return, that it was a necessity to have such a system as a defense system." General Dwight D. Eisenhower echoed these sentiments and was perhaps the most significant military man influenced in this way. Referring first to his role in the 1919 transcontinental truck crossing, he wrote in his memoir, "The old convoy had started me thinking about good, two-lane highways, but Germany had made me see the wisdom of broader ribbons across the land." He further reflected, "After seeing the autobahns of modern Germany and knowing the asset those highways were to the Germans, I decided, as President, to put an emphasis on this kind of road building. . . . This was one of the things that I felt deeply about, and I made a personal and absolute decision to see that the nation would benefit by it."[9]

Just one year into his presidency, Eisenhower assembled the President's Advisory Committee on a National Highway Program, which united military and business interests behind the interstate program. The committee's members were largely industry experts with direct stakes in the plan's passage. They included executives from Bankers Trust, the civil engineering contractor Bechtel, the earthmoving equipment manufacturer Allis-Chalmers, and the Teamsters union. Heading the committee was Lucius Clay, a retired general turned board member of General Motors, which had recently entered the earthmoving equipment field through its purchase of the Euclid Company. Frank Turner served as executive secretary of what became known as the Clay Committee. Often credited with making the interstates a reality, Turner was a longtime BPR employee (he eventually became its director) and a wartime veteran of both the Alcan Highway project and postwar Philippine highway reconstruction work. The committee's recommendations helped lay the final groundwork for the Federal-Aid Highway Act of 1956.[10]

A critical feature of the 1956 act was its strong federal-funding provision. This authorized an initial fund of $25 billion for a forty-one-thousand-mile, ten-year building program. The program duration, roadway length, and fund size all increased over time. One constant, however, was the federal government's 90-percent sponsorship of interstate construction costs, versus the fifty–fifty cost sharing with the states that characterized primary roadways. The impact of this funding was significant. Before the act's passage, the federal government had spent just over a billion dollars, cumulatively, on urban highways. By 1960, that figure was up to one billion dollars *per year*. Douglas Fugate, the former commissioner of the Virginia State Highway and Transportation Department, voiced an opinion shared by many of his fellow engi-

In January 1955, President Dwight Eisenhower accepted the Clay Committee's report on the national highway program, which laid the groundwork for the interstates. Committee members included representatives of related business interests, including (*l.–r.*) retired army general Lucius Clay, Frank Turner of the Bureau of Public Roads, Steve Bechtel of Bechtel Corporation, Sloan Colt of Bankers Trust, Bill Roberts of Allis-Chalmers Manufacturing, and Dave Beck of the International Brotherhood of Teamsters. (Courtesy the Dwight D. Eisenhower Presidential Library and Museum)

neers when he concluded that, if they had been left up to the individual states, the interstates probably would not exist today.[11]

Federal involvement also extended to the highways' form. As one Colorado engineer recalled, "I certainly felt [the federal engineers'] influence a lot more on the interstate system than I did on the secondary and primary [roads]." Civilian and military interests—from the BPR to the AASHO and Department of Defense—jointly established highway design standards. These standards endorsed straight roads with modest grades. By contrast with the two-lane roadways that dominated primaries of a previous era, more than 90 percent of interstates consisted of four lanes or more. Lanes were

This 1967 aerial view of a stretch of I-70 in Colorado shows the character of interstate highway construction, as well as the large-scale land clearance and earthmoving necessary to accommodate this infrastructure within the existing landscape. (Steve Larson/Denver Post/Getty Images)

also widened, increasing from nine to ten–foot widths in earlier iterations to twelve to thirteen–foot widths on the interstate. Broad medians and wide shoulders further expanded the interstates' footprint, making the paved area per mile 2.75 times wider than that of predecessor roadways. The total associated right-of-way acquisition area was roughly the size of the state of Delaware. The accompanying excavation was greater as well, totaling 9 times as much as that required for roadways of the preceding generation. According to one boosterish BPR publication, the interstates would involve "moving enough material to bury the whole State of Connecticut knee-deep in dirt."[12]

The construction industry enabled the relatively rapid and economical achievement of these physical feats. The same kinds of powerful equipment

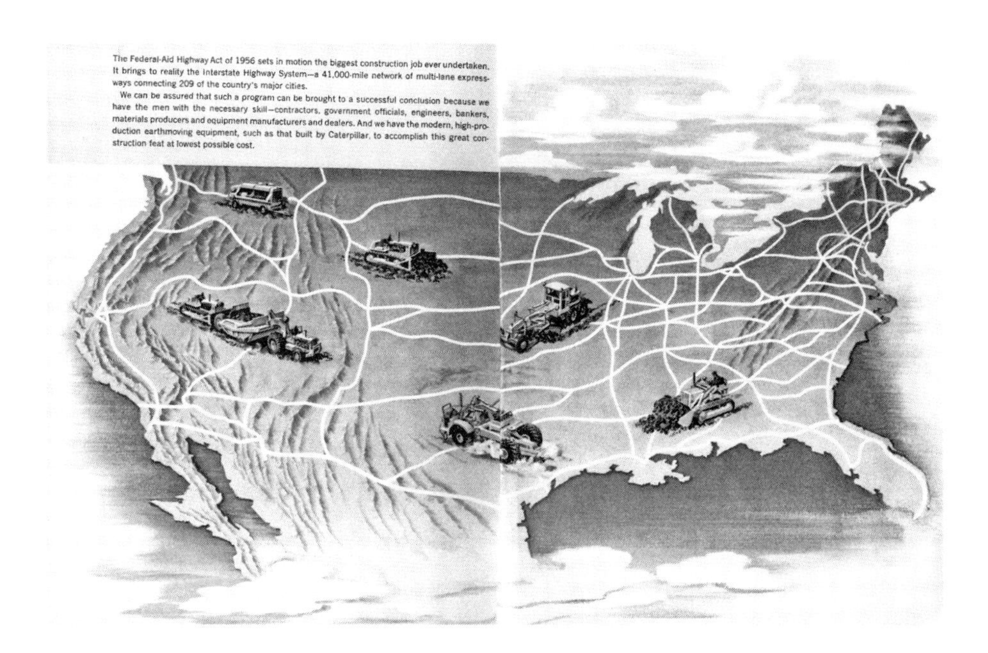

This depiction of the interstate roadmap in Caterpillar's 1956 annual report portrays the diverse kinds of earthmoving equipment necessary to realize these highways. (Reprinted courtesy of Caterpillar, Inc.)

that moved earth for postwar suburban development proved similarly useful in highway work. In particular, the commercial launch of the rubber-tired bulldozer in 1947 and the vibrating roller in 1948 greatly assisted the road-building industry. One 1949 article in *Life* also chronicled the all-important loaders, carriers, rock crushers, and "treedozers that pluck out good-sized trees like radishes. These Paul Bunyan-esque devices have not only moved mountains but have made highways much safer, since their capacity permits leveling of dangerous hills and moderation of curves on a scale heretofore unfeasible." Descriptions like this aptly emphasized that the equipment's technical prowess advanced efficiency, capacity, and—importantly—safety. Further, the machinery accomplished all of this while simultaneously taming the landscape. The annual reports of manufacturers like Caterpillar even likened their highway equipment's dispersal across the country to the settling of the frontier. In this case bulldozers, scrapers, and other construction machinery took the place of stagecoaches in leading the latest wave of development. These were themes that were repeated throughout postwar promotional literature. Not only the technical capabilities but also the sheer volume of this equipment facilitated the grand scale of interstate work. Prior to passage of the 1956 highway act, the American Road Builders' Association inventoried the industry. The group counted 331,000 pieces of equipment, including trucks, tractors, and scrapers. Combined with recent investments in new manufacturing plants, these promised ample mechanized supply to initiate the project.[13]

Despite such coordination across public, private, and military organizations, reports that overpass vertical-clearance standards would not guarantee safe passage of all military vehicles led some to cast doubt on the sincerity of at least the last of these connections. Some observers even suggested that the defense appellation was mere "window dressing," as one California transportation consultant put it, appended at the last minute to tip the scales in favor of the much-debated program. Yet in early 1957, representatives of the army, navy, and air force all agreed that fourteen-foot clearances would meet their needs. Their assessment was pragmatic, a recognition that such specifications would accommodate the majority of military vehicles. Exceptional cases would simply require special provisions, such as routing vehicles around existing overpasses, partially dismantling them, or transporting them by rail. As equipment increased in size, and particularly after the Russians' launch of *Sputnik* in 1956 elevated the significance of missile defense systems, some members of the military began questioning the long-term viability of these workarounds. Thus, when Congress held hearings in 1960 to address discrepancies between vehicle height and roadway clearance, the members were neither illuminating an overlooked issue nor doubting the connection between highways and military needs. Rather, they were investigating whether there had been "negligence or carelessness" in fully appreciating the need for higher clearances.[14]

Given significant overlap between military and civilian highway usages, satisfying defense demands rarely required military-specific designs. According to the army's assistant chief of transportation for transportation engineering, the "defense highway" purposes of the interstates included "goods movement" by repurposed civilian vehicles in order to connect more than one hundred highly dispersed defense-related industries; "flexibility and control over the movement of large volumes" of vehicles, particularly under national emergency conditions; and troop movement on controlled-access roadways.

Most of these functions could be accomplished within established height limits. In the early 1960s, the military revised its minimum overpass clearance expectations to sixteen feet (in rural areas, with special provisions for urban interstates). Over time, engineers made height adjustments on roughly 350 existing structures that served 95 percent of major military installations. These investments affirm the tangible significance of the interstate highway system to evolving military objectives, even as a combination of economy and practicality restricted the extent to which those objectives explicitly shaped overall highway design.[15]

Engineers and the Bulldozer Mind

With the relevant policies in place, civil engineers led the execution of interstate development. Populating both state highway departments and a growing number of private contractors, they implemented plans on a scale previously unrealizable. In fact, the *New York Times* declared 1956 "the age of the engineer." Also referencing advances in communications, atomic weapons, and bridge building, the newspaper concluded, "All of these have been formidable projects, and if on many an occasion the doctrine of eminent domain has been brought forward to cancel out the castle rights of man—well, the greatest good for the greatest number. Speeding along the [New York State] Thruway, it is inevitable to think of the engineer as riding high, wide and handsome."[16]

These postwar engineers earned both power and prestige through their distinctive "engineering mentality" or, as Lewis Mumford put it, the "bulldozer mind." That is, engineers generally prioritized the technical logic of efficiency and economy over social and environmental concerns. While the engineering mentality was not born during the postwar era, it flourished then. The public initially embraced engineers' technocratic expertise as being removed from political influence. Many engineers shared this view. As the California engineer Chuck Pivetti recalled, "The staff believed they were building the public's highways. That they were building them the way the public wanted them and where they wanted them." At the same time, in practical terms, massive federal funding and road-building power enabled their grand plans to become built realities.[17]

The unfettering of the engineering mentality served several functions. It offered clear priorities for executing the massive undertaking of highway work in a coordinated manner across the nation. It also lent this work prestige, which served as both a recruiting tool and a psychological validation for its practitioners. But, as Pivetti also acknowledged, "There was a very strong sense that those highways were ours, they belonged to the division of highways, and if the public had enough sense they would realize we were putting them in the right places and doing the right things with them. We tended to bully and steam roller opposition to our plans for the highways." Similarly, a Colorado engineer paraphrased evolving public sentiment as "You guys just bowl your way through." Implemented to excess, the engineering mentality eventually spurred critical feedback that helped make the interstate project more humane and environmentally sensitive.[18]

The engineering ranks of postwar state highway departments swelled during the 1950s and 1960s. Many senior engineers had worked in these departments for a decade or more (albeit perhaps punctuated by wartime service), but lower-rank positions were harder to fill. While states like California and New York paid well for the profession, private-sector engineering salaries generally exceeded those for public-sector highway work. In Kansas, for example, one engineer estimated that only 10 percent of the civil engineering graduates in his class chose state highway departments for their profession. He further recalled that, of his six job offers, the Kansas highway department's compensation "was the lowest offer I had, by a considerable amount." The development of active recruiting and training programs helped fill this void. Recruiting materials sometimes tried to compensate for financial shortcomings by stressing the immense scale of the undertaking and the unique opportunities to move massive quantities of earth and play with great machines. California also sent recruiters to technical schools as far away as Mississippi. By contrast, states like Kansas and Colorado focused on the local labor force. As one engineer recalled of the new recruits, "Most of them were returning veterans. A lot of them were—a lot of them did pretty good." While state departments trained these men in their own schools, over time, local college graduates also began adding to the ranks.[19]

While about half the states increased their internal personnel to staff interstate development, the remainder hired outside consultants. As in the cases of urban renewal demolition and suburban earthmoving, private engineering and heavy construction firms necessarily grew in response to this increased demand. One Kansas engineer recalled, "Certainly at the time, we had firms that came into existence or maybe they were small and they got real big during the interstate. We had lots of firms—lots of grading contractors. Now [by the 1980s] that's died out, for instance, we don't have many grading contractors." These contractors were predominantly local, and this same engineer estimated that in-state contractors accomplished three-quarters of the highway work.[20]

Both new recruits and established highway personnel recognized the nationally significant public benefits derived from their postwar highway work. As a Kansas engineer reflected, "There is a sense of accomplishment, that whole group. There [are] a lot of people who were World War II veterans that came back and were members of the various highway organizations that got to be a part of the interstate system. Look back on their lives and their work—the work aspect of it was very satisfying. Very satisfying. And I think it made them better people." He elaborated, "It was such an exciting time that everybody just enjoyed it. If you hear people talking about all the hard times of the interstate, you look them in the face and they are smiling, because they enjoyed it. The challenge was out there to do something that was going to really—everybody believe[d it] really was going to help the country, whether it is economically or just the fact that it makes my vacation easier." Such characterizations incorporate the postwar engineers into the mythology of the Greatest Generation. Ultimately, he concluded that the interstate era produced a sense of "wonderment" in engineers and "gave us, I think, a good feeling."[21]

Accompanying these public benefits was the corollary appeal of personal power. Bertram Tallamy recognized his almost dictatorial control: "I had the authority to build the road anywhere I wanted to make it. I had authority to buy property and sell any. . . . I myself could do it, or my signature on the whole thing, for 680 miles of highway." A photograph of Tallamy appearing in *Life* just months after his appointment as federal highway administrator makes visible the technocratic power of this "exacting and practical engineer." Standing suit-clad beside a portion of the completed New York State Thruway, Tallamy motions with his hands to explain one of the roadway's

The inaugural federal highway administrator, Bertram Tallamy, appears alongside a stretch of recently completed New York State Thruway in 1956. By contrast with this serious portrait of technocratic power, outtakes from the *Life* photo shoot depict Tallamy smiling contentedly beside the roadway and peacefully enjoying the outdoors while canoeing. The published image illustrates the description of Tallamy in the text as "an exacting and practical engineer." (Paul Schutzer/The LIFE Images Collection/Getty Images)

technical details. While the waist-level cropping of the image obscures his overall stature, his serious pose pairs with the long roadway stretching behind him to suggest his larger importance.[22]

The bulldozer mind revealed itself most clearly in the establishment of highway routes. After federal officials identified broad highway corridors across the nation, state highway departments translated them into specific locations. Traffic-flow studies primarily directed their selection process, with engineers generally favoring the shortest route and straightest alignment for maximum savings. According to the engineer Edward Haase, of Colorado, "The cost-benefit ratio to the traveling public was the thing, and you could prove cost-benefit simply by way of the shortest distance. I mean, that was a very critical factor because [of] the saving of mileage, the saving of gas to hundreds of thousands of motorists and so forth." In mountainous terrain, straight alignments often meant maximum earthmoving. As Haase put it, "To get standard alignments, that is, standard speed alignments through the mountain, the mountain doesn't necessarily know that that's what you want, and sometimes it required big cuts and big fill and so forth. I thought that was just part of the game." A Kansas engineer later acknowledged that such technocratic decisions were not necessarily the most accommodating of affected residents or environments. He concluded, "I would say as a group that probably engineers haven't been overly sensitive. . . . If they had a choice between building a road in a certain place or preserving some species of frog, . . . this wasn't a hard decision for the engineer to make."[23]

Highway design practices had not always followed this approach. Before the war, engineers in places like Iowa, Missouri, and West Virginia tended to construct crooked or steep roads. By contrast, when engineers in prewar Kansas faced the choice to either "go through the countryside with the least possible disturbance to the natural drainage/grades" or "establish a point here and a point there and go through there," they took the second option. As one Kansas engineer recalled, he and his colleagues "didn't hesitate to disturb the natural." The added "benefit," he argued, was the displacement of large quantities of earth that were then put to use in filling valleys for the creation of more than a hundred dams and lakes. During the interstate era, the Kansas way became nationwide practice. As a Georgia engineer noted,

As shown in this 1970 *Denver Post* image, I-70 cuts through Colorado's Glenwood Canyon. Despite the incorporation of terracing on adjacent slopes, controversy arose regarding overall route selection through this natural landscape. (Floyd H. McCall/Denver Post/ Getty Images)

"In the early years you got to go from point A to point B, on an interstate highway and boy, let's get there and move dirt and put down pavement."[24]

The interstates' dramatic imprint upon the landscape derived from both this preference for point-to-point routing and states' attempts to maximize their individual shares of federal highway funding. When deciding between widening an existing roadway or forging a new alignment, state engineers generally chose the new route. As one California engineer put it, "I think a lot of the perception was, hey, get all you can while you can, and it only costs ten cents on the dollar." Maximum land clearance resulted. Interstate 5, in southern California, illustrates this phenomenon. Multiple options existed for routing the corridor between Mexico and Canada, including incorporating the existent state highway 99. "But since there was 90 percent money to build it, and the people who were at the top figured that they needed another highway over against the hills," as another engineer explained, "they designated the Route 5 as being over against the hills. 99 wasn't an interstate." While some have questioned why particular freeways were not interstates, the California engineer explained it simply: "Well, it's not an interstate highway because we already had a lot of four-lane construction out there. The perception was that we could continue to build it with our state resources and we would take the new windfall of interstate money and go build new things. Get as many miles of new roadway as possible." In total, nationwide, engineers forged new routes for roughly 80 percent of interstate highways.[25]

Advances in photographic and computational technologies enabled engineers to locate these routes precisely. Although photogrammetry, or aerial photographic surveying, began in the late nineteenth century, it improved significantly during the world wars. In the postwar decades, the engineer Douglas Fugate recalled, "We went into aerial photography in a big way." The Federal-Aid Highway Act of 1956 authorized the use of photogrammetry for mapping, and highwaymen deployed the method to compile topographic maps that would enable the selection of "the most economical or most feasible" interstate alignments. This saved time and money over previous manual land surveys. Eventually, the method's accuracy was so well recognized that in some places before-and-after photogrammetric surveys provided the main mechanism for calculating earthwork payments. This was a sharp break with previous practices. As one Colorado engineer described, "To determine earthwork quantities, people were sitting over desks, humped over desks, moving a plastic strip up and down and across the cross-section to determine the end areas and day after day after day. Just awful. Surveying methods were archaic. Nothing had changed for many, many years." Thus, new tools beyond construction equipment alone made engineers' shape-shifting designs for the vast interstate network more achievable. They also further distanced engineers from the affected landscapes, lessening the visibility of their social and environmental impacts.[26]

Right-of-Way Acquisition and Rural Roadways

Following route selection, right-of-way acquisition could prove both challenging and expensive. In the prewar years, the federal government would not assist with legal or procedural obstacles related to this important step. With the Defense Highway Act of 1941, however, it first began reimbursing right-of-way expenses. The Federal-Aid Highway Act of 1944 even subsumed these under the banner of "construction" costs, which the federal government then covered in part (up to 50 percent in the case of primary roadways, and at 90 percent for interstates after the 1956 act). Also significant was this later act's allocation of funds to states based on need, rather than a fixed apportionment. Previously, in order to save their limited federal funding for actual construction, many states had relied on counties and cities to furnish rights-of-way free of charge. After 1956, however, they no longer had to apportion funds in this limited way. Greater support for rights-of-way enabled the acquisition of more land for more new roadway construction.[27]

Many states began right-of-way acquisition in rural areas. This was the path of least opposition and also the easiest terrain on which to make quick, visible progress. As one Kansas engineer recalled, "The decision was made early that for the most part we would build the rural interstate first, and get as many miles under construction as we could for the—with the money that was available to give the people in the state a chance to really see what the interstate was all about. Where if we had started in the urban areas, we would have spent a lot of money and really wouldn't have very many miles to show for it." The Federal Highway Administration (FHWA) endorsed this rural-first strategy, which many other states also followed.[28]

The long, wide roadways they plotted caused hardships for some. Rather than interrupting traffic flow with entrances for every driveway, farm, and crossroad along the highway's path, as many prewar roadways had done, the interstates had limited access. This stranded some local businesses that would no longer be easily accessible. As the Colorado engineer Charles Shumate recalled, "Well, you've got a guy whose [*sic*] got a filling station here peddling gasoline to the people and you've got to tell him that they're not going to come by here anymore, it's pretty hard to convince him that he's not going to get hurt." For many farmers, impassable roadways divided their property, impeded cattle passageways, and imposed diagonal obstructions that disrupted efficient planting and farm equipment operation. A Kansas engineer recalled an uncomfortable incident that occurred while he was out

In this photo-worthy case from 1962, Colorado's I-25 briefly jogs to the west to avoid the costs and hassles of relocating the irrigation structures that lay in its original path. (Lowell Georgia/Denver Post/Getty Images)

surveying plans for a roadway: "A farmer met me with a shotgun. And he said, 'You are not going through my land.' And I didn't. He took it to court, and I had to be a witness!" More commonly, however, when separation damages exceeded costs of construction, engineers tried to accommodate farmers with underpasses and other work-arounds. Only rarely did the interstates interrupt their straight trajectories to allow for special impediments.[29]

Some speculative individuals sought to profit from right-of-way acquisition. They might purchase property expected to lie in the path of future land assembly with the goal of reselling it at a higher price. One group, for example, developed plans for a shopping center in Ohio and then complained about the millions of dollars in damages they would experience due to construction of the highway and its interchange. As one engineer recalled, "I said, 'You'll be glad to know that we changed the design and we are not going to touch it.' Before they could think, one guy says, 'My God, what will we do with it now?'! They had it all set out where they could make a real killing." Congressional hearings would uncover similarly questionable practices among private developers who were paying off government employees for advance notice of planned interchange locations. In these cases, their payoff would be a jumpstart in acquiring and building upon land in what would soon be the most desirable locations.[30]

To minimize both the disruptions of highway construction and the economic burdens of speculation-induced compensation, engineers aimed to acquire land early and abundantly. But even in exceptional places like San Diego, where discretionary funding and strong public support for highways enabled an active right-of-way preservation program, the results were limited. As one California consultant observed of his state's policy, "I was proud of it, but it wasn't all that great. It wasn't really enough to do the kind of job if we knew how to do it." Rather than using the funds to plan ahead, the money primarily went toward "preventing somebody from putting something in that we'd have to buy sooner or later." Essentially, it was a "putting out fires fund." Advance property acquisition also posed practical challenges. Highway departments had to prove near-term necessity in order to satisfy private-property rights and minimize tax losses. Then, following acquisition, states often struggled to hold on to property for too long. After California began parting with excess rights-of-way in the face of opposition, one engineer lamented, "If I had my druthers, I would very rarely sell off any of the right-of-way. I would keep it for at least ten years and find out how the climate changed."[31]

Such practices supported a larger criticism levied by some engineers against the interstate development process: that it impeded truly big thinking. Greater right-of-way acquisition would have provided buffers for future development and created opportunities for eventual roadway expansion. Instead, engineers generally designed for twenty-year volumes and no more. As the former district director of the California Department of Transportation later reflected, "The only mistakes we ever made were building to standards too low. . . . We never plan big enough, that's been my point. Political problems made it impossible to plan big enough." Despite the grand scale of completed interstate work, full fruition of the engineering mentality might have yielded even greater expanses of property acquisition, with greater potential for future construction.[32]

House Moving and Demolition for Urban Interstates

The greatest battles over property acquisition occurred not in rural areas but in cities. Urban interstates more dramatically disrupted the built environment and physically relocated greater numbers of residents and businesses. Most states experienced their first urban highway development as a result of the interstates, often coordinating these routes with urban renewal activities. With the eradication of blight as a corollary goal, demolition became the dominant clearance method. House moving could be an alternative approach when the built fabric was still salvageable, yet cost and logistical challenges, combined with conflicting objectives among those funding, managing, and executing interstate development, often led to demolition instead.[33]

Although many locales started their interstate work in rural areas, states like Florida and Illinois began in the cities. Rather than seeking to make a quick impact, they prioritized urban areas for their greater need and the high acquisition values at stake. In Florida, for example, the urban 10 percent of the system would carry more than half the traffic. Officials also estimated that right-of-way costs would increase two- or threefold if they waited a few years longer to acquire the necessary rights. Even in states that delayed urban roadway development, cities soon became a priority.[34]

Many of the parcels to be acquired contained physical "improvements," such as buildings, that required clearance before highway construction could commence. The federal government aimed to dispose of those obstacles at the least cost to the project. The state highway departments offered fair-market-value compensation to owners of such property. Following acquisition, the buildings would be demolished. But when the structures were in a condition to be reused elsewhere, and when available relocation sites existed nearby, house moving could offer a viable alternative. The original property owner might negotiate for adjusted compensation that allowed for the move. Or an outside individual might purchase a recently acquired property in order to relocate it. This worked out, however, only if the house-moving hassles were worthwhile and if the timing accorded with the highway-building schedule.

While house moving in the United States dates back to early Americans relocating log cabins, the industry grew substantially during the postwar years, alongside other clearance-related businesses. The major trade group, the International Association of Structural Movers, did not form until 1983, but local house movers and house-moving organizations preceded it. In 1920, for example, Los Angeles experienced an "epidemic of house moving," according to the *Los Angeles Times,* as individual homeowners relocated to the suburbs. In 1948, Pete Friesen started Advanced Moving in Chicago, eventually building the company into the largest structural-moving firm in the state. He recalled that house-moving activity in Chicago peaked during the 1950s and 1960s. At that time, he was moving four hundred structures per year, versus only a quarter as many by the late 1970s and early 1980s. Over that same period, the number of house-moving companies in the Chicago area declined from about thirty to five.[35]

House moving was a technically straightforward process. The contractor passed steel beams through a series of holes drilled in the building's foundation. Four-wheeled hydraulic dollies positioned underneath these supported the structure for the move. An attached tractor provided the motive power. Reconnecting the foundation and utilities at the new site completed the process. All the while, the contents of the home remained intact inside, eliminating the typical relocation activities of packing and unpacking one's pos-

The home of Mrs. Alice Blackburn awaits its move in southern California in 1948. According to *Life,* on the evening of the relocation, Blackburn hosted a party that ended only when it was time for the move to commence. (Loomis Dean/The LIFE Picture Collection/Getty Images)

sessions. One contractor boasted that thanks to hydraulic lifts, the houses he moved stayed sufficiently level throughout the process that a glass of water set atop a table would remain standing and spill-free. Some residents even traveled as passengers inside their relocating homes. *Life* highlighted the seamlessness of integrating a house moving into one's social calendar. A 1948 issue included a photo-essay documenting the move of Los Angeles resident Alice Blackburn's two-story house. Blackburn invited friends up the ramp to a party in her jacked-up home. The gathering ended at midnight, when the structure began its three-mile, five-and-a-half-hour move across town to make way for a freeway.[36]

It was in the spaces outside the relocated building that greater challenges arose. Contractors often had to reposition utility wires. They might also remove the roof or sever the structure to facilitate its passage down narrow pathways. Or they would lift parked cars out of the way with a crane. As Blackburn's story illustrates, house movers conducted many such trips at night—both to minimize disruptions and to comply with municipal regulations. But there were local limitations on how far houses could be moved, and multiple permits might be required before the relocation could begin. As one contractor testified in a congressional hearing about highway land disposition in Florida, "A building can be moved any distance that is economically feasible. Now by that I mean the removal of wires, trees, and so forth, in the way." Logistical concerns usually trumped the technical, and this same practitioner concluded, "The paperwork is actually harder than moving the building."[37]

Experienced professionals could move large numbers of modest-size buildings relatively quickly and cheaply. In 1949, the *Saturday Evening Post* profiled Abb Wilson, southern California's "foremost virtuoso" of house moving. His biggest job was clearing many of the buildings in the path of Los Angeles's hundred-mile freeway network. Wilson estimated that he had moved an average of one building per week over his forty years of experience. But during the interstate era, many companies worked at an even more aggressive pace. One Indianapolis house mover boasted that he could move three to five single-family houses per day, while a Florida contractor said he could do up to ten houses per day, provided he was "tooled up" for it.[38]

Perhaps the greatest impediments to the house-moving approach to clearance were the divergent objectives of the public and private actors involved. While the government sought to reduce costs—sometimes by finding buyers, rather than wreckers, for structurally sound properties—road-building contractors typically viewed existing structures as obstacles in the way of the more profitable construction parts of the job. Thus, they valued the rapid removal of these obstacles over any potential economic upside to be realized by their slower relocation and resale. Congressional investigations into improprieties uncovered some of the specific challenges. As one contractor testified, "We were not looking to the house removals for that profit. They were a nuisance item and we wanted to get rid of them as fast as we could. Like having a cancer or a disease. We wanted to get rid of them so we could proceed with construction." On the projects to which he was referring, contracts totaled $1.5 million, on top of which the potential incremental upside from the sale of moved houses represented less than 1 percent. Unsurprisingly, the contractor often chose demolition instead.[39]

Other contractors found ways to profit from house moving, but at the expense of the road-building project. As uncovered during congressional hearings, at least one Florida contractor preyed upon the inexperience of state right-of-way cost estimators. It bid one price for a building-removal job, and then subcontracted it out for a lower price and split the profit between the two companies. While legal, this practice incited charges of "highway robbery" as interstate costs spiraled upward. With these incremental profits predicated on the existing buildings' resale, some structures remained on-site longer than intended while awaiting purchase and relocation. Meanwhile, grading and fill work commenced around them, with embankments of dirt rising three to four feet up against the homes. The belated removal of the houses slowed the project and increased expenses—such that the state ultimately had to suspend this contract in order to proceed more quickly with building removal.[40]

Increasing the profitability and efficiency of building removal would require a more measured pace of property acquisition that preserved the subject buildings for longer. As one contractor advised the Florida Highway Department, an ideal scenario would be to release small groups of five to ten buildings at a time for public auction. This would facilitate direct sales to individuals, rather than glutting the market or creating bundles of buildings too sizable for individual purchase. If the sales were pursued as soon as the roadway route had been confirmed, he argued, the relatively appealing environment of a still functioning neighborhood would probably create a context more conducive to them.[41]

Vandalism and looting further limited resale value. In the ninety days between the acquisition of thirty-four buildings in Florida and their transfer to a house-moving contractor, for example, the structures lost substantial quantities of windows, plumbing equipment, and other fixtures. As the contractor observed, "The attitude of the buyer, I imagine, is considerably changed when he looks at a building like that. Beyond what it would cost to repair the building and put it back in its original condition, the property would lose even more if someone sees a house damaged three or four thousand dollars' worth that they are not apt to pay $2,000 [for]. . . . It just knocks your sale out completely." While the efficiency of this vandalism often suggested organized activity, individual departing residents could cause substantial damage on their own. As another contractor noted, displaced residents "don't like it, and they're not happy, and they move out at night and take whatever out of those houses, and we don't know about it. Maybe it is them or their neighbors, the night they move out, or maybe it is organized vandalism. We don't know." After such property damages, demolition would become the most economical course. Thus, even in the face of possible alternative approaches, clearance for interstate highways wrought a largely physically destructive path through the urban landscape.[42]

Interstate Impacts and Opposition

Given the many challenges of implementing the interstate system, it is not surprising that the initially celebrated program soon incited criticism. Opposition developed differentially across the country in response to variations in the kinds of building practices, their impacts on residents, and the level of mobilization within the affected communities. But a first step in changing both practice and policy was to reform the engineering mind behind the endeavor.

When World War II began, a young Kansan named Edward Haase enrolled at the University of Colorado at Boulder to study civil engineering. After he graduated, he was commissioned as an ensign in the Civil Engineer Corps and shipped out to Guam with the Seabees. His first assignment was to build a B-29 airfield. After the war, he became an office engineer with the Highway Department in Colorado's Durango District. He recalled, "In rural areas, we were almost heroes." Everyone there "thought highways were the greatest thing since sliced bread." His only problems came from individual property owners who felt they were being unjustly compensated for their losses. In the mid-1960s, Haase transferred to Denver and encountered a wholly different environment: "Quite frankly, when I was dumped into this thing in Denver, I was surprised and shocked, and wondered what I had gotten into. I didn't understand what it was. I had to learn for myself." In developing I-70 through North Denver, what Haase called "people problems" came first: "It was the people adjacent to the highway or who would be taken by the highway. The environmental thing wasn't that strong. I think it came along as sort of an afterthought, in my opinion." Haase's experience points to the changing public responses to the interstates across both space and time.[43]

Dealing with "people problems" was a new challenge for many engineers. They had learned the technical elements of highway design in school and from AASHTO Road Test data. But training in the sociology and politics of highway development was harder to acquire. Professional meetings offered some opportunities to acquaint engineers with the social obstacles likely to come their way. As Haase noted, "The annual meetings . . . of AASHTO were invaluable in that way because things were happening more quickly along those lines in other states, more populated states, than were happening here. So we could take our clues from California, for example." Indeed, a California engineer recalled his state's early attention to the social impact of highways: "In the '50s, we had lots of studies of relocation, disruption, business relocation—how a community fared by having a freeway built through or around it. I had a man named Stew Hill, in our right-of-way department, who turned out report after report. We're talking early '50s here. This is a long time before environmental law and relocation assistance acts and things like that."[44]

But California was an outlier, and advance warning from engineers working there and in the northeastern states did not insulate Coloradans from troubles of their own. When asked if it would have been helpful to have staff members who were trained specifically on the people problems, Haase was skeptical: "When we first saw the problem, I don't think any of us would have even thought of bringing someone on staff at the time. We wouldn't have believed that anybody had the expertise. . . . We just had to learn slowly. We didn't understand how anybody could be learning faster than we because we were on the cutting edge, we thought." As one Virginia engineer reflected, "Well, I think city planners, county planners, people who planned a broad range of public services, had a greater appreciation of the impact than the highway engineers. The highway engineers very shortly learned of this impact when this massive highway program started, however."[45]

Public hearings offered a formal avenue for community input. The Federal-Aid Highway Act of 1956 required one public hearing for every routing. The impact of these forums, however, could be more administrative than substantive. As the journalist Grady Clay wrote early on in the interstate

District engineer Edward Haase discusses highway plans at a Colorado public hearing in 1970. (David Cupp/Denver Post/Getty Images)

era, "I cannot speak of the thousands of hearings I have not seen, but from some personal observation I am forced to conclude that the public hearing is a carefully staged performance designed to show the audience why the route officially agreed upon in private cannot be changed." An Ohio engineer corroborated Clay's theory, reflecting, "When we went to a hearing, we all really basically knew what we were going to do, because we knew what it was the public needed. And they didn't. . . . And we'd go through this exercise, but unless they showed us something that we had forgotten, because we were so intelligent they rarely do, we are going to go ahead and build a straight line between A and B, basically. Clearly speaking. And that wasn't quite right." As hearings increased in frequency, however, highway opponents became more adept at using the process. As a Virginia engineer recalled of his own experience with public hearings, "We were scared of them, frankly, and, as it turned out, with some there was some reason to be. As time went along, it proved to be a fairly useful tool in bringing to our attention things that we had overlooked, but also, it turned out to be a real weapon in the hands of people who wanted to delay or stop the highway program."[46]

The substantial scope of highway-related takings and displacement increased the opposition. Highway departments acquired more than 750,000 properties across the entire interstate system. At peak interstate construction, approximately 70,000 housing units were demolished per year. Opponents of highways in general, and of specific routes in particular, protested the destruction of homes, the construction of highways in too close proximity to nondisplaced homes, and the loss of temporarily empty lots that had been appropriated for public use.[47]

The organization of this opposition into larger-scale freeway revolts soon helped galvanize broader resistance to urban and environmental destruction—for highways and otherwise—across the country. San Francisco initiated the freeway revolts in the late 1950s following the construction of the waterfront Embarcadero Freeway. Critics opposed the city's extensive highway plan in large part on aesthetic grounds. Over time, local oppositional movements to freeway development emerged in other cities as well. In the 1950s, Jane Jacobs and other "park mothers" successfully thwarted Robert Moses's proposal to build a highway through the sacred space of New York City's Washington Square Park. The 1960s and 1970s saw a flurry of grassroots efforts with even broader bases of opposition. In Baltimore a diverse coalition of neighborhood and community groups organized Movement Against Destruction (MAD) to oppose further highway construction in the city in favor of neighborhood revitalization. In Boston, the complementary efforts of Urban Planning Aid, the Boston Black United Front, and the Greater Boston Committee on the Transportation Crisis worked successfully to remove 343 miles from the interstate system. They lobbied not only against highways but also in favor of democratic and participatory decision making. Many other such battles were waged across the nation, albeit with varied results. In general, sustained opposition led by well-connected white populations tended to have the greatest success in halting or transforming individual projects. As these local revolts raged on, they benefited from changes in urban and highway policy that offered alternative development approaches. Less influential, minority-led opposition often proved less materially successful. Yet these groups simultaneously harnessed representational forms ranging from murals to poetry and fiction that give enduring cultural voice to their dissent.[48]

Although some engineers later claimed to be unaware of the injustices they were causing, many who did recognize them believed at the time that the gains justified the means. One Colorado engineer noted, "For the most part [highway alignment] was for the benefit of the highway user, for the benefit of the highway. Not too much thought was given to the effects on the people and I think everybody thought that they would be compensated if their property was taken, and that was enough." But compensation was not always sufficient. The former commissioner of Virginia's highway department reflected, "In many cases the locations of the interstate highway caused the property owner great difficulty. I think encroachment on property and damage to the property sometimes wasn't fully compensated for. There are ethical values and family values that you can't put a good value on." An Ohio engineer similarly recalled, "I think that the thing that struck me as being strange and different was the way that we divided farms and communities and . . . it seemed maybe in a way unfair to the local people to have to suffer for what the Interstate did to them. But yet, it was progress, and we just accepted it as progress, you know."[49]

The community burdens and the psychosocial costs wrought by the interstates were similar to damages then also being inflicted in the name of urban renewal. The victims of each were likely to be poor ethnic or racial minorities living in urban areas. Both politics and economics shaped the uneven geography of displacement. A former inspector general from Kansas observed, "When you first start to look at a new location for a highway, whether it is an interstate or anything else, one of the things you always look for is low land cost. The other thing you look at is service to traffic." In other words, property values and proximity to downtown directed route selection. He continued, "And of course, then, if you are going to go through land that is developed, the underprivileged areas of town which would have a lower unit cost as far as the dwellings were concerned, always [catch] the engineer's eye ahead of trying to go through Snob Hill." For such inequities, urban renewal became known as "Negro removal," while highway opponents decried the government's building of "white men's roads through black men's bedrooms." From the lead-dust exposure and empty lots of urban renewal to the lingering damages of exhaust fumes and noise pollution produced by highways, environmental injustices characterized both processes.[50]

What distinguished the displacements created by urban renewal and those wrought by interstate highways was the scope of reparations. In both cases, the government compensated property owners for the market value of their property. But whereas residents displaced by urban renewal were guaranteed new accommodations and moving assistance, highway victims were not. As one Kansas engineer bluntly put it, "In the acquisition of rights of way, in the earlier stages you just acquired the right of way. You had no provision allowed in law to pay for relocation of people or businesses; to move people, to move businesses. You just bought them out and said, 'Here, you get out of here now. It's your job to move.'" The engineer was also careful to note, however, that they were just following procedures for what they were required and able to do: "You can't always pay for all the inconveniences of somebody when they're put out of their home, but they weren't necessarily mistakes made. They were things for which there was no authority and it took a little experience and then the movement through Congress to get the authority to do these things."[51]

The Federal-Aid Highway Act of 1962 began rectifying this issue by requiring states to provide relocation advisory assistance to displaced residents. But benefits were still uneven. Participation was optional by state and varied between renters and owners. These geographic distinctions became particularly obvious where roadways crossed state borders and neighboring displacees received different treatment. By the mid-1960s, only two-thirds of states were authorizing residential-relocation payments for highway displacees. Moving expenses for displaced business owners were capped at three thousand dollars for highway displacement, versus twenty-five thousand for urban renewal. The Uniform Relocation Assistance and Real Property Acquisition Policies Act of 1970 finally mandated more even relocation assistance across the nation.[52]

Although social impacts provoked some of the loudest early opposition to highway construction, environmental dissent built up over time. Environmentalists decried, for example, the paving over of undeveloped natural space, the spread of urban sprawl, damage to local wildlife and streams, and the aural, visual, and air pollution produced once highways were in use. In addition to local protests at public hearings and on job sites, these highway opponents wrote angry tracts, including A. Q. Mowbray's *Road to Ruin* (1969), Helen Leavitt's *Superhighway—Superhoax* (1970), and Ben Kelley's *The Pavers and the Paved* (1971).[53]

Many engineers had initially been blind to these viewpoints. As a Virginia engineer observed, "I am afraid highway people, administrators and engineers, didn't appreciate the effect that the expanded highway program was going to have on the environment, and we were so anxious to get ahead with it that we didn't pay sufficient attention to environmental ill effects." Another engineer, from Illinois, went so far as to characterize this group as "the world's worst enemies of the environment, prior to the mid-1960s." Even when engineers did recognize the collateral damages these critics pointed out, they felt that the benefits outweighed the costs. "It's just a balance of choices that had to be made," one North Carolina engineer reflected. "I'm comfortable with the fact—and I think I always will be—that we had to do what we had to do. That we had to cut trees, and we had to put pipes in where there were beautiful little streams with waterfalls. I didn't like to do that! But I had to do it."[54]

Over time, many state and federal officials attempted to soften—if not necessarily eliminate—the environmental offenses caused by the highways. The same North Carolina engineer, for example, contrasted changes in building practices between Interstates 40 and 26, built more than a decade apart. Reflecting on the frequent rock slides on I-40, he noted, "I-40 was built in an era when you didn't worry about what was left; you just blew the mountainside to smithereens and hoped that you didn't have to haul too much, that it would just blow it away. So there was a lot of collateral damage—residual damage—to what was left over there on the slopes. . . . We're paying the price for doing it real cheap. It was not so much of a cut-rate thing back when they did it. That's all they could afford when they did it, but it goes more to the heart of the design. I know that I-26 was much more carefully planned out." Edward Haase also recalled a BPR official who "gave me a lot of hassle about one piece of alignment that I had designed through a mountainous area there. He felt the cut that I had designed was too high, unnecessary, and so forth. I was kind of surprised by that. But I conceded it to him, and we were highly complimented after the thing was built as to how nicely the thing fit into the countryside."[55]

These engineers suggest that highwaymen tended to be receptive to more sensitive and collaborative building techniques once they were introduced to these alternative approaches. A California engineer recalled, "All during the

'60s, there was growing concern for aesthetics in highway design and for errosion [*sic*] control. It was the FHWA that kept insisting that we terrace our excavation slope so they would revegitate and wanting us to do things like use variable slopes so that the landform would have an appealing look. I don't think any of us opposed those things, it just hadn't occurred to anybody." A Colorado engineer echoed this point, "I think part of it was a lack of knowledge on our part. We hadn't thought about wetlands or wildlife or anything like that." The former Virginia commissioner Douglas Fugate felt that the extra time spent on this additional training was worth the hassle. He noted, "As time went along, why, the program was slowed down considerably by environmentalists who were attempting to teach us, and did in a rather drastic way, that we had better pay more attention to the land through which the facility was being constructed and the people that lived on the land. This was a good development. I hope we have reached a good accommodation with the environment now."[56]

Others were less sanguine, resisting the irritations caused by outside pressures and seeing the environmental backlash as overly idealistic, lacking in pragmatism, and unnecessarily driving up costs. Reflecting upon his experiences on the Citizens Advisory Committee to Colorado's Glenwood Canyon project, one construction company owner recalled, "I guess I had a great amount of disappointment with a lot of the environmental groups. They—they helped, without a doubt—it was important that the attitude that they had forcibly portrayed, because with that we were able to do what you see being done. And yet, if everything would have been turned over to them, there would not be a highway, which, in my mind, just has to be proven to be wrong. You know, the state would have been totally choked off." One Kansas engineer was more critical: "I think maybe we moved too far in saving some things because we didn't have enough knowledge." Decades after a segment of I-470 in Colorado was removed from the interstate system, the engineer Charles Shumate still felt the frustration of that loss—particularly as the segment was eventually built, but as a primary project (rather than an interstate) that cost the state an incremental $18 million. He dismissed the environmentalists' argument against the roadway: "It was going to introduce urban sprawl and there was urbans sprawled all over there before we got started." Fueling his frustration was his perception of highway opponents' failure to offer a positive alternative: "They just wanted to kill everything. They were against everything."[57]

By the late 1960s and early 1970s, new legislation introduced regulatory instruments for safeguarding the landscape. These laws increased the time, expense, and deliberation associated with highway development. The National Environmental Policy Act (NEPA) of 1969 required environmental assessments and impact statements for any project involving federal funds. In tandem, the Wilderness Acts of 1964 and 1970 offered protection to public lands from a variety of human influences. Establishment of the Environmental Protection Agency in 1969 further cemented federal oversight of these growing regulatory programs. Celebration of the first Earth Day on April 22, 1970, aptly recognized the growing ferment of environmental advocacy while also helping to launch further environmental legislation and institutional efforts in the decade ahead. The National Historic Preservation Act (NHPA) of 1966 offered similar protections for the built world. It provided tax credits to incentivize increases in the number of properties protected by registry listing. In addition, like NEPA, the NHPA explicitly required consideration of the impacts of federal projects on properties listed (or, eventually, eligible to be listed) on the National Register.[58]

These policy changes' impact on highway clearance was significant. An Ohio engineer observed that NEPA "created the greatest upheaval . . . throughout the nation than any other single piece of legislation." He contrasted the situation before and after the act: "Up to that point, [smaller agencies] theoretically had power, but politically they couldn't enforce that. But now they can enforce it." Compounding the situation were "the historic preservation people and, here again, they can tie you up so long on delays—they can't stop anything, but they can delay you so long that you might as well say forget it." For some of these engineers, the interventions of outside agencies replaced much of the excitement of highway building with frustration. Many even doubted that the interstate system could have been built any later than it was. Bertram Tallamy was clear on this point: "No way in hell would [we] be able to go from NY to Albany, across the whole state. Why? Because of the interferences and each one can hold you up . . . [for] many, many months, years. Interferences include local governments, local citizens groups, environmental groups, other forms of competing transportation, private groups, industrial groups, unions, etc." Reflecting back in the late 1980s, the Ohio engineer put it bluntly: "In fact, you couldn't build the interstate system today with the restrictions that we've got. It's impossible." On the flip side, another Ohio engineer recognized the greater certainty that came from a more expansive process. He reflected, "You know, we now go through all of the environmental concerns—water quality and air quality, the noise, historical data, Indian relics and things like that. We check these things out so thoroughly now that we feel assured that we are going in the right direction, and it has lessened the possibility of court cases stopping us. . . . It takes more time and money, but we are more certain that we are doing the right thing, or the acceptable—acceptable."[59]

By the mid-1960s, the climate surrounding highway construction had changed. Public opinion was transformed from an atmosphere of broad support, grounded in the engineers' singular expertise, to a more subjective understanding of the expansive nature of the highway-making process. Accompanying this shift was an increase in the incorporation of social and environmental concerns into what had previously been viewed as a primarily technical exercise. The mindset of the engineers also changed as older personnel aged out of the profession, replaced by those who had grown up in a new environmental era. Later, NEPA process requirements also created new positions within government and industry organizations that expanded project development beyond engineers alone. The presence of biologists, archaeologists, ecologists, hydrologists, and civil rights specialists increasingly working alongside engineers in the office and on the job site led to a broader understanding of highway-building work.[60]

Rapidly escalating costs further diminished public support. The 1956 act had authorized $25 billion for the interstate project. In 1958, the BPR estimated the system's total construction cost at $37.6 billion. That figure was up to $41 billion by 1961, just one year after the start of hearings, led by Congressman John Blatnik, uncovered several questionable highway-building practices. As one of the hearing participants summarized in a *Saturday Evening Post* article, suspect practices included "fraud or carelessness involving right-of-way acquisition in 24 states, shoddy or deliberately dishonest practices in 21 states, [and] payola accepted by highway-department employees in seven states." Even though such improprieties tarnished the project in public opinion, they were the exception. Total estimated interstate costs more than doubled between 1961 and 1972 to reach $89.2 billion. Yet most of these increases derived from inflation, normal overruns, or other errors—not from corruption. Another sixth of the difference came from categories of costs not previously recognized as part of the highway construction process. Some $3.7 billion came from "increased social, economic, and environmental requirements," and another $3.8 billion supported incremental right-of-way expenses for relocation reimbursement. The shift toward more equitable and sustainable highway development came at greater expense of time and money.[61]

It also became increasingly possible to redirect interstate funds toward other uses. The Federal-Aid Highway Act of 1973, for example, permitted the withdrawal of interstate routes and the reallocation of Highway Trust Fund sums to mass-transit options. Although the mileage eliminated was relatively small, the significance of this policy change lay in the federal government's willingness to fund alternative development approaches to clearance and new construction. This was a change that echoed concurrent broadening of policies surrounding urban revitalization, from clearance alone to rehabilitation. While highway spending continued, most funding through the end of the decade went to maintenance, rather than clearance for new routes.[62]

By the end of the 1970s, the interstate highway network was nearly complete. The era's relatively unfettered implementation of the engineering mentality had yielded dramatic physical and social impacts. It produced not only the largest single public-works project in history but also wide ribbons of demolished and relocated buildings, trammeled countryside, and displaced residents across the nation. Despite positive early depictions in the popular press, clearance for the postwar highway system ultimately proved less straightforward than the expertly choreographed battleground victory described in early reports. Although originally the outgrowth of a purely technical approach to infrastructure creation, road building ultimately proved to be a complex undertaking. Like its contemporary processes of urban renewal and suburban development, the highway construction process was more complicated, costly, and contentious than initially imagined.

Part Three

Bulldozers of the Mind

(*Overleaf*): Catherine Danner, *Buster Bulldozer,* illus. Mary Alice Stoddard (Racine, Wisc.: Whitman Publishing, 1952), cover.

Chapter 6

Unearthing "Benny the Bulldozer"

Children's Books and Tonka Trucks

In 1947, the author Edith Thacher Hurd and illustrator Clement Hurd published *Benny the Bulldozer.* This thirty-three-page children's book, filled with colorful images and simple text, tells the story of a moody bulldozer named Benny whom road builders persuade to help clear the way for a new highway. Without Benny's assistance, the project supervisor laments, the workers will never finish the road in time for the Fourth of July celebration. Benny soon hauls away trees, digs through rocks, and evens out the soil. Thanks to his help, the crew meets its deadline and rewards the bulldozer by putting him at the head of the opening-day parade. Leading a procession of cranes, dump trucks, earthmovers, and mixing and paving machines, adorned with American flags, Benny rumbles through town to the exuberant cheers of well-wishers. The townspeople wave to him, set off firecrackers in his wake, and cheer with delight: "'Hurrah!' . . . 'Hurrah for the new highway!'"[1]

In the postwar years, the simultaneous prominence of the bulldozer and "bulldozer books" was no accident. Illustrating parallels between the landscapes of fact and feeling, these books reflect the social, political, and cultural preoccupations of their era. They helped the younger generation make sense of the world around them as their environment underwent massive destruc-

Benny the Bulldozer leads a Fourth of July parade in commemoration of the completion of the new roadway in Edith Thacher Hurd's *Benny the Bulldozer,* illustrated by Clement Hurd (1947). (© Clement Hurd, 1947. Reprinted with permission.)

tion. At times, the books move beyond reflection to prescription. By means of positive associations, the authors make the process of real-world demolition and development seem like a natural progression. The historian David Nye has called these kinds of tales "technological foundation stories." His work on nineteenth-century America explores narratives that depict the clearing power of the ax as progress. Bulldozer books update this story for a more modern age. In place of potentially scary but realistic images of earthmoving machines, the books offer cheery pictures. By putting a happy face on demolition for their readers, the books subtly endorse the culture of clearance.[2]

Although relatively few postwar writers have written specifically about contemporary clearance work, children's literature has long given bulldozers their due. A spate of children's picture books about demolition and clearance began appearing on library shelves in the 1940s. The books feature images and text with an informational tone and explain how the world works. Their subjects are the machines and operators behind excavation and wrecking,

Smiling and strong Buster Bulldozer, the hero of Catherine Danner's 1952 book, is one of many cheery depictions of the machines of demolition and clearance in postwar children's books.

and their audience consists of young readers, generally between four and twelve years old. While some of these books had a relatively small readership, others, published by major trade houses like Houghton Mifflin and Scribner's, were best sellers. Popular series like Little Golden Books and Book-Elf included bulldozer titles, and the *New York Times* and *Publishers Weekly* reviewed them. Some of their authors and illustrators—including Virginia Lee

Burton, David Macaulay, and Clement Hurd—received the prestigious Caldecott Medal and Boston Globe–Horn Book Award. And despite their age, many of these books remain in circulation today.[3]

Although children's literature may initially seem an unlikely venue for a discussion of demolition, its status as an "inferior" literature held practical appeal for demolition boosters. Children's books have historically escaped the scrutiny that society applies to most products of commercial culture, and they have been subjected to far less critical examination than other genres. As the literary scholar Lois Kuznets notes, the books' characters can express desires that "evade individual and societal censors." Women dominate children's book writing, publishing, and library services, another reason for its positioning as a stepchild within the larger field of "literature." The derogatory term "kiddie lit" suggests its second-class status. Moreover, nonfiction children's books received even less attention and fewer prizes than fictional stories. Overall, authors could make use of the clarity of children's books to promote the virtues of clearance without critics questioning their agendas.[4]

Children's literature also offered direct access to a large, significant audience: young children and the adults who shared story time with them. Scholarship on children's literature suggests that young readers generally approach books with open minds, using them to explore new ideas rather than substantiate existing ones, as adults tend to do. Moreover, the critic Paul Hazard has observed that whereas all readers find and respond to the socializing norms within books, child readers are particularly susceptible, picking up attitudes and values that affect their behavior. The works can offer youngsters their first exposure to cultural knowledge and attitudes on morality, imparted with the sanction of the parents and respected adults who read the stories with them. Schools and libraries facilitate the access of authors to this malleable young group. And the writers recognize the power they wield: the acclaimed children's author Virginia Lee Burton has articulated her own belief that the books are "amongst the most powerful influences in shaping [children's] lives and tastes." This combination of an unregulated creative space and a large, impressionable, and accessible audience made children's literature an ideal platform for the generally uncritical, pro-clearance texts that emerged in the postwar era.[5]

It should come as no surprise that some businesses recognized the public relations potential of bulldozer books and supplied authors and publishers with photographs and information. The companies' involvement explains the sense of outright propaganda characterizing some of the works, although most were more circumspect. In her books on building wrecking, Jean Poindexter Colby benefited from the assistance of the John J. Duane Wrecking Company, Controlled Demolition Company, Central Building Wrecking Company, International Harvester, Bay City Shovels, Unit Crane and Shovel Corporation, and the editors of *Wrecking and Salvage Journal*. Similarly, *Contractors and Engineers* magazine, the Portland Cement Association, Ford Motor Company, and Standard Oil supported David C. Cooke's *How Superhighways Are Made,* while a Caterpillar Tractor Company employee, according to the Acknowledgments page, "researched much of the information, supplied photographs and reviewed the final copy." Although Cooke was the book's official author, Caterpillar might well have claimed the honor. The company also lent its brand name to small-scale models of its machines, which became popular as children's toys. Mound Metalcraft Company introduced Tonka trucks in the same year the Hurds published *Benny the Bulldozer.* In combination with the bulldozer books, miniature earthmovers by Tonka and other manufacturers helped make bulldozers a prominent part of the postwar child's play world.[6]

The Children's Literature of Clearance, 1940s to 1960s

As children's literature entered the postwar era, one creative tendency was giving way to a new one. The fairy tale had dominated since the early twentieth century, and it remained a favorite genre. As the influential librarian Anne Carroll Moore argued, imaginative fairy tales could protect the innocence and free-spirited exploration of childhood. But not everyone shared Moore's perspective. In 1921, the progressive educator Lucy Sprague Mitchell pioneered a new, reality-based approach to children's literature with her *Here and Now Story Book*. This was not a fairy tale but an information book filled with the experiences of everyday life in the modern world. The city featured prominently. The literary critic Margery Fisher has argued that information books offer more than "a mere collection of facts"; they "contain fact, concept, and attitude" as well. Or as Mitchell herself stated, they aim to give readers "not so much new facts as a new method of attack." Through these means, they help children understand the world around them. Although Mitchell's conception of childhood realism broke with the fairy tale tradition, it still recognized "children's ability to imagine possibilities beyond the everyday." Animated machines fell within the bounds of her "realistic" children's book world. Postwar bulldozer books built on this tradition, developing four common themes to subtly advance the culture of clearance: the promotion of technological progress, the involvement of readers in the work of clearance, the masculinization of bulldozer operators, and the heroicization of anthropomorphized machines.[7]

The most obvious theme uniting the bulldozer books is the celebration of growth and progress enabled by modern technology, with scarce acknowledgment of its negative aspects. This triumphal spirit coincided with the nation's glorification of science and engineering in the areas of space travel and atomic warfare. It also accorded with support for education in science and technology for early grade students as mandated by the National Defense Education Act of 1958. The act was a response, most immediately, to the Soviet Union's launch of *Sputnik*. Just as in the 1959 Kitchen Debate, during which Vice President Richard Nixon countered Russian advances in rockets with discussions of U.S. growth in home construction, so too were Cold War American children's book authors showcasing the construction industry. While progress often remains an unstated subtext of these books, David Cooke invokes the word itself in *How Superhighways Are Made:* "Of all the stories that can be told about progress in modern America, none is more exciting than that of how superhighways are made!" Replacing the word *superhighway* with *high-rise* or some other physical construction reveals a theme characteristic of many of these works.[8]

Some authors promoted the cause of demolition-based progress by equating it with the advance of civilization. Jean Poindexter Colby's *Tear Down to Build Up: The Story of Building Wrecking* historicizes machine-based wrecking within a paradigm of natural evolution. Just as dinosaurs once foraged the earth with their teeth, and as early man struggled to tear down buildings by hand, Colby suggests, today's dinosaur-sized cranes bring a new era's power to an age-old process. By rooting the machines in this larger history, narratives like Colby's identify "the work of constructive, creative demoli-

Jean Poindexter Colby's *Tear Down to Build Up,* with illustrations by Joshua Tolford (1960), depicts the work of bulldozers and cranes as part of "constructive, creative demolition." (Courtesy Hastings House)

tion"—as she calls it—as a natural and enduring fact of life. Although the illustrations can convey a gloomier tone, Colby's overall message is positive: "Of course, these cranes are there for the purpose of destruction—the tearing down of buildings—but they are also, in many ways, a sign of progress, of the world moving forward toward better living everywhere."[9]

Authors further demonstrated technologically based progress by depicting bulldozers taming nature. The excavating machines in these stories dam rivers, cut through mountains, flatten hills, and fill valleys, bringing order to the seemingly haphazard natural world. Virginia Lee Burton's *Mike Mulligan and His Steam Shovel* demonstrates this tenet pictorially by juxtaposing the previous development of homes atop rolling terrain against the straight and narrow order of a modern roadway. Norman Bate's technology series even better exemplifies this phenomenon. In *Who Built the Dam?* Bate gives waterways defiant personalities. He describes a river that enjoys racing down the mountain, selfishly washing away everything in its path: "'I'll smash this mountain,' it hissed. 'I'll tear it apart! I'll wash it away!'" When the engineers productively harness the power of the river for energy generation, they also tame the whims of a willful river. Books like this suggest that machines are not destroying nature, but rather optimizing its use.[10]

Acts of omission offer another means for promoting the authors' single-minded view of progress. In many of the works, authors skip quickly past the losses caused by new construction. By ignoring the victims of progress, they demonstrate the selective nature of their "information" books. *The Busy Bulldozer,* for example, begins with a bulldozer putting the finishing touches on a city lot slated for new post office construction. The text avoids the question of what building or which people previously occupied the site and how the land in this prime corner location became available in the first place. In another story, Carla Greene's *I Want to be a Road-Builder,* young Beth and Jim are delighted to learn of a new road being built that will link their family farm with the town. Yet they fail to consider the destruction of other farms that construction of the new road probably required. Moreover, while agriculture admittedly generated its own environmental costs, the book overlooks the fact that completion of the roadway added new automobile-based pollution while taking away open space.[11]

Even when some books include mentions of preexisting structures on a site, they often do so while glossing over the building-removal stage or oversimplifying what happens to them when the bulldozer comes along. The removal of houses in *Let's Go to Build a Highway,* for example, receives just one sentence of attention. The authors avoid accountability, writing vaguely and passively, "Any buildings on the land are removed while the engineers plan the highway." Colby more directly addresses the reality of building demolition in *Tear Down to Build Up* by calling "amusing" the idea that some building owners fight to save their lone structures. By contrast, in *How Superhighways Are Made,* Cooke introduces the prospect of house-moving, noting, "When land is being bought by the government for the new highway's right-of-way, the owner may decide that he wants to keep his house or building instead of having it torn down." Yet Cooke omits images or fuller descriptions of the more common demolition solution. In a similar way, Virginia Lee Burton emphasizes the mobility of the house at the center of her fictional story *The Little House.* Burton's house begins life out in the country, surrounded by fields. Development grows up around her until the house eventually finds herself beneath an elevated train track, surrounded by skyscrapers. Saddened by the loss of her former pastoral existence, she finds solace through an "escape syndrome" of relocating back to the country. The cultural historian Joe Goddard has classified this work, with its preference for the countryside, as symptomatic of an emerging environmentalism. Yet even a pioneering environmentalist like Burton, through the limits inherent in her localized solution, underestimates the necessity for larger-scale countermovements to the advances of developers' bulldozers.[12]

Frank Tashlin's *The Bear That Wasn't* teaches a similar moral about the benefits of accommodating demolition and development, rather than taking a stand to manage this change. While a bear sleeps in his forest cave one winter, a construction crew arrives, bulldozes the trees around him, and erects a factory on top of his cave. The awakened bear tries briefly to alter the course of the construction, but ultimately he returns to his lair for the next seasonal sleep. Just as he did before the factory's arrival, he can rest happily while dreaming the sweet dreams of hibernation. In a manner similar to Burton's, Tashlin's resolution reinforces the viability of physically or mentally transporting oneself back to a less developed world without confronting the harsh reality of the present day. Instead of reclaiming his forest, the bear closes his eyes and goes to sleep. This strategy of avoidance suffices for the short term, but it is not a solution to the problems posed by the spread of sprawl and losses of homes, rural land, and open space enabled by a culture of clearance.[13]

The second, less dominant, tactic used by authors of bulldozer books for promoting the idea of clearance is not only to educate readers about why progress is good but also to invite them to take part in it. This strategy helps transform the books from static texts in the imaginative world to calls to action in the real one. *Let's Go to Build a Highway,* for example, concludes with a list of nine activities for readers that will complement their newfound knowledge. They might learn how to mix cement or "write a letter to a newspaper in your town explaining where you think a highway should be built and why." Through techniques like this, young readers learn about the road-building trade while also receiving a broader education in aspects of citizenship and civic participation in support of highway development.[14]

Career-oriented books take a longer-term perspective in their call to action. *I Want to Be a Road-Builder* articulates that objective right in its title. The dust jacket for Herbert Zim and James Skelly's *Tractors* reads, "The authors discuss tractor skills and tractor jobs, suggesting how a young person can get both." *Vermont Roadbuilder* is dedicated "to all roadbuilders and the children who want to become roadbuilders." Finally, authors promote the jobs of machine operators and road builders by positively depicting the training, camaraderie, and compensation the workers enjoy. Colby emphasizes the work's profitability, interest, and variety, as well as the tradition of family ownership characteristic of many wrecking firms. To a similar end, Zim and Skelly valorize the skill the work requires. They tempt their readers with the allure of one day securing a job in the construction industry: "Not everyone is quick enough and has good enough control to run a giant crane. Perhaps someday you can be the man who sits at the controls of one of these powerful machines."[15]

A third theme in support of demolition and clearance is the books' inspirational portrayal of the manly machine operators. Bulldozer books celebrate the all-American masculinity of smiling, strong, selfless construction workers. This association is particularly relevant in light of the Red-baiting and the Lavender Scare that also characterized this era. Since conservative anti-Communists believed sexual and political unorthodoxy went hand in hand, they targeted government employees who appeared to be insufficiently masculine. The bulldozer books helped educate readers in the characteristics of the increasingly important public portrayal of manliness.[16]

Few of the men in the books exhibit the brawn characteristic of older depictions of manual laborers. The drawings of the crew in Vera Edelstadt's *A Steam Shovel for Me!* from 1933 typify this earlier aesthetic. Her muscular workers share the iconography of proletarian art. Their chiseled features, giant skulls, and massive volume suggest seriousness and strength. To a lesser degree, this aesthetic also characterizes photographs of construction workers from the era, such as Lewis Hine's images of Empire State Building construction, originally published in the children's book *Men at Work* (1932). In general, however, the muscular, male image characteristic of construction workers and bodybuilders of the interwar years fell out of favor in the postwar era as the audience for such images became associated with homosexuals. The imagery of children's books seems to have followed suit.[17]

By contrast, the masculinity represented in postwar children's books is friendlier and derives more from the workers' skill and their association with machines than from their physique. These characteristics become clear in the title alone of Elisabeth MacIntyre's *The Affable, Amiable Bulldozer Man* (1965). Illustrations in these works frequently depict operators smiling and waving to the children who watch them at work. Some even invite their young admirers to join them for lunch, give them a construction site tour, let them make an examination of their machines, or offer them a turn at the equipment's controls. In *Vermont Roadbuilder,* the authors write about one construction worker, "Fred likes his work. Fred likes children, and the children like him." By association, Fred and his machine seem so innocuous that the children happily spend time playing in the bulldozer's open jaws. Some books take the man and machine association a step farther, suggesting that operator and bulldozer essentially become one—creating "bulldozer men," as one book puts it. In *The Busy Bulldozer,* the relatively scrawny and clean-cut Joe and his powerful bulldozer exhibit this joined-at-the-hip quality as they amaze a little boy watching them. In one scene, as Busy Bulldozer gets ready to dig up a hill, the boy says, "Maybe the hill will be too steep for you." Joe responds defiantly, laughing, "Ha! Ha! Ha!" while the bulldozer emits a "GR-R-R-R-R-R!" and pushes away a pile of dirt. This pattern repeats throughout the story, with man and machine vocally mimicking each other as they grunt and "grrr" their way through the task.[18]

Alternatively, when authors feminize the machines, the male operators demonstrate their masculinity through their roles as female guardians. Virginia Lee Burton's *Mike Mulligan and His Steam Shovel* works this way. When steam shovel Mary Anne outlives her usefulness, Mike Mulligan feels unable to send her to the junk pile. Instead, together they begin one last task of excavating the cellar for a new town hall. Once finished, they realize they have dug themselves permanently into the hole. Mike therefore converts Mary Anne into a furnace for the building and becomes her caretaker. They live out their days together in the cellar, like an aging husband and wife. While female machines are characteristic of Virginia Lee Burton's books, girls typically play relatively minor roles in these works. Males dominate, either as operators or as admiring observers. The boys seemingly rehearse their own protective roles as they accompany token female companions to view the machines in action. The example of *Vermont Roadbuilder,* a later work (1975), proves the exceptional nature of female protagonists. In this self-proclaimed "nonsexist book," girls equally balance out boys, and one of them, Sarah, corrects two boys who claim that "girls never drive bulldozers." Sarah replies, "Girls can do anything that boys can do." Then she climbs up onto the backhoe and drives it herself.[19]

Children's books portray construction men and their jobs as not only appealing in their exercise of masculine power but also virtuous and well-intentioned. As authors Zim and Skelly write in *Tractors,* "Even though the work is rough and the hours long, volunteers always are found among the bulldozer men." *Tear Down to Build Up* describes the steel helmets the men wear for protection and quotes one workman who observes, "If you're really hit, nothing will save you. That's the chance you take." As another demolition worker notes, "I'm glad to have a hand in pulling down some of these old tenements. Someday there'll be a beautiful building there." The men's good work extends from removing old tenements to making the roads safer. As David Cooke writes of highway builders: "The crews of road builders have labored hard to finish the job, often working in pouring rain and blistering sun. It has been a struggle every inch of the way. They know that their high-

In Elisabeth MacIntyre's *The Affable, Amiable Bulldozer Man* (1965), a satisfied-looking equipment operator sits at the controls of his bulldozer after demolishing a home so thoroughly that little memory remains of its previous presence. (Elisabeth MacIntyre, *The Affable, Amiable Bulldozer Man* [New York: Knopf, 1965], n.p.)

The books often depict friendly equipment operators, such as in this exchange of smiles and waves between a bulldozer operator and a young boy and girl and their dog. (Carla Greene, *I Want to Be a Road-Builder,* illus. Irma Wilde and George A. Wilde [Chicago: Children's Press, 1958], n.p.)

As depicted in Constance Cappel and Raymond Montgomery's *Vermont Roadbuilder* (1975), after spending time with a personable road builder, this group of children happily climbs into the mouth of his bulldozer blade to play. (Courtesy Constance Cappel and Raymond Montgomery)

way will make driving easier for countless people for many years to come. They know that their highway will carry trucks taking all kinds of products to market, that it will take families to vacation areas faster and thus allow them to enjoy longer holidays. But perhaps most important of all, they know that they have built a safe highway." Despite the difficult, tiring, and oftentimes risky nature of their work, the operators impress by willingly shouldering the burdens of self-sacrifice for society's greater good.[20]

The books' authors dramatize both the blue-collar laborers who do the work of clearance and the white-collar professionals who plan and manage the projects. In this way, they offer readers multiple role models, from laborers to draftsmen and middle-class businessmen, while broadening the parameters of masculinity. A picture of three young boys gazing smilingly up at a white-collar worker in Zim and Skelly's *Hoists, Cranes, and Derricks* (1969), for example, suggests their admiration. *Let's Go to Build a Highway* begins with a drawing of suit-and-tie-wearing businessmen chatting happily at a contractor bidding session. Project overseers and civil engineers appear in other works. This embrace of the middle class is consistent with the consequent real-world rise of the "Organization Man" and other white-collar bureaucrats. By linking the men in the office with those physically operating the machines, the books boost the masculinity of white-collar workers as well.[21]

The ultimate heroes in these books are not the men but the machines. By anthropomorphizing bulldozers and shovels, the authors confer valor and even humanity on the operators' mechanical counterparts. Anthropomorphism in children's literature is not new, but mid-twentieth-century authors extend the nineteenth-century example set by Hans Christian Andersen and others to the trains and earthmoving equipment of their own day. The animation of the equipment creates a personal connection between the reader and the fictional characters, making the machinery seem safer. As one critic has observed, "The machine that has a soul is not dangerous anymore; it is reasonable." Consequently, the authors transform construction equipment from potentially scary, mechanical monsters into valiant, all-American friends.[22]

The fictional machines' many heroic exploits demonstrate their civic-mindedness. Benny the Bulldozer saves the day by helping complete the roadway on time. In Virginia Lee Burton's *Katy and the Big Snow,* snowplow Katy (a tractor equipped with a snow-clearing blade) rescues the city of Geopolis from a debilitating snowstorm. The illustrations of Katy establish her friendly nature: alert eyes replace headlights; her plow looks like a heart-shaped mouth; and red paint completes her loving, Valentine-esque appearance. But her actions truly endear her to the town. She clears off roadways to rescue the police, the postmaster, the electric company, the telephone company, and the water department. In the process, she also delivers a patient to the emergency room. All are saved, "thanks to what Katy did." The title character of *Buster Bulldozer* further proclaims that being helpful both pleases himself and serves others. Heading off on a new task at the end of the book, Buster sings, "Clankety clank! Clankety clank! / And soon the work is done! / Clankety clank! I'm helping now, / And this is lots of fun!" As the unheralded workhorses who keep the town running and the populace out

The anthropomorphized character of Katy the tractor (equipped, in this case, with a snowplow) appears smiling, helpful, and unthreatening. (Illustration from *Katy and the Big Snow,* by Virginia Lee Burton. Copyright 1943 by Virginia Lee Demetrios; copyright © renewed 1971 by George Demetrios. Reprinted by permission of Houghton Mifflin Harcourt Publishing Company. All rights reserved.)

of trouble, machines like Buster and Katy exemplify service. The choice of the Fourth of July as the milestone in *Benny the Bulldozer* also suggests a civic-oriented connection. Many of the books close with patriotic celebrations, their pages brimming with flags, government officials, and ribbon-cutting ceremonies. Through stories of public-spirited work, the books promote patriotism and civic responsibility.[23]

Such one-dimensional personifications of these industrious machines, however, do not offer young readers a range of qualities for emulation. As one scholar has argued, in their nearly singular focus on work over play—eschewing adventure, while toiling ceaselessly at demanding jobs—the machines can function as dangerous role models. In addition to conveying a positive ethic of selflessness, such character traits place standardization over individuality and the common good over personal need. These trade-offs are also consistent with the philosophy underpinning eminent domain. Prioritizing a broadly defined public good over individual property owners' rights facilitated the demolition of much construction that stood in the way of redevelopment projects. *How Superhighways Are Made* explains this doctrine this way: "In some cases these landowners may not wish to sell their property. However, a special law makes it possible for the government to buy their land anyway, because highways are considered to be so important to everybody. This law is called the 'Law of Eminent Domain.'" Through such explicit language and the behavior of their mechanical characters, these children's books offer examples of some of the sacrifices that postwar citizenship increasingly required.[24]

Tonka Trucks and Model Toys

Accompanying the rise of children's books about construction equipment was the emergence of toys that offered a three-dimensional counterpart. Like the books, these toys invited their young owners to participate, physically and imaginatively, in the postwar construction and clearance trades. In 1946, the brothers Charles and Fred Doepke began making small-scale steel versions of trucks and construction vehicles, which they called Model Toys. They were motivated in part by their individual experiences in World War II. Charles, an engineer, had spent the war on the home front, manufacturing ordnance for the army in his machine shop. When those contracts dried up at the end of the war, he identified the toys as a new civilian good to take their place. Meanwhile, Fred served in the Army Air Corps, an experience that, he recalled, exposed him to the impressive construction feats of the Seabees and Army Corps of Engineers. Noting that millions of other soldiers had also observed this work, he speculated that familiarity with construction equipment would probably lead both operators and their observers to recognize scale models of the real thing—and that, he hoped, would induce them to make a purchase.[25]

Through close study of construction vehicles in person, photographs, and sometimes even manufacturers' working drawings, the Doepke Manufacturing Company developed a series of toys that mimicked their full-scale counterparts at a 1:16 scale. The company's product line was extensive and included bulldozers, dump trucks, bucket loaders, scrapers, clamshell cranes, and many other machines. Doepke branded its individual toys according to the products on which they were modeled, from Euclid dump trucks to Caterpillar D6 tractors and bulldozers. The constituent parts made these miniatures even more lifelike, with tires purchased from manufacturers that produced the real thing and delicate controls that made the machines' attachments fully functional. The designs were so good, in fact, that one construction equipment manufacturer purportedly solved a design challenge he was facing by adopting the solution he had observed in Doepke's smaller-scale execution. More commonly, however, distributors would bring the models on sales calls to help them sell the full-scale machines. With such uses in mind, some manufacturers specifically requested that Doepke develop a model of their product next. Even if the toy maker could make only a single working model for demonstration purposes, it was typically happy to do so as long as the real-world equipment manufacturer footed the bill.[26]

Although the company's close attention to materials and design drove up the toys' prices, they sold well. The brothers entered the market by shopping an initial mockup around the annual Road Builders' Convention in 1946, mimicking the marketing techniques of the full-scale vehicle manufacturers whose machines were also on display at the show. The manufacturers and dealers they met showed interest, and by the early 1950s the company was producing about 270,000 units per year—priced between twelve and twenty dollars apiece—for its dual customer base of businesses and children. Doepke continued production through the end of the decade, when escalating steel prices effectively put the company out of business. Available alternatives like wood and plastic, both of which were becoming popular among other toy manufacturers, did not offer the requisite precision for their designs.[27]

A more mainstream participant in the children's construction toy market also emerged alongside Doepke. In 1947, Mound Metalcraft introduced its first toy steam shovel and crane. It added trucks the next year, and a full line of earthmoving machines followed. The company branded its products with the name Tonka, after nearby Lake Minnetonka, in Minnesota. Like Doepke, Tonka sought to make its toys durable, operable, and realistic enough to invite children to play with them. While it did not copy specific real-world designs or the precision of their controls, the company used a strict 1:18 scale relative to the generalized proportions of full-size machinery. It also constructed the toys using accurate materials, including rubber tires, automotive enamels, and heavy gauge steel. Accessories ranged from dual headlights to hydraulic lifts. When the company acquired a manufacturer of hydraulic components in 1963, it even entered into the business of producing some of these parts for full-scale vehicles.[28]

Tonka resembled real-world construction equipment manufacturers in its business strategies and performance. Production volumes soared over the postwar period, from 500 toys per week at the outset to 125,000 per week by 1962. By 1965, it was producing nearly as many vehicles per year as the entire American automobile industry. Heavy demand led to the steady growth of its physical plant: the company increased its manufacturing footprint by a third in 1962, and again by more than half in the three years that followed. In 1963, Tonka even expanded internationally by completing a licensing agreement to produce its toys in Australia and New Zealand. Next, it broadened product lines, introducing Mini-Tonka in 1963 to target lower price points (down to three and five dollars) and Mighty-Tonka in 1965 to capture an older demographic (shifting the upper age limit of their target consumers from twelve to fourteen years old). The company's financials positively reflected this performance, with 20 percent annual sales growth and relatively steady 6-percent profit margins during the early 1960s. This put Tonka in an advantaged market-share position, as average sales growth among all toys at the time was only 12 percent.[29]

Through the efforts of both Tonka and Doepke, small-scale construction vehicles quickly became valued mainstays of postwar children's toy collections. Parents and children alike wrote to the manufacturers to express their appreciation. In a *Saturday Evening Post* article, a New York doctor even noted the positive effect the toys had on his cerebral palsy patients, writing,

---

(*Opposite*) This 1952 *Popular Science* advertisement for one of Doepke's Model Toys, modeled on the Caterpillar D6 Tractor and Bulldozer, invites children to imagine themselves at the controls of a powerful real-world earthmoving machine. (Courtesy Estate of Charles William Doepke)

EVER WATCH A BULLDOZER WORK?
Breaking through earth, brush, trees, it's a demonstration of mechanized might that fires the imagination of every red-blooded boy, turns his interest from cops and robbers to the he-man activities of a construction engineer.
Here, in the newest Model Toy, is an authorized, working reproduction of the famous
CATERPILLAR*
D6 TRACTOR & BULLDOZER
*Reg. U. S. Pat. Office
1 Made of heavy-gauge steel to outlast other toys 3 to 1. 2 Scraper blade adjusts from driver's seat. 3 Heavy-duty Tenite treads operate independently for sharp turns, dig into rough ground, yet will not scratch floors. 4 Tow hitch for pulling other Model Toys.
With the blade lowered, bulldozer levels roadways and airports.
With blade in up position, tractor is used to tow other Model Toys.
Realistic "Diesel" engine looks just like the powerful "Cat" engine.
Operating lever adjusts scraper blade to three positions.
Separate sections are securely hinged to form continuous tread.
$12.95
O.P.S. Ceiling Price
11 Western States—$13.75
Write for free catalog describing all Model Toys.
THE CHAS. WM. DOEPKE MFG. CO.
ROSSMOYNE 1, OHIO
Model
TOYS

"The children seem to experience a sense of power in duplicating a major construction operation in miniature." Decades later, the adult science fiction and nonfiction author Michael A. Banks recalled the combined appeal of these toys and books like Stephen Meader's young adult novel *Bulldozer:* "I first read Bulldozer in 1959, at the very beginning of the Space Age. . . . I was constantly exploring how things worked and trying to do more with my chemistry set and microscope than was possible. I built models, I read books, magazines and newspapers. With my friends I even built miniature Interstate highways on the dirt floor of an abandoned barn, using our Doepke and Tonka bulldozers, graders and other toys."[30]

Doepke tried to stimulate such enthusiasm through its marketing. One advertisement depicts a young boy pushing his toy bulldozer in the dirt and seemingly imagining his real-world counterpart in the distance behind him. The text reads, "Breaking through earth, brush, trees, it's a demonstration of mechanized might that fires the imagination of every red-blooded boy, turns his interest from cops and robbers to the he-man activities of a construction engineer." An article in a period business magazine forecast even greater ramifications for this toy trend, noting, "As many children today are playing with toy tractors, power shovels, trucks and scrapers as with cowboy pistols and toy airplanes." The net effect, the author concluded, was not only additional entertainment, but also "a new occupational outlook for youth."[31]

The related introduction of small-scale, rideable bulldozers during this period represented the most direct example of children's participation in the work of clearance. In 1950, *Popular Mechanics* reported on the development of a child-size Caterpillar bulldozer. Other than the substitution of pedals for engine power, the article noted, the tracked machine "looks and works like the real thing." The magazine further advised that youngsters could put the bulldozer to practical use in clearing snow. In 1966, the same publication featured a photograph of a youngster operating an even more powerful Cat-branded machine. Two marine sergeants had created this "pint-size bulldozer" out of a small automotive transmission, a seven-horsepower engine, and salvaged parts. As the magazine noted, the bulldozer "not only serves as a toy but does real landscaping chores" as well. Like growing numbers of operable toy construction vehicles from the postwar period, this product straddled the line between childhood plaything and real-world training device.[32]

Conservation and Criticism, 1960s and Beyond

Despite the celebration of the bulldozer embedded in many children's books and toys from the early postwar period, by the mid-1960s and early 1970s, more critical voices had emerged. In the world of fact, both urban and suburban residents vocalized disapproval of large-scale clearance work. They organized into movements supporting open space, environmentalism, building rehabilitation, and historic preservation, helping to bring about new legislation. Just as children's books from the 1940s through the 1960s mirror the rise of the bulldozer during that earlier period, so too do texts from this later era convey these changed public attitudes toward clearance that characterized subsequent decades.

This dissent within the literature did not appear out of nowhere. There had always been counternarratives to the celebration of progress, even in books that initially seem consistent with the dominant trope. Virginia Lee Burton conveys this productive tension particularly effectively, as she buries Mike Mulligan in the same modernity he created and leaves the technotopia of Geopolis dependent on a single machine to survive the snow. Similarly, just as *The Little House* suggests the limitless nature of pastoral land, it also gloomily depicts the creative destruction of the city. Gray and black urban buildings contrast with the house's pink pastel, while building demolitions make way for taller structures that sadden and overshadow the Little House. Don Freeman's *Fly High, Fly Low* (1957) is similarly despondent about urban change. The book recounts the love story of a pigeon and dove that make their home together in the letter "B" of the San Francisco Bay Hotel sign. When the pigeon comes home one day to discover that a crane has removed the sign, the reader feels the triple loss of his missing partner, their unhatched eggs, and the partially dismantled old building.[33]

Animals of the forest, rather than the city, are some of the most frequently depicted fictive victims of earthmoving machines. In Roald Dahl's *Fantastic Mr. Fox* (1970), for example, the fox family digs as fast as it can to get away from an onslaught of tractors outfitted with mechanical shovels. Similarly, the namesake machine of Catherine Woolley's short story "Butch the Bulldozer" (1953) threatens the woodland habitat of birds, squirrels, rabbits, and frogs when he arrives to clear land for the construction of new houses. By the story's end, though, the animals have successfully diverted his path by explaining that he will be destroying their own houses in the process.[34]

During the 1960s and 1970s, such critical depictions became less exceptional. Authors increasingly introduced children's books that presented fuller alternative perspectives on landscape destruction. In 1970, Suzanne Hilton published *How Do They Get Rid of It?,* a children's book about finding reuses for old things. Although the author demonstrates an interest in recycling, she stops short of advocating a reduction in demolition itself. She even describes the process of building wrecking as an art on par with that of building construction. Still, Hilton differentiates her work from the earlier celebrations by even considering alternatives to disposal of the old in favor of the new. In another book, Pearl Augusta Harwood's *Mr. Bumba and the Orange Grove* (1964), Bill, Jane, and Mr. Bumba learn of the impending clearance of a neighboring orange grove. Unlike earlier protagonists, they lament, rather than celebrate. Then they develop a plan. Mr. Bumba convinces the grove owner to preserve small ribbons of trees in order to justify naming his new subdivision Green Belt Manor. Noting that "people would pay more money for a place like that, with a name like that," the grove owner agrees. This compromise resolution manages emerging aesthetic and environmental aspirations within existing economic constraints.[35]

The change in times also tempered the tone of some otherwise celebratory texts. Although Jack McClellan and Millard Black devote the bulk of *What a Highway!* (1967) to a detailed description of road-building work, they also touch briefly on the removal of residents and disruption to daily life that the process entails. In two scenes about boys who get stuck while playing on the machines, they also try to teach respect for the dangers of heavy equipment. In *Tractors* (1972), Zim and Skelly momentarily interrupt their celebration of earthmoving machines to recognize some of their problems. While the authors never hinted at such sentiments in their two previous books on related subjects, they concede toward the end of this later text: "Men and tractors are changing the world we live in rapidly. They are turning forests into farms, cultivating large areas of land, building roads and bridges. Until a few years ago, all of this work was considered to be good. It was part of our progress. Now we know that the world pays a high price for this kind of growth. When changes are too great, both land and water suffer." To remedy the potential damage to fish, wildlife, and natural ecosystems, they entreat each of their readers—not just the machine operators—to consider the best path forward. In so doing, they attempt to involve readers in a quite different role from that of identifying future road-building sites. Nevertheless, the book concludes on a positive note: "These men are proud of the skill with which they operate their machines. They much prefer to work where they will help the world, because they and their tractors have done a good job."[36]

Most authors who truly broke with the celebratory past tradition tended to target a slightly older audience than the elementary school demographic. While pictures remain, text dominates the multiple brief chapters of their young adult works. Perhaps the more mature sensibilities of the intended readers, combined with the texts' longer length, better suited the more nuanced perspectives they present on the remaking of the landscape. The characters in *Andy's Landmark House* (1969) respond variously to the planned demolition of a street of brownstones: from excitement over the new construction to frustration about being depicted as a roadblock to progress and finally to resignation about the futility of opposition. Only idealistic Andy remains committed to the belief that "no one had the right to bulldoze a good neighborhood. It would be like stepping on an anthill and wiping out a way of life." By researching the history of one of the buildings, he rallies the neighborhood into united opposition that saves both the structures and the community. By contrast, the lead character of another book, *My House Is Your House* (1970), fares more poorly in her bout against the bulldozer. In this story about an urban renewal project on Manhattan's West Side, young Juana and her family are some of the last to leave a neighborhood in forced, clearance-based transition. On her final walk along the block, she sees some of the dirty undersides of urban clearance: "Dust in the air and yellow mud in the streets from the dump trucks. Empty lots strewn with rubble and beer cans alternated with rows of condemned buildings." Unlike optimistic Andy, Juana feels that nothing can save her own building. Yet the story concludes more positively, with Juana finding solace in a new home on Long Island. Rather than channel her criticism of urban destruction into physical opposition, Juana, like the Little House, looks for a new place to call home.[37]

What *My House Is Your House* does highlight, however, is the intimidating appearance of many wrecking and earthmoving machines. This nightmare potential is what makes the celebrated friendly depiction of bulldozers in many postwar books seem so surprising. Recall how the children in *Vermont Roadbuilder* played happily inside a bulldozer blade. The title page of *Johnny's Machines,* a Little Golden Book, similarly depicts a smiling toddler sitting inside a raised shovel bucket. *Are You My Mother?* goes a step farther. The baby bird in this story hatches from his egg and falls out of his nest while his mother is away. After asking various passersby if they are, improbably, his mother, he is saved by a power shovel named Snort. Snort lets the bird climb into his bucket and then lifts him back into his nest, where the mother bird arrives home happily to meet him. But not every child might look as favorably as these young characters do on large, loud, construction machines. Alongside the excitement stimulated by heavy equipment at work lay the equally viable prospect of its evoking fear. *My House Is Your House* taps into this other perception. During her last night in her West Side apartment, Juana has a terrible nightmare: "She dreamed that there were crosses on her forehead, white crosses like the ones on the windows of the condemned buildings. In her dream she was running down a long, dark hallway trying to escape a bulldozer that was clanking after her. As she tried to run faster, the bulldozer came closer and closer. But her feet would not run. Suddenly, Juana felt wet with perspiration, and wet below. She forced herself to awaken to end the nightmare and found that she had wet her bed." Readers might look past this traumatic scene as a fiction from which they and Juana need only awaken to move on. Juana even enacts this solution by finding salvation in the sunlit suburbs. Yet Robert Weaver's illustration, using dark colors and lines to create claustrophobic space, cements the text's overall negative depiction of the 1970s-era bulldozer.[38]

(*Opposite*) As depicted in a nightmare scene illustrated by Robert Weaver, the bulldozer that will tear down Juana's apartment building comes chasing after her instead in Toby Talbot's *My House Is Your House* (1970). (Courtesy Toby Talbot)

Other authors play on the menacing fear lurking behind clearance work to craft an even less positive story. One example is David Macaulay's *Unbuilding,* published in 1980. Through his more critical, ironic, and complex point of view, Macaulay breaks with his predecessors, who had largely cast an uncritical eye on the human reshaping of the physical and natural landscape. By eschewing the happy ending, the book forces readers to consider more direct responses to unwanted environmental changes. It also helps distinguish between construction and destruction. The author trades in his admiration of the building process, as depicted in his earlier works, for a satirical tone in describing the process of structural undoing.[39]

Unbuilding is the fictional story of the physical deconstruction of the Empire State Building. It is the near future—1989—and an oil baron named Prince Ali Smith has decided to purchase, dismantle, and relocate the landmark building to the Arabian desert as his new company headquarters. Many New Yorkers initially cringe at this news. But in typical American fashion, Macaulay notes, people quickly forget about it and abandon their protesting. Once the prince offers the city regular supplies of oil for taxis and buses, he is able to seal the deal. The demolition firm Krunchit and Sons spends the next three years taking down the building. Then a tanker loaded with the structure's dismantled elements sets sail, sinking a short distance from its final destination. The entire cargo is lost. Even though he stops short of stating it explicitly, Macaulay makes his disdain evident. His narrative suggests the wastefulness of dismantling a functioning building that is less than sixty years old. The story concludes with the opening of Empire State Park in place of the now lost building, but Macaulay does not end his epic on this relative high note. Instead, he offers a more ominous image: a nighttime view of Midtown Manhattan, with a black, rectangular void where the Empire State Building stands half-disassembled. In the last lines of the text, he returns to Prince Ali on the morning after the park's opening. Macaulay writes, "Prince Ali left for home. As his airplane rose swiftly into the sky he settled back on his reclining throne and turned his thoughts to the next . . . meeting, at which he would present Krunchit's estimate for the dismantling of the Chrysler Building."[40]

Unbuilding shows how far children's bulldozer books have come since the early postwar years. It also demonstrates advancements in demolition techniques, which were increasingly incorporating deconstruction and recycling into physical destruction. Macaulay's detailed drawings depict the work as more complex than earlier grunting and "grrr-ing" machines suggested, following a trend among his recent antecedents to emphasize the equipment's technological feats over animated faces. The text also reflects other cultural preoccupations of its age, from fears surrounding the late 1970s oil crisis to anxiety over the invasion of foreigners and foreign goods. In many other respects, however, *Unbuilding* follows the example of its predecessors. For Macaulay, clearance offers a lens onto technological progress; he calls on readers to get involved in the decision making surrounding clearance and uses his characters as reflections of the inherent value of that work. Yet the book's overall tone is changed. *Benny the Bulldozer* celebrates the bulldozer and the singularly positive value of its work. It describes how the little machine clears a forest to make way for a roadway and then shows readers everyone's enthusiasm with the result. Although Macaulay's detailed drawings and text offer a similar—and even greater—education in how the machines of demolition and clearance work, he shuns triumphalism for a more thoughtful attitude about the role of demolition in society. He makes this distinction clearest on the book's dedication page. Recall that the earlier works were often dedicated to future road-builders everywhere. By contrast, Macaulay dedicates *Unbuilding* "to those of us who don't always appreciate things until they are gone."[41]

Perhaps it was in the safe and simple world of children's literature that the celebratory culture of clearance expressed itself most clearly. Using happy illustrations and enthusiastic text, the authors of postwar bulldozer books offered multiple reasons for embracing the destruction of the natural and built environment. They argued that clearance was bound up in the forward sweep of technologically driven progress. They presented it as something that even young children could support, albeit in modest ways. Most compelling, these books depicted equipment operators as inspirational examples of virtuous masculinity, and powerful earthmoving machines as their buddies, accomplishing their work with a friendly smile. In these ways, bulldozer books—along with construction equipment toys—put a pleasing face upon a potentially upsetting topic.

With the passage of time, however, the darker undersides of clearance made their way into new kinds of stories. While these counternarratives remained in the minority within children's literature, the transformation that occurred within this literary subgenre reflected changing attitudes in the wider world. At this same moment, oppositional points of view were finding increasingly complex expression in other cultural forms. The art world offered a potent venue for this more critical, sophisticated, and enigmatic discourse. As groundbreaking conceptual artists took up the bulldozer as their paintbrush and the landscape as their canvas, they demonstrated how widely the machine had enmeshed itself into varied facets of postwar American life.

Chapter 7

Bulldozers as Paintbrushes

Earthworks and Building Cuts in Conceptual Art

In May 1960, the artist Walter De Maria wrote an essay, "Art Yard," describing a fictional project he might someday build. Sounding more like a scene out of *Benny the Bulldozer* than a proposal for serious artwork, De Maria's text combined the era's interest in "happenings" with burgeoning thinking on earth-based sculptural art. He proposed creating an art installation far outside the gallery's confines on an empty piece of dirt-covered ground. There, he imagined, spectators would gather, dressed in their best: "Then in front of the stand of people a wonderful parade of steamshovels and bulldozers will pass. Pretty soon the steam shovels would start to dig. And small explosions would go off. What wonderful art will be produced. . . . As the yard gets deeper and its significance grows, people will run into the yard, grab shovels, do their part, dodge explosions. This might be considered the first meaningful dance. People will yell 'Get that bulldozer away from my child.' Bulldozers will be making wonderful pushes of dirt all around the yard. Sounds, words, music, poetry. (Am I too specific? optimistic?)." While this scene had a fanciful quality, De Maria concluded with a more practical request: "And if this paper should fall into the hands of someone who owns a construction company and who is interested in promoting art and my ideas, please get in touch with me immediately. Also if some one owns an acre or so of land (preferably in some large city . . . for art . . . thrives there) do not hesitate."[1]

During the 1960s and 1970s, De Maria and his peers—Robert Smithson, Michael Heizer, Robert Morris, and Gordon Matta-Clark, among others—put the fantasies of children playing with toy bulldozers into practice in the art world. The works they produced were often large, site-specific projects consisting of reshaped earth and carved-up buildings. These young men were members of the generation that had grown up reading bulldozer books. As adults, they corralled construction equipment—including shovels, tractors, bulldozers, chainsaws, and backhoes—to dig earth and crush buildings as part of their work. As Smithson aptly observed about one of his fellow artists, "Instead of using a paintbrush to make his art, Robert Morris would like to use a bulldozer instead."[2]

Unlike the prototypical contractor, who employed these machines to accommodate the fast pace of progress, these artists tried to slow time down. Contractors cleared the landscape to erase it, seeking to create flat, blank slates to cover with new construction. But late-postwar artists dug and cut to produce mounds, voids, and gaping holes that would last long enough to allow public inspection. They invited viewers to stop, look, and ponder the complex meanings of clearance. Recognizing their sculptures' implicit ephemerality, they often prolonged the projects' existence by capturing them in still photographs and film. All of these activities improved the visibility—and the possibility for interrogation—of the bulldozer's increasingly normalized activities and impacts.

Ultimately, these artists' cooption of the bulldozer for creative use suggests the expansive reach of the postwar culture of clearance. In addition to influencing journalism, films, photography, toys, and fiction, this culture even inspired new forms of conceptual art. That these explorations were occurring in the realm of high art, rather than for mass consumption, does not diminish their significance to the larger culture. Using the language of sculpture, these artists articulated changing ideologies. By repurposing the bulldozer's destructive possibilities—and creatively removing select parcels from the capitalist landscape—they gave material form to the late-postwar period's emerging critique of the culture of clearance. Eventually, these artists' material influences began to transcend the art world. As some expanded their goals from cultural representation to social and environmental rehabilitation, they helped reuse sites that clearance had despoiled. Working cooperatively or responsively, some architects, planners, and social activists effectively mirrored the examples set by these artists.[3]

Earthworks Artists Make a Studio of the Land

In October 1968, New York's Dwan Gallery hosted an extraordinary exhibition of a new kind of art: earthworks. The organizer, Robert Smithson, included fourteen artists exhibiting earth-based works, or the documentations of such works (since the actual art resided either out in the landscape or as yet only in the artist's imagination), within the seemingly incongruous environment of an art gallery. Gallery owner Virginia Dwan ran bold advertisements in *Artforum* magazine to build interest in the event. Like a ground-level perspective on one of William Garnett's Lakewood, California, landscapes, Dwan's ads foregrounded dirt. The soggy ground of a construction site nearly filled the frame of an ominous black-and-white photograph. The tracks of a heavy vehicle traversed the ground. This cryptic image suggested that the event would explore not only the earth but also humankind's mechanical imprint upon it. At the bottom of the ad appears the show's title, *Earth Works.* "Earth art," "land art," and "environmental art" have all been used to describe these projects, but Dwan's exhibition gave this informal category of art its enduring "earthworks" label.[4]

Although this work was a product of the 1960s, it continued an artistic tradition dating back to at least the early twentieth century. In the teens and twenties, Marcel Duchamp had created his Readymades out of a snow shovel, a urinal, and a bicycle wheel. His projects expanded the bounds of art objects and materials, while connecting art with everyday life. In 1930, Gutzon Borglum took large-scale artistic sculpture to the landscape when he blasted and reshaped the Black Hills to create Mount Rushmore National Monument. Borglum's former student, the Japanese-American sculptor Isamu Noguchi, later developed proposals for earth-based projects of his own. Inspired by World War II battleground photographs, Noguchi conceived of a project—never realized—to be called *This Tortured Earth,* in which he would bomb a parcel of land. The resultant slashes and scars would symbolize war's de-

A full-page advertisement in *Artforum* for the Dwan Gallery's *Earth Works* show, in 1968, features an ominous image of a construction vehicle's tracks. This photograph suggests that the event will engage not only with art made of earth but also with our mechanical imprint upon that terrain. (Photograph by Virginia Dwan, *Artforum,* October 1968. Used with permission.)

structive terrestrial impacts. During the 1950s, Jackson Pollock and Robert Rauschenberg extended the scope of artistic media inside the gallery. They intermingled paint, dirt, and other found objects on their canvases. In 1955, Herbert Bayer created *Earth Mound.* This grassy, forty-foot-diameter mound set down in an Aspen, Colorado, meadow was perhaps the first instance of earth-based sculpture. It was not until the 1960s, however, that greater numbers of artists began taking up these broad interests and more commonly turning to the landscape as their canvas. By the end of the decade, a critical mass had arisen, warranting its first group show.[5]

While earthworks was always less an explicit movement than a loose collection of artistic productions by like-minded individuals, its practitioners were united in the use of common methods and materials. These elements blurred the lines between the artworks and actual construction. Dirt was a common building block, and construction equipment the means by which many of these artists molded it. As the *New York Times* critic Grace Glueck explained in an exhibition review, "The medium (and message) is Mother Earth herself—furrowed and burrowed, heaped and piled, mounded and rounded and trenched." Just as real-world construction projects were expanding in scale, so too were these artistic works. As one earthworks practitioner noted, "Artists have always been frightened by things like the Grand Canyon. These forms may be impossible to duplicate or rival, but they are important. Now we have to take them on in their own ballpark." Accordingly, artists co-opted the tools of the professional builder to tackle these new forms. They used aerial photographs both to search out building sites and then to exhibit the finished projects. Heavy equipment made their work practical, while ubiquitous materials made it economical. Echoing the sentiments of a suburban developer, one artist noted, "Dirt is reasonably cheap, so you can use a lot of it."[6]

Even as earthworks artists mimicked the practices of the Bulldozer Man, they also invited criticism of him. These artists broke most significantly with their construction peers in the ethic they offered about the environment that served as site and substance of their art. As one art critic noted, "So the earth workers are posing questions about how and where we live. How do we deal with our natural resources? What kind of environment do we live in? Where have all the flowers and the open spaces and the beautiful vistas gone?" As earthworks makers blurred the line between art and life, they invited viewers to see the landscape in a new light. They questioned the era's march toward progress. They challenged the commodification of art and land, seeking to make both sacred instead. In an era in which the tyranny of the blank slate threatened to wipe out distinctive landscapes and impose flat sameness instead, earthworks artists argued for the value of place. As they called attention to humankind's ongoing consumption of the environment, their appreciation for entropy countered the order being sought in professional clearance and construction practice.[7]

A brief survey of some prominent earthworks will show the range of artistic approaches underpinning this movement, as well as the varied relationships between the art and construction worlds. Dennis Oppenheim, for example, drew upon both his art school training and his past work in construction to tackle a 400-acre plot of Napa Valley land. There, in the mid-1960s, he deployed earthmoving equipment to reshape cliffs and valleys into a massive mound of earth and sod. After moving to the East Coast, Oppenheim collaborated with city planners and engineers to enliven the vast expanses of space abutting superhighways. On a gravel pit near Exit 52 of the Long Island Expressway, he produced *Landslide* (1968) by placing a series of parallel boards down the slope. Perhaps inspired by this project, the Yale School of Architecture invited the artist to collaborate on an "environmental highway project" in New Haven. In the resulting *Contour Lines Scribed in Swamp Grass* (1968), Oppenheim used a mower to cut grass in the shape of surrounding contour gradients on a 150-by-200-foot area of highway-side swamp. The Dwan exhibition included maps and photographs of the scene. In *Directed Seeding—Cancelled Crop* (1969), the artist applied mowing equipment to a large agricultural field. Although the associations may have been unintentional, the juxtaposition of neighboring housing with a field traversed by a giant "X" imports the shorthand of demolition into an agricultural context. Just as the crossed-out windows of Juana's New York City apartment building signaled its impending demise, Oppenheim's notation suggests the comparable clearing of farmland.[8]

Robert Smithson, the most prominent figure behind the earthworks movement, received his introduction to the field while working as a contractor. In 1967, the engineering firm of Tibbetts, Abbott, McCarthy, and Stratton (TAMS) hired Smithson as an artist-consultant on their proposal for a new Dallas–Fort Worth airport. As he later recalled, "I found myself surrounded by all this material that I didn't know anything about—like aerial photographs, maps, large-scale systems, so in a sense I sort of treated the airport as a great complex, and out of that came a proposal . . . [for] sculpture that you would see from the air." Although their collective proposal was not chosen, the experience opened Smithson's eyes to new representations of space and new possibilities for sculptural scale.[9]

Smithson's critical perspective distinguished him from his professional construction colleagues. In contrast to the contractor who lauded American mechanical might or the engineer who proudly proclaimed, "You ought to see my 'cats' move that dirt!," Smithson called these same machines "dumb." He wrote, for example, of "grim tractors that have the clumsiness of armored dinosaurs, and plows that simply push dirt around." The overall impact of this equipment, he believed, was to "turn the terrain into unfinished cities of unfinished wreckage." Indeed, "with such equipment, construction takes on the look of destruction; perhaps that's why some architects hate bulldozers and steam shovels." Such disdain might also arise from the architect's emphasis on the completed building, rather than on the enabling clearance work. But Smithson's preferences were for the latter. As he concluded, "These processes of heavy construction have a devastating kind of primordial grandeur, and are in many ways more astonishing than the finished project—be it a road or a building."[10]

Walter De Maria shared Smithson's perception of heavy construction as a process of destruction. Although earthworks artists rarely wore their politics on their sleeves, De Maria's contribution to the Dwan show was unusually explicit. From his base in Germany, the artist sent instructions to his friend Michael Heizer to create a large stretched canvas, measuring roughly seven feet tall by twenty feet long, covered entirely in "Caterpillar yellow" oil paint. When the specified hue was unavailable, De Maria settled for the shade of yellow used on John Deere earthmoving machinery instead. A small stainless-steel plaque, mounted in the middle of the work, announced its title: *The Color Men Choose When They Attack the Earth* (1968). By equating construction equipment with violence to the land, the work expresses a clear environmentalist statement. In keeping with the subtleties of the art movement's messages, however, the words are relatively small in size. Only by moving in closer—and stopping to look—does the viewer catch the argument.[11]

Some artists bridged the divide between the real and art worlds by building earthworks in the gallery. Smithson dug up dirt and rocks in the New Jersey Pine Barrens for display in the Dwan show. He filled metal receptacles with his materials and plotted their original locations on an aerial map. He called the work *Franklin Non-site* (1968). Similarly, in *Earthwork* (1968), Robert Morris deposited a six-foot-square, twelve-hundred-pound pile of dirt and peat moss on the floor of the Dwan Gallery. He mixed these materials with pieces of copper, aluminum, brass, zinc, felt, grease, and bricks he had found on a Manhattan construction site. De Maria contributed not only his yellow canvas but also a photograph of Heiner Friedrich's Munich gallery, which he had filled two feet deep with potting soil. He titled the piece, variously, *Munich Earth Room* and *50 M^3 (1,600 Cubic Feet) Level Dirt/The Land Show: Pure Dirt/Pure Earth/Pure Land* (1968). Visitors to the Dwan Gallery could only experience the work photographically. Yet this viewpoint was not much different from those of the visitors to the German exhibit itself, whose entry to the room was blocked by a waist-high barrier.[12]

Michael Heizer distinguished his work by its dramatic scale and its enduring presence in the landscape. His father, Robert Heizer, was a prominent archaeologist at the University of California at Berkeley. The younger Heizer reflected that his teenage experiences recording technical excavations "might have affected my imagination" for his future medium. "My personal associations with dirt are very real," he said. "I really like it, I really like to lie in the dirt. I don't feel close to it in the farmer's sense." Political currents also framed his decision to work with earth. He recalled that he "started making this stuff in the middle of the Vietnam war. It looked like the world was coming to an end, at least for me. That's why I went out in the desert and started making things in dirt." Heizer completed his first earthwork, *Nine Nevada Depressions,* in late 1967. Deploying manual and mechanical labor, he dug "artfully shaped holes" in the hardened earth and lined them with metal. In less than a year's time, he had moved twelve tons of dirt. The Dwan exhibit included a transparency depicting the piece, titled *Dissipate #2* (1968). *Double Negative* (1970), Heizer's most famous completed project, dwarfs even these volumes. On a square mile of land located eighty miles northeast of Las

Vegas, a crew armed with bulldozers and dynamite cut two ramplike notches opposite each other on the edges of a mesa. The massive trench that was created between them reached fifty feet deep, thirty feet wide, and fifteen hundred feet long. It also displaced 240,000 tons of material. The Dwan Gallery exhibition of *Double Negative* consisted entirely of photographs. Yet the original work resided in the landscape, and visitors can still experience it there today.[13]

Other earthworks artists shared Heizer's grand ambitions. In 1970, for example, Smithson leased ten acres around Utah's Great Salt Lake and hired a contractor to move 6,650 tons of dirt and rock from a nearby hillside onto the lakebed. The earthen deposits created a fifteen-foot-wide, fifteen-hundred-foot-long dirt roadway that extends from the shoreline into the lake before spinning into a spiral. *Spiral Jetty* endures today, although changing water levels continue to alter its appearance over time. *Three Continents* was the title of one of De Maria's most ambitious—but ultimately unrealized—projects. He planned to carve two lines and a square into the surfaces of Indian, African, and North American deserts. Once those were complete, he would superimpose satellite photographs of the etchings on top of each other. For the African portion, he spent over a week in January 1969 bulldozing a line a mile long and ten feet wide across the Sahara. Unfortunately, the project is largely lost to us today. When his equipment acquisition and site selection led local authorities to fear that he might be an oil speculator, they arrested him. Their confiscation of the precious film he had already produced destroyed any record of the completed work.[14]

These artists' works implied several criticisms of the culture of clearance. Perhaps most obvious, they were ambivalent, at best, about the era's march toward "progress." Creative destruction—or the capitalist tendency to clear away in order to build anew—underpinned this pursuit of progress. By contrast, although many earthworks artists employed similar practices of landscape removal, they often left the voids they had created unfilled. This emphasized the environmentally destructive nature of the act, rather than concealing it under new construction. They left the absences visible for further contemplation.

In a 1966 interview, while earthworks were still in their formative stage, Carl Andre expressed an interest in commissions that would allow him to dig

The dump truck and loader of a local contractor appear here moving earth and rock to create Robert Smithson's *Spiral Jetty* (1970) in the Great Salt Lake. (Robert Smithson, *Spiral Jetty,* Great Salt Lake, Utah, 1970/unidentified photographer; Robert Smithson and Nancy Holt papers, Archives of American Art, Smithsonian Institution; Art © Estate of Robert Smithson/Licensed by VAGA, New York, NY)

and create "negative sculptures." Soon afterward, in 1967, Dennis Oppenheim made his first earthwork by carving a five-foot wedge out of the side of a mountain in suburban Oakland. He then lined the hole with translucent Plexiglas. As Oppenheim later recalled, "I was more concerned with the negative process of excavating that shape from the mountainside than with making an earthwork as such." Similarly, Robert Smithson pursued sculpture as absence, or "voids that displace the solidity of space." Perhaps most obvious in this regard was Michael Heizer's work. He explicitly referred to *Double Negative* as a "negative sculpture." As he further elaborated, "I make something by taking something away." Clearance was not just these artists' means; it was their end.[15]

Along with their critiques of the blind pursuit of "progress," earthworks artists challenged the postwar period's culture of consumption—particularly as it applied to the landscape. The "consumer's republic" required both commodification and possession. By creating ephemeral artworks, earthworks artists resisted both. Excluding selected pieces like *Spiral Jetty,* many earthworks endured for only a short period of time. Oppenheim's *Accumulation Cut* (1969) exemplified this tendency. The artist chainsawed a two-hundred-foot-long trench in frozen Beebe Lake, in Ithaca, New York. Within twenty-four hours, however, the lake had refrozen and the work was gone. Even works composed of more durable materials suffered a similar fate, disappearing through the natural decay caused by weather and time. Human influences sometimes exacerbated these effects. Only a few years after the completion of *Double Negative,* for example, a reporter visiting the project noted the degradation of the previously sheer walls, as well as the presence of graffiti on them. Urban projects faced comparable risks. Several years after Heizer's excavation of *Munich Depression*—a circular pit measuring a hundred feet across and fifteen feet deep in a vacant lot on the outskirts of Munich—the work was gone. Apartment houses now reside on the site, marking a somewhat fitting ending to a project that was intended to reflect nature itself.[16]

It was not only their ephemerality but also the projects' remoteness and scale that contributed to their resistance to commodification. Ironically, these aspects of the artworks also increased the costs of their creation, while simultaneously diminishing their economic payback potential. Even though dirt was cheap, moving so much of it was expensive, especially when factoring in property acquisition costs and the rental fees for heavy equipment. De Maria estimated, for example, that a proposed project consisting of two twelve-foot-high, mile-long walls would cost him five hundred thousand dollars. A small group of dedicated patrons funded much of the early work. These included gallery owners Virginia Dwan in New York and Heiner Friedrich in Munich, as well as the New York taxi fleet owner and pop art collector Robert Scull. Since patrons would rarely be able to possess the artistic output in a traditional sense, their financial support bought them "research" or "interesting activity," rather than a tangible collection. At best, patrons might visit the works or acquire photographs or models. Sometimes artists even dispensed with documentation altogether, on the premise that the absence of a material record should not detract from the overall work. As long as someone had seen the final product, they argued, that should be enough.[17]

The decommodification of art also had political ramifications, since the art "could not be bought and sold by the greedy sector that owned everything that was exploiting the world and promoting the Vietnam War," as one critic put it. Heizer speculated that the new relationships these artists were forging between capital and earthworks might signal a change in the consumption of art more broadly. "One of the implications of earth art," he noted, "might be to remove completely the commodity status of a work of art and allow a return to the idea of art as . . . more of a religion." By nature of this association, these artworks also suggested corollary critiques of the commodification and desacralization of the larger landscape.[18]

Ultimately, earthworks artists reasserted the power of place, particularly from an environmental perspective. Developers, builders, and clearance crews viewed nearly any site—no matter how varied in terrain or previously populated by peoples and buildings—as a potential space for new building. All that was necessary was to level it and begin the new construction. Canvas-based artists worked in an analogous manner, initiating their projects upon empty, white cloth terrain. Most of their finished products would end up hanging on the white walls of galleries. In stark contrast, earthworks artists adapted their works to their messier starting points on land. It was also there, out in the landscape, that their completed projects remained on display. These artists' unique creations criticized both the real world's and the art world's fascinations with blank slates.

The *Earth Art* show, organized by Willoughby Sharp at Cornell University's Andrew Dickson White Museum of Art in February and March 1969, exemplified these artists' connection to place. The show's premise was to assemble artists who would use the materials and sites in the surrounding town of Ithaca, New York, for the creation of new works. Smithson contributed what he called a "non-site," consisting of materials from the nearby Cayuga Rock Salt Company. Richard Long arranged rocks from a nearby quarry on the gallery's front lawn. Dennis Oppenheim reworked the ground at an abandoned Ithaca airport. David Madalla dug up several tons of earth from behind the museum, dumped it on the building's south side, and turned the pile to mud by watering it with a hose. These works all derived from Ithacan geography, with the only limitations set by climate (cold temperatures made some proposals impossible) and the availability of local materials. In the gallery, the maps that sometimes accompanied these artworks located the projects' ingredients and outputs geographically.[19]

The significance of place to the *Earth Art* show was typical of the larger movement. As Oppenheim put it, "Rather than isolate a form, I prefer to tie in with forms that already exist, as in a huge piece of landscape." Consequently, the artists selected their sites with great care. Smithson made multiple treks to the quarries and Pine Barrens of New Jersey to develop his nonsites—indoor works consisting of materials from outdoor sites. Heizer spent over a year surveying the Swiss Alps, Nevada, Wyoming, and Montana from the air in search of a location appropriate for a vertical piece he hoped to execute on the face of a granite cliff. As he noted, "The work is not put in a place, it is that place." Even as they altered existing landscapes, earthworks artists argued by example that place still mattered. Whereas the construction professionals behind the commercial culture of clearance tended to erase place, earthworks artists made specific places the starting points of their work.[20]

Earthworks as Land Reclamation

The connections between earthworks installations and construction practices became more tangible during the 1970s. Private companies and public officials often called upon artists to bring public art to sites that had been despoiled by landfills, strip-mining, other industrial uses, or abandonment. Although this was not the first time that artists had participated in land reclamation, such activities flourished during this period. This was a consequence of the capabilities of earthmoving equipment, the interests of the artistic community, the obligations increasingly falling upon corporations to clean up their environmental ruins, and the funds set aside to help pay for that work. The Surface Mine Control and Reclamation Act of 1977, for example, instituted federal regulation of coal mines and created a fund to pay for the cleanup of abandoned mine lands. On some of the locations where contractors had recently sent out bulldozers to clear the earth, artists took up those machines for aesthetic and environmental repair. Through these practices, artists expanded the applications for which large-scale earthmoving could—and even should—be employed.[21]

Robert Smithson was a vocal proponent of this practical approach to art making. After developing one of his earthworks on the land and lake of a former quarry, he began seeking out land-reclamation projects. In the early 1970s, for example, he submitted proposals to Hanna Coal, in Ohio, and Minerals Engineering, in Denver, for reclamation of former strip-mining sites. He also developed plans for the reclamation of a massive, three-mile-wide mining pit near Bingham, Utah. Although Smithson died before making much headway on these projects, his writing on the subject influenced later artists. He saw reclamation projects as both opportunities and obligations, in which earthworks could be used to "mediate between the ecologist and the industrialist." They would improve upon sites "that have been disrupted by industry, reckless urbanization, or nature's own devastation." With the help of the artist, he argued, "such land is cultivated or recycled as art." Yet Smithson's was also a self-serving position. While the exclusion of most earthworks from commodity circulation posed challenges for project funding, land reclamation—as Robert Morris put it—was "the key that fits the lock to the bank" of local, state, federal, and industrial funding. The potential value of that sponsorship could be measured in access to hundreds of thousands of acres of sites, not to mention millions of dollars.[22]

Not all artists agreed with Smithson. Dennis Oppenheim felt no obligation to repair the environmental damages he found on the "ravaged sites" upon which he worked. Morris also disagreed with Smithson, but on political grounds. In an article published in *October* magazine, he argued that while art always exists in the service of others—from museums to consumers—the patronage of this particular public-private alliance threatened to put it in the service of landscape loss. "Insofar as site works participate in art as land reclamation," he argued, "they would seem to have no choice but to serve a public-relations function for mining interests in particular and the accelerating technological-consumerist program in general." Morris further posited that "when the U.S. Bureau of Mines contributes to an artist's reclaiming the land . . . art must then stand accused of contributing its energy to forces which are patently, cumulatively destructive." He concluded his argument with a tongue-in-cheek proposal: "Should the government/industry sponsorship of art as land reclamation be enthusiastically welcomed by artists? Every large strip mine could support an artist in residence. Flattened mountain tops await the aesthetic touch. Dank and noxious acres of spoil piles cry out for some redeeming sculptural shape. Bottomless industrial pits yawn for creative filling—or deepening. There must be crew out there, straining and tense in the seats of their D-8 caterpillars, waiting for that confident artist to stride over the ravaged grounds and give the command, 'Gentlemen, start your engines, and let us definitively conclude the twentieth century.'"[23]

Others viewed earthworks—as produced for reclamation or otherwise—as inherently detrimental. One critic posited, "Earth Art, with very few exceptions, not only doesn't improve upon the natural environment, it destroys it." Others saw reflections of the "way developers insensitively force their will on the land." Some even likened the work of both Heizer and Smithson to symbolic rape. When asked whether he saw any destructive elements in his own work, Smithson demurred, "It's already destroyed. . . . It's a slow process of destruction. The world is slowly destroying itself. The catastrophe comes suddenly, but slowly." By taking the long perspective on the entropy of geological time, Smithson sidestepped his own relatively insignificant impact upon the landscape. He also suggested that the motivation behind the work was more significant than its physical impact, arguing in *Artforum* for "the possibility of a direct organic manipulation of the land devoid of violence and 'macho' aggression." This parsing of intent and impact could also be applied to postwar suburban development, where developers seemed to be suggesting that the progress of development outweighed any environmental damages incurred as a means to that end.[24]

Some potential allies felt differently. In 1972, the painter and environmental activist Alan Gussow assembled a collection of American landscape paintings in the volume *A Sense of Place: The Artist and the American Land.* Gussow included works ranging from the Hudson River School to more recent abstract art. While lauding the inherent beauty of this long-lived tradition, he also contrasted the lyrical poetics of these landscape painters with "the earth works artists who cut and gouge the land like Army engineers." Smithson, however, defended himself and affiliated actors: "The farmer's, miner's, or artist's treatment of the land depends on how aware he is of himself as nature; after all, sex isn't all a series of rapes. The farmer or engineer who cuts into the land can either cultivate it or devastate it."[25]

The gendered nature of the earthworks movement invited the environmentalist critiques of masculine aggression that were also characteristic of construction-related land clearance. Despite the participation of female artists like Nancy Holt, Ana Mendieta, Mary Miss, and Mierle Laderman Ukeles, men overwhelmingly dominated the earthworks genre. This gender imbalance partially resulted from the old boys' network of patronage, which constrained access to funding, and the physically laborious processes necessary to bring many of the works to fruition. The art historian Susan Boettger has termed the image of the strong earthworks artist "cowboy bravado." Feminists have noted that these works "seemed to be all about getting out behind the wheel of a big ditch digger." Several artists reinforced the "cowboy bravado" image by gathering at a New York City watering hole called Max's Kansas City dressed in their cowboy boots. In 1969, Heizer and De Maria also made a short film, *Hard Core,* which depicts the two men donning cowboy hats and engaging in an old-fashioned shootout in the Black Rock Desert outside Reno, Nevada. This western Bulldozer Man character further came alive upon the landscape.[26]

The contributions of earthworks artists to land reclamation helped combat the negative, Bulldozer Man image. In 1979, for example, the Arts Commission and Department of Public Works in Kings County, Washington, teamed up to convene the project *Earthworks: Land Reclamation as Sculpture.* With funding from the Percent for Art program, the U.S. Bureau of Mines, and the National Endowment for the Arts, the county retained artists for reclamation of a landfill, an eroded creek bed, gravel pits, a former naval air station, and the area surrounding the Seattle-Tacoma airport. The funders selected Robert Morris for the pilot project. His design, which he executed at the 3.7-acre Johnson Pit, was for a bowl-shaped, terraced amphitheater carved out of a gravel quarry. As a reminder of the site's past history, he preserved the stumps of some fir trees that had been growing along the rim. Morris's modest approach provoked negative reactions among some area residents, due in part to their distaste for his design, early problems with erosion, and the perception that the project preserved destructiveness. Yet by reclaiming the location in a manner that conserved and commented on its previous

Robert Morris's *Earthwork* (1979) for the Johnson Pit in Kings County, Washington, was an early example of artistic land reclamation. In addition to beautifying the site, Morris also retained reminders of its past usage though the preserved stumps and terraced bowl of the pit. (ARTonFILE.com; © 2014 Robert Morris/Artists Rights Society [ARS], New York)

exploitation, Morris made a more visible statement than a picturesque park that covered over the area would have done. As much as was possible, he froze the cleared landscape in time for all to see. Morris's intervention and the larger county project suggested the potential alliances to be formed among earth art, private development, and public projects.[27]

Buildings and Chainsaws as the Vocabulary of Artistic Critique

While rural landscapes inspired the majority of earthworks artists, contemporaries like Gordon Matta-Clark pursued a corollary approach back in the city. Where Smithson and Heizer gouged earth for their projects, Matta-Clark severed and punctured buildings. By exposing the destruction of urban and suburban built environments, Matta-Clark's art disrupted the logic of the real-world building demolition that was then increasingly attracting popular criticism. Working in a variety of locations, but especially in and around New York City, Matta-Clark turned to aged and abandoned buildings for his canvases. Sometimes he even worked in consultation with wrecking companies, since their impending arrival made it easier for him to get permission to explore these sites. He dissected the structures to dramatic effect. Matta-Clark's work challenged the solidity of architecture, blurred the boundaries between inside and out, and like earthworks, pushed the limits of what we categorize as "art."[28]

Matta-Clark's creations spoke to social and political themes beyond the art world. Although his process, at a surface level, seemed to mirror that of professional demolition companies, the artist's practices and products differed notably. First, whereas demolition companies came in with a wrecking ball and bulldozer, Matta-Clark frequently used only manual tools and a power saw. In Genoa, Italy, he even cut through the foot-and-a-half-thick concrete walls of an old factory with a hand chisel. This craft approach slowed the pace of his work, replacing the swift sweep of the wrecker with the careful removal of the artist. A slower process allowed for closer observation and introspection about the building cuts he was executing. Second, the two activities diverged in the scope of their destruction. While the wrecker cleared an entire site, Matta-Clark practiced selective removal, often cutting out only portions of a building's facade and leaving unexpected openings in their place. Rather than clearing everything away, the artist used the demolition process to open up previously unseen spaces. Finally, the artist and demolition crews treated the detritus differently. As the wrecker struggled to find suitable dumping sites for burying rubble, Matta-Clark photographed and even displayed the cut-out remains for all to see. For wreckers, demolition was a means to a practical end. For the artist, the process and products of demolition were the creative ends in themselves.[29]

Matta-Clark's interest in the built environment can be traced to his training in architecture. He earned his undergraduate architecture degree from Cornell in 1968 and remained in Ithaca following his graduation. He was still there when the *Earth Works* show came to town. While it was there, Matta-Clark introduced himself to many of the participant artists. He also had the opportunity to contribute to the execution of some of the artworks, including

Smithson's *Mirror Displacement (Cayuga Salt Mine Project)* (1969). Shortly after the show ended, Matta-Clark moved to a SoHo loft and integrated himself into the burgeoning conceptual art scene. The art he ultimately produced combined architecture, sculpture, film, and performance to reconfigure disused metropolitan spaces. He often applied the term "anarchitecture" to his approach, in recognition of its anarchist relationship with the professional practice of building design.[30]

In one of Matta-Clark's earliest site-specific works, *Fresh Kill* (1972), the artist incorporates a bulldozer into his performance. This thirteen-minute, mostly silent film begins with Matta-Clark driving his red pickup truck along an empty roadway in Staten Island, New York. Just a minute and a half into the journey, a bulldozer appears. It waits menacingly and silently at Fresh Kills Landfill, the destination toward which the truck is approaching. In one violent instant, the truck crashes into the bulldozer's waiting blade. The crash is then repeated in close-up to increase the effect. While the immediate cause of the collision remains unclear, its consequences are not. The bulldozer springs to life, crawling backward and circling around the damaged vehicle like an animal examining its prey. Then the dozer's sharp blade attacks. The machine hits the truck in the front, crumples the roof, pushes the wreckage down the road, and flips and rolls over it. When a second bulldozer arrives, they smash the automotive skeleton between them before a dragline's clamshell bucket loads it onto a dumpster truck. A tractor then pulls the dumpster to the landfill, where it deposits the debris on the edge among the rest of the rubbish. When another bulldozer covers the remains of the pickup with refuse, the job is complete.[31]

Fresh Kill portrays the process and implements of wrecking in a violent light. At the first appearance of the bulldozer the soundtrack goes silent, signaling the doom to come. The machines' ensuing movements seem primitive, yet endowed with terrible strength. In the space of less than ten minutes, they turn a carefree red truck into a mangled, immobile, and unidentifiable pile of trash. The fate of the truck's driver throughout the mauling of the vehicle remains disconcertingly ambiguous. In these ways, the film refutes the friendly and helpful view of heavy equipment characteristic of postwar children's books. The title of the piece, a play on the name of its geographic site, speaks to the violence of the work the machines perform. Each new object they encounter becomes their latest "kill." By depicting the seemingly wanton destruction of a productive vehicle, Matta-Clark critiques the unchecked power of demolition machines. In tandem, he draws attention to the culture of disposability—a point he emphasizes by using a landfill as his performance site. This theme of refuse and disposal would continue to resonate in his future projects.

At the same time that he was creating *Fresh Kill,* Matta-Clark initiated his "building cuts," in which he transposed his interest in wrecking from vehicles to the urban and suburban built environment. In 1970, during the early stages of his thinking, Matta-Clark contacted several New York City officials to inquire about renting, or otherwise acquiring, some of "the many condemned buildings in the city that are awaiting demolition." He intended to put them to artistic use, writing to officials, "I am interested in turning wasted areas such as blocks of rubble, empty lots, dumps into beautiful and useful areas." He speculated that his work might involve cleaning up the designated buildings, using them as galleries or community spaces, and generally "teaching + working with young people." In reality, however, his art was far more complex than he let on.[32]

Without waiting for permission, Matta-Clark began to seek out, appropriate, and excavate selected structures on his own. For him, blighted neighborhoods were the sites of artistic inspiration. He recalled driving around New York City in his pickup truck, "hunting for emptiness, for a quiet abandoned spot on which to concentrate my piercing attention." There, "the omnipresence of emptiness, of abandoned housing and imminent demolition gave me the freedom to experiment." Once he had selected a building, often in the deteriorating South Bronx, Matta-Clark and a small band of co-conspirators would raid the structure and use generator-powered chain saws to cut out large geometric sections from the walls and floors. They photographed the newly opened spaces and sometimes transported the excavated layers of plaster, linoleum, and wallpaper back to a gallery. Fellow artist Ned Smyth recalled the excitement and danger of these outings: "This was always scary, with blocks and blocks of empty, boarded-up buildings, haunted by junkies who would steal copper wire and pipe to sell as scrap to get money for drugs. You never knew what you might run into, and if anything happened, no one would ever know. But it didn't seem to bother Gordon." Matta-Clark saw not

In *Splitting* (1974), Gordon Matta-Clark cut a vertical slice through an abandoned suburban home at 322 Humphrey Street, Englewood, New Jersey, in order to open it up to the light. Although the building soon fell to demolition, the artist relocated the four roof corners to a gallery. (Gordon Matta-Clark, *Splitting,* 1974, gelatin silver print; 16 in. × 20 in. [40.64 cm × 50.8 cm]; San Francisco Museum of Modern Art, Gift of The Estate of Gordon Matta-Clark; © 2014 Estate of Gordon Matta-Clark/Artists Rights Society [ARS], New York)

just the buildings but also their broader social landscapes in a wholly different light.[33]

In 1973, Matta-Clark shifted to the suburbs and expanded the scale of his building cuts. In the first such project, he performed radical surgery on a structurally sound single-family house located in a predominantly African American neighborhood in Englewood, New Jersey. In the film *Splitting* (1974), in which he documents this six-week surgical process, Matta-Clark uses a saw to cut through the structure from roof to foundation. He fissures walls, window frames, staircases, and all else that crosses his path. Next, the bare-chested artist bevels down part of the concrete foundation and tilts one of the two separated building halves, creating a V-shaped slit. Two feet wide at the top, this slit admits light into the empty interior. Matta-Clark concludes his film with a brief screen of text that describes the ultimate fate of the house: "The building lasted three months before being 'demolished' for urban 'renewal.'" Since this renewal project, like so many others, was never completed, the clearance process turned both a former home and, in this case, an artwork into an empty lot.[34]

In his second house-cutting project, Matta-Clark more clearly contrasts the laborious and specific removals of the artist with the rapid, careless destruction of the bulldozer operator. In a film titled *Bingo/Ninths* (1974), he documents the final days of a red-shingle two-story house in Niagara Falls, New York, that had been slated for demolition. The film begins with Matta-Clark drawing the grid of a Bingo game (or a tic-tac-toe board) on the house's rear facade. Then, over the course of ten long workdays, he and five crew members methodically measure, cut, remove, and crate the roughly five- by nine-foot half-ton sections of the facade. This process exposes the interior profile of the building within. The one piece he leaves attached is the center square, which Matta-Clark called the "bingo."[35]

Soon, the demolition company's bulldozer arrives. It enters from the right side of the screen, led by the teeth of its blade. The machine proceeds to gouge the roof, knock off the remaining central piece, and reach inside the house to collapse the second floor. The entire frame and roof swiftly topple over upon themselves. Again the artist managed to preserve some of his work through both the documentary film and the crated pieces of building facade that he deposited, like a debris slide, on the banks of a nearby river gorge. In this way he suggested the gorge's natural reclamation of the building materials. As with *Splitting,* other remnants of the building found their way into museum collections.[36]

Future projects in the building cuts series explored the aesthetic and creative possibilities of building destruction in various settings. In *Day's End* (1975), Matta-Clark commandeered an abandoned warehouse on Manhattan's West Side Pier 52, cutting channels in the floor and ceiling and using an acetylene torch to open a crescent shape in an end wall. His objective was to turn a largely forgotten space into an industrial cathedral. Later, Matta-

In this still from the film *Bingo/Ninths* (1974), the "bulldozer" begins demolishing the house that Gordon Matta-Clark and his crew have spent the previous ten days carefully cutting and crating. The machine's brute, unselective approach topples the structure in mere minutes. (Gordon Matta-Clark, *Bingo/Ninths,* 1974. Courtesy Electronic Arts Intermix [EAI], New York; © 2014 Estate of Gordon Matta-Clark/Artists Rights Society [ARS], New York.)

Clark moved from two-dimensional lines and cuts to more visibly three-dimensional excavations that illuminated what he called "dynamic volume." In *Conical Intersect* (1975), he bored a spiraling, circular opening into the walls of two townhouses abutting the construction site of the Centre Pompidou in Paris, creating views both out of and into the buildings. In *Office Baroque* (1977), he cut boat-shaped openings in multiple floors of a waterfront commercial building in Antwerp, producing overlapping voids. In *Circus* or *The Caribbean Orange* (1978), he inserted a circular slice into a brownstone building adjacent to the Museum of Contemporary Art in Chicago. This project was particularly unusual in that the artist was invited to do the work by the museum—rather than completing it covertly—and also because

In *Conical Intersect* (1975), the hole carved by Gordon Matta-Clark into the wall of a Paris townhouse opens up the city to the building, and vice versa. The multilayered, spiraling shape of the building cuts he has produced contributes to the sense of blooming new vistas and textures revealed through his openings. (Gordon Matta-Clark, *Conical Intersect,* 1975, gelatin silver print, 10⅝ in. × 15⅝ in. [26.99 cm × 39.69 cm]; San Francisco Museum of Modern Art, Accessions Committee Fund: gift of Frances and John Bowes, Collectors Forum, Pam and Dick Kramlich, and the Modern Art Council; © 2014 Estate of Gordon Matta-Clark/Artists Rights Society [ARS], New York)

the building-turned-canvas was going to be renovated, not destroyed, after Matta-Clark was through. Nevertheless, the renovation would cover over Matta-Clark's work, as had occurred with most of his previous projects.[37]

These dramatic artworks earned Matta-Clark not only attention but also criticism for the violence they seemed to perpetuate. As the artist recalled, one architect drew on romantic notions of the life cycle of the built environment when he claimed that Matta-Clark "was violating the sanctity and dignity of abandoned buildings by interrupting their transition to ruin or demolition." The artist also recalled another person who "saw what I did as out and out rape." This conflation was consistent with contemporary criticisms of all kinds of destructive acts upon the landscape, from bulldozers clearing forests to earthworks artists digging in the desert. One student explicitly connected the artwork with urban renewal–style clearance. In Matta-Clark's recollection, she criticized his work as being in "collusion with the forces of destruction and the renewal, and so forth." This student was commenting on Matta-Clark's building cuttings in Paris, at a time when housing was still a dire need in that city. He never intended the project as a housing proposal. But as with similarly violent practices among earthworks artists, it could be difficult to separate an artist's intention from his artwork's material effect.[38]

Matta-Clark was quick to debunk these associations. He distinguished his art from urban renewal practices by arguing that he worked on dilapidated buildings for pragmatic reasons. Those were the sites available for his use. He would, apparently, have been interested in cutting apart occupied buildings in order to observe the impacts of his cutout openings on daily life. Thus, for Matta-Clark, the excavations of his wrecking work were projects of exploration, rather than obliteration. The artist also expressed his moral objection to destruction pursued for nonartistic purposes. He noted, "I don't try to make destruction into a beautiful experience by any means. I think of it as being part of an immensely wasteful condition." He identified this wastefulness in both natural and built environments. Matta-Clark described suburbia as the product of a "brutal uncivilized war waged by the insulated ruling homes against the forests of antiquity (defoliating the primeval forest to clean its floor of cannibals) / The garden waging war against the ancient jungle halls." He offered an environmentalist critique of the chemical poisoning and clearance of the natural environment, as well as of the militaristic undertones behind such acts.[39]

For Matta-Clark, however, chain saws were not base instruments of physical destruction but the vocabulary of his social critique. The artist suggested that political dissent—rather than an endorsement of large-scale clearance—motivated his work. As he noted, "By undoing a building there are many aspects of the social conditions against which I am gesturing." He viewed the built environment as comprising "both a miniature cultural evolution and a model of prevailing social structures. Consequently, what I do to buildings is what some do with languages and others with groups of people: I organize them in order to explain and defend the need for change." While he recognized the violence embedded within his approach, he intended that violence as a means toward a greater end. As he once reflected, "The first thing one notices is that violence has been done. Then the violence turns to visual order and hopefully, then to a sense of heightened awareness."[40]

Yet urban renewal was clearly a focus of his criticism. Matta-Clark had professional experience with the program and its planning principles. After Cornell, he got his first job at the firm of his former professor Werner Seligman, during which time he worked on an urban renewal project for the city of Binghamton, New York. While the plan was never implemented, the work gave him firsthand experience with declining postwar cities and the demolition approach to addressing their problems. By turning away from architecture toward art after he left the position, he implicitly rejected this practice. Matta-Clark spoke derisively about urban renewal policy, noting, "So what I am reacting to is the deformation of values (ethics) in the disguise of Modernity, Renewal, Urban Planning, call it what you will." Urban renewal also instigated the demise of many of his own projects, with the bulldozers arriving hours, days, or, at most, a few years after the completion of his works. As his goal was the production of enduring—rather than temporary—artworks, urban renewal was more of an enemy to Matta-Clark's practice than its accomplice.[41]

Thus, urban renewal's great symbol, the bulldozer, would become the object of the artist's scorn. Not only in *Fresh Kill* but also in *Bingo/Ninths,* Matta-Clark portrays the machine as the enemy of both landscape and art. He later recalled the arrival of the machine on the *Bingo/Ninths* project site in disparaging terms, noting, "Even before the last section to be removed had been crated, the bulldozer crew arrived tearing up buildings and trees to either side of the project." The machine's depiction in the film supports his critical description of "violent bulldozing." The destruction wrought by the machine is wanton and total. Its quick demolition of the suburban home also contrasts sharply with the ten days of meticulous labor that produced Matta-Clark's more artistic, partial deconstruction. The bulldozer, he recalled, squelched "the life of the project" in only a few minutes. Even the bulldozer's end products were trampled and splintered building materials, rather than the artist's carefully crated pieces of history and art.[42]

Ultimately, Matta-Clark's cuttings were modern, urban archaeological investigations with preservationist undertones. As the artist noted, "This work reacts against a hygienic obsession in the name of redevelopment which sweeps away what little there is of an American past, to be cleansed by pavement and parking. What might have been a richly layered underground is being excavated for deeper, new building foundations. Only our garbage heaps are soaring as they fill up with history." In many ways, his building cuts revealed more than they destroyed. In contrast to the wrecker, who deconstructed in order to create a uniform surface founded upon erasure, Matta-Clark cut and sliced to shed liberating light upon the varied layers of a more complex history. As he once noted, "Better to let light pass through a space that still had some of its wits left before its identity would finally be pulverized in the cause of higher efficiency, higher densities."[43]

Embedded within this approach is an effort to interrupt the utilitarian logic of capitalist development through architectural punctures. Matta-Clark's work slowed the process of destruction, thereby exposing "the layering, the strata, the different things that are being severed." His creations prompted others to pause and look at the built environment's constituent parts as archives of history, rather than obstacles to progress. "The openings I have made," he concluded, "stop the viewer with their careful revealing." Even his photographic practice, in which he sometimes used collages of pictures to create more complete views, privileged revelation over erasure. Many of those who saw the work agreed. As one writer noted in the artist's *Artforum* obituary, "He gave me room to see."[44]

Over time, Matta-Clark became interested in moving beyond political critique to more enduring physical and social interventions in the lived landscape. In his earlier projects, he emphasized his art's lack of use value by defining it more as sculpture than as architecture. By the mid-1970s, however, Matta-Clark identified "more direct community involvement" as a focus for future work. A project in Milan expanded Matta-Clark's conceptions for his art, awakening him to the possibility of doing his work "not in artistic isolation, but through an active exchange with peoples' concern for their own neighborhood." While scoping out a site for a piece, he met a group of radical youths who had been occupying an abandoned factory complex. They viewed their occupation as a form of resistance to real estate development, and they hoped to use the site for a community center instead. Their approach inspired

Matta-Clark to extend their example to places like the South Bronx, where he might work with neighborhood youths to convert an abandoned building into a social space. "In this way," he reasoned, "the young could get both practical information about how buildings are made and, more essentially, some first-hand experience with one aspect of the very real possibility of transforming their space."[45]

These aspirations finally positioned him to deliver the socially engaged art he had described in his letter to the New York City official several years before. He began putting this philosophy into practice in the building cut for the Chicago Museum of Contemporary Art. There, he replaced artists with laborers from Manpower, an employment agency, to create his work. In reflecting upon this decision, he noted the possibilities of community ownership, participation, and execution that such an approach suggested—not only for art but for architecture as well: "Instead of most architectural situations where you dictate the plan or you dictate the situation, it would be a situation constantly subverted, a dictatorship constantly subverted by the people who were investing their time and energy in making it happen." Matta-Clark died of cancer only six months later, leaving the full potential of this vision unrealized. At the same time, however, other built environment professionals—including urban planners and landscape architects—were embarking upon projects similarly grounded in social engagement. These actors blurred the lines between art, architecture, and urban social life, suggesting the interactive relationship among these categories in the dynamic cities of the late postwar decades.[46]

The Art and Architecture of Human Renewal

In an urban analogue to artistic land-reclamation projects at former landfills and mining quarries, a group of designers began looking for ways to revitalize the spaces of damaged cities and put them to new uses. Their plans incorporated explicit goals aimed at social and environmental regeneration. By the early 1970s, roughly eighty community design centers across the country were pursuing this practice, using the empty lots and discarded rubble of urban renewal detritus to build new community projects. The landscape architect Karl Linn was a pioneer in the field, and his experience demonstrates the ways alternative art and architecture practices came together to critique and change postwar urban clearance on the ground. While the landscapes he produced were in part products of professional design practice, they could equally be described as urban environmental art.

As one journalist put it, Linn saw "latent beauty in urban blight." His collaborative design interventions brought that latent beauty to light. Linn emphasized elements missing from the urban renewal process. He engaged communities in reshaping their neighborhoods by treating as precious, rather than disposable, the building materials and abandoned lots that wreckers often left in their wake. Through his work, Linn posited a broader range of socially inclusive, technologically democratic, and place-specific possibilities for the renewal of the urban landscape.[47]

Beginning with his 1959 appointment as a professor at the University of Pennsylvania, Linn brought the skills of landscape architecture to the lower-income communities of Philadelphia. While his inspirations included Jane Jacobs and Herbert Gans, past personal experiences also informed his outlook. He grew up on an orchard in Germany, where thoughtful cultivation of the land nourished his family's livelihood. During World War II, he observed the development of community gardens and makeshift playgrounds in the bombed-out urban lots of European cities. Later, as a child psychologist, he utilized plants and neighborhood parks in horticultural therapy. He built on these experiences as a landscape architect by developing a therapeutic approach to repairing the landscape. Linn's projects enriched local environments while also orienting his profession toward more urban-focused, socially responsible, environmentally sustainable, and mentally healthy goals. Through community-based neighborhood-design projects, he taught an approach he called "community design-and-build service education."[48]

Working in Philadelphia, Linn and his students followed a fairly standardized process to turn abandoned urban lots into "neighborhood commons." They worked within the structures of urban renewal, rather than in opposition to the policy. They began by coordinating with the urban renewal authority to acquire bricks, marble steps, flagstones, and other materials from buildings slated for demolition. Project design and implementation were more collaborative affairs. Volunteers, including Penn architecture students, architects at local firms, and art students from the Philadelphia College of Art, assessed the needs of the community in order to develop a plan. The plan's execution relied upon the neighborhood. Laborers, ranging from settlement-house workers to area youths, gang members, and the unemployed, learned new trades while investing in their local environments. The acquisition of skills in horticulture, masonry, carpentry, and welding offered them opportunities for long-term professional advancement. As Linn later reflected, "We had a lot of bricks, and we didn't have money for machines, but we did have a lot of unemployed people. How much nicer it is to have a sidewalk made out of bricks than out of concrete." In a play on the name of the newly formed Peace Corps, Linn called his initiative the Neighborhood Renewal Corps.[49]

During the spring and summer of 1960, the group created Melon Neighborhood Commons, its first major project. After a local Quaker social service agency, the Friends Neighborhood Guild, acquired a few abandoned lots in northern Philadelphia, Linn agreed to partner in enhancing them. From an initial site of just a few parcels of land, ongoing demolition eventually expanded the area to include a half block of open urban space. Although the Guild had originally planned to build a simple playground in the area, Linn set his sights higher. He envisioned a broader commons that would serve the needs of the neighborhood's many age groups and constituencies, providing recreational space as well as a social and political forum.

Linn dissuaded the Guild from pursuing further clearance—in this case, bulldozing the site's varied terrain—in favor of building on existing topology. While the Guild did bring in a bulldozer to remove debris, it left the existing mound of earth intact. The mound soon became a site of creative play in which "kids made all kinds of contraptions and used them to roll down and play King of the Mountain." The designers supplemented this space with an amphitheater, a patio, an area with trees for relaxing, and sections with woodchips and sandboxes for children. Finally, they embellished the site with found objects, including a cable reel that served as a merry-go-round and concrete cylinders that became a bench.[50]

Melon Neighborhood Commons was just the beginning of Linn's work. Over time, he expanded his efforts to Washington, D.C., where he founded the Neighborhood Commons Nonprofit Corporation. The Washington- and

Philadelphia-based groups produced six commons in each city. Linn's work inspired similar organizations in Baltimore, New York, Chicago, and Columbus, Ohio, among many other cities. At the same time, his former students brought his teaching to the design schools where they eventually became faculty members, including the University of California at Berkeley and Davis, the University of Oregon, and the University of Toronto. The activists, teachers, and professionals involved in these ventures helped turn Linn's local efforts into a nationwide alternative renewal practice.[51]

Even without a direct connection to Linn, some urban renewal agencies also began pursuing projects along these lines. Recognizing the emergence of a critical moment "between clearance and renewal," these bureaucrats looked to vernacular, temporary parks as potential antidotes to the vacant lots produced by demolition. The corollary benefit, of course, was to provide recreational areas to members of the community who were directly affected by urban renewal. In April 1971, for example, the San Jose (California) Redevelopment Agency invited the city's Youth Commission to participate in a project to transform the former site of a downtown hotel. Area businesses donated materials and equipment, while community members contributed ideas, labor, and additional supplies. The completed facility consisted of a fountain, tables made out of cable reels, kiosks for art displays, flowers, trees, and open space. When Curtis Place Park opened in February 1972, the mayor was on hand for the dedication ceremony.[52]

Linn continued to develop his work in community design as he moved from Philadelphia to academic positions at the Massachusetts Institute of Technology, the Massachusetts College of Art, Drexel University, Antioch College, and the New Jersey Institute of Technology. He arrived at the last of these institutions, in Newark, after the race riots of the late 1960s had ravaged the city and created an unplanned clearance of their own. As in Philadelphia, without endorsing destruction, Linn recognized the potential held within these lots. He recalled, "Rather than seeing the abundance of vacant lots as a detriment, I saw the possibility of using this land to cultivate community gardens and market farming." He aspirationally called Newark "the Garden City of the Garden State." Linn studied one of the city's residential neighborhoods with a high number of vacant lots, hoping to repopulate it with rejuvenating wildflowers. Although these designs were never realized, he summarized his proposals in a publication, *From Rubble to Restoration,* which continues to be relevant to cities like Detroit today.[53]

Not surprisingly, Linn's work did not always achieve all of the designers' material and social goals. At Melon Neighborhood Commons, for example, problems began during the building of the project. Tensions developed between black and white laborers, energy levels waned in the execution of too-difficult designs, and poor weather dealt its own blows. After the construction was finished, it was left up to area churches, neighborhood organizations, and public agencies to devise a way to "awaken the commons" on the finished landscape. Continued building clearance undermined their efforts. As ongoing urban renewal displaced additional residents, it eliminated the lifeblood of the area. More clearance increased the size of the Commons, making it harder for the remaining population to maintain and secure the space amid escalating neighborhood violence and drug use. When gang members and energetic youths further damaged the property, the city was unwilling to take responsibility for a project it had not initiated. Ultimately, city leaders decided to bulldoze the entire site and replace the commons with their favored open space, a vest-pocket park. Without an enduring ownership stake by either residents or city officials, the design concept never realized its full potential.[54]

Given his project's mixed outcomes, perhaps Linn's greatest significance lies not in his built legacy but in his projects' social impacts on the individual community members who became involved. As he later reflected, "Those who participated in building and using the projects will always remember it as an empowering experience." Moreover, programs in Washington, D.C., and Columbus developed the infrastructure to support local residents' ongoing cultural and vocational development beyond the limited timeframes of particular projects. The construction workspaces that these groups instituted offered places where participants could work regardless of weather, and where the refuse of demolition, including large stacks of used bricks, could be stored for future use. Some local residents found opportunities to create sculptures out of these salvaged materials and then to sell them to galleries. Thus, the landscape architect helped to make artists out of area youths.[55]

Art as Protest

In January 1970, Kent State University invited Robert Smithson to take up a weeklong residence, during which he would develop landscape-based artwork on campus. Smithson accepted. After cold temperatures made his original mud-pour concept impossible, he and some students came up with the idea for *Partially Buried Woodshed* instead. The work consisted of a wood-and-stucco structure, upon and into which a backhoe dumped twenty truckloads of dirt. When the building's center beam began to crack from the weight of the dirt, Smithson deemed the work complete: weather and time would take care of the rest. In effect, the project combined the earthmoving propensities of the earthworks artists with the building-wrecking practices of Gordon Matta-Clark; viewed differently, the project spilled Walter De Maria's *Earth Room* out of the gallery and into a wood shed. The significance of this unusual piece increased when National Guardsmen shot and killed four students in and around the site of an antiwar protest on the Kent State campus later that year. Sometime after that event, someone painted the words "May 4 Kent 70" on the building, linking the physical and metaphorical breaking points represented by the tragedy and by the art. In 1975, arsonists burned part of the building. Groundskeepers later used another backhoe to remove constituent pieces as they fell to the ground. Less than a decade later, through gradual deterioration and removal, the work was essentially gone.[56]

Partially Buried Woodshed was, in many ways, the quintessential artwork of the period. Piles of earth and an abandoned built structure constituted its main materials. The means of production included construction equipment, community participation, and the manual labor of removal. The project was ephemeral, while owing to the happenstance of proximity and the subsequent actions of one graffiti artist, it also became part of one of the major protest movements of the period. Smithson embraced this association. He called the work both "prophetic" and "intensely political" and created an antiwar poster that included an image of the work. Thanks to these unintended, but obvious, political connotations, *Partially Buried Woodshed* made explicit the often implicit connections between artworks and resistance movements.[57]

The work of the varied late-postwar artists discussed here demonstrates the dramatic breadth of the culture of clearance's reach. Further, it reveals a

A backhoe piles dirt atop the roof and inside a wood-and-stucco shed on the campus of Kent State University in 1970, in the creation of Robert Smithson's *Partially Buried Wood-shed.* The modern building rising behind this decrepit structure suggests the iconography of urban renewal, even as this artwork operates in a different social and cultural register. (Robert Smithson, *Partially Buried Woodshed,* 1970. Art © Estate of Robert Smithson/ Licensed by VAGA, New York, NY. Copyright © the Holt-Smithson Foundation, Licensed by VAGA, New York. Courtesy of James Cohan Gallery, New York and Shanghai.)

subtle but evolving critique of the shape-shifting development of the American landscape. Walter De Maria wrote an infant manifesto for the field. He then extended his provocations by filling a gallery with dirt and covering a canvas in construction yellow. Michael Heizer worked at a larger scale and outside the gallery's confines, deploying earthmoving equipment to excavate notches and holes out in the desert. Robert Smithson became not only the voice of the earthworks artists but also a prominent earthworks practitioner, piling dirt, rocks, and minerals to create new forms in and out of the gallery. Along with Dennis Oppenheim and Robert Morris, Smithson also conceived of some works in the service of industry, infrastructure, and land reclamation. In these ways, earthworks art straddled a line between provocation and practicality. Gordon Matta-Clark offered an earthworks analogue for the urban built environment. Like Heizer, he produced cuts and holes on a large scale. Like Smithson's, his prolific writings helped explicate the political messages embedded in his unconventional works. And over time, like so many of his earthworks peers, his practice evolved from one motivated exclusively by artistic concerns to one that incorporated social factors as well. As a design professional invested in the community, Karl Linn took the social landscape as his starting point. Yet he too, like Matta-Clark, reached a similar outcome of creating beauty, utility, and visibility out of the abandoned landscapes of the changing city. With increasing clarity, the many built projects of these artists and designers articulated a critical counterargument to the previously celebratory clearance culture of the day.

While earthworks and related projects may seem less political than both the activism then taking place outside the art world and the intensely activist art that followed, these artists were indeed political actors. Their critiques made important links between art and the built environment. Sometimes these links were purely representational, while in other cases they actively participated in social and environmental acts of renewal. More commonly, however, these artists used cultural forms to subtly comment upon a host of concerns, from urban renewal to environmental degradation and even to the waging of war. Outside artistic circles, protest movements were increasingly organizing around these causes. Rather than take to the streets, these artists decamped to more diverse urban, suburban, and rural landscapes. In these spaces, they proclaimed their concerns in the language of dirt, debris, and artistically inspired destruction. As a counterpoint to the dominant clearance trends of their day, they sought visibility over erasure, slowness over speed, and a more thoughtful engagement with the natural and built world around them. Through their written and representational voices, they subtly declared their support for an end to the culture of clearance. At the same moment they were creating these artworks, other actors were more openly, violently, and explicitly amplifying their message of dissent.[58]

Conclusion

Toward a Culture of Conservation

In 1975, Edward Abbey published *The Monkey Wrench Gang,* a novel about a group of anarchist environmentalists working to disrupt the development of the western American landscape. The book's protagonist, George Washington Hayduke, is a veteran Green Beret trained in demolition during the Vietnam War. When he returns home to the United States, Hayduke is dismayed by contemporary builders' degradation of the environment. In response, he and a crew of accomplices set out to sabotage construction projects. Their methods include damaging both the equipment and the built products—from dams to roadways—of the earthmoving trades. Notably, the Monkey Wrench Gang's aim is not to injure construction workers but rather to upset their machines and the landscape-scarring projects their work has produced.

One of the group's attacks occurs in the Arizona desert. After a road-building crew has locked up its equipment and departed for the evening, Hayduke and his friends begin their work. Their first target is a twenty-seven-ton Hyster compactor with a Caterpillar engine and sheepsfoot rollers. They cut wires, fuel lines, and hydraulic hoses before pulling the dipstick from the engine block and pouring sand into the crankcase. Finally, they dump four quarts of Karo syrup into the fuel tank. When they are done, they stand back in wonder: "All were impressed by what they had done. The murder of a machine. Deicide. All of them, even Hayduke, a little awed by the enormity of their crime. By the sacrilege of it." Afterward they tackle a Caterpillar D9A—"world's greatest bulldozer, the idol of all highwaymen"—and damage it in a similar manner. "Having done all they could to sand, jam, gum, mutilate and humiliate the first bulldozer, they moved on to the next." The evening proceeds in this fashion: "They worked over the Cats, they operated on the earthmovers, they gave the treatment to the Schramm air compressors the Hyster compactors the Massey crawler-loaders the Joy Ram track drills the Dart D-600 wheel loaders not overlooking one lone John Deere 690-A excavator backhoe, and that was about all for the night." The following morning, the road-building crew awakes to a terrible surprise. After they start the engines, "those little particles of sand, corrosive as powdered emery, began to wreak earth's vengeance on the cylinder walls of the despoilers of the desert."[1]

The Monkey Wrench Gang offers an extreme illustration of emerging environmentalist concerns over land clearance in the 1960s and 1970s. The landscape destruction wrought by large-scale building projects became the target of the protagonists' aggression. At the same time this critique was evolving, awareness of the damaging impacts of war was also growing. The Vietnam War had left Hayduke psychologically unhinged. Instead of preparing him for productive postwar building trades, his war experience had manifested itself in violent ecological resistance. His was a more radical version of the larger environmental movement gaining traction across the country. Through acts ranging from small-scale inconveniences to larger publicity stunts, Hayduke and his friends try to transform the culture of clearance into a culture of conservation. The politicians and builders behind the big projects they disrupt play the dupes to their commitment to fighting the Man. Even the equipment itself—while respected for its power—reveals its inherent weaknesses as it is quickly and easily undone by something as simple as a wrench. The quest to save the environment ultimately becomes a cause worthy of martyrdom, and at the novel's end Hayduke dies defending his principles.

Hayduke's opposition to the culture of clearance is similar to the sentiment expressed in a *New Yorker* cover cartoon from 1971. The drawing, by the cartoonist Mischa Richter, depicts a gallant white knight preparing to battle a hungry power shovel. His shield—shown in color as a red cross on a white background—references Saint George, the dragon slayer. In this mythical tale, which dates back to the Middle Ages, Saint George fights a dragon that has been terrorizing a city. The people temporarily appease the monster with animal and human sacrifices. When a princess falls into its grasp, Saint George arrives to save the day. By invoking this story in the *New Yorker* image, Richter likens powerful construction equipment to modern-day dragons. These machines devour sacrificial lambs until someone strong enough to slay them comes along. The scene reflects a transition in popular perceptions of clearance during the late postwar years.[2]

Richter's depiction of the instruments of clearance differs markedly from the images of these machines created during their heyday. His monster power shovel calls to mind Virginia Lee Burton's "Mary Anne"—except for the two rows of sharp incisors where there should be a smile. It also recalls the more aggressive and anthropomorphized World War II construction equipment of Boris Artzybasheff—except that the victims are no longer Axis leaders, but everyday citizens. This shovel stands in for the bulldozer, the scraper, and every other construction-based instrument of destruction then being criticized by the public. These implements were so interchangeable in the popular mind that when Mary Anne Guitar referred to the *New Yorker* image in her 1972 *Property Power: How to Keep the Bull-Dozer, the Power Line, and the Highwaymen Away from Your Door,* she mislabeled the machine, calling it a bulldozer. In 1963, the economist Stuart Chase had published an essay in the *Saturday Review* in which he explained how San Francisco residents were organizing to "turn back the bulldozers." Chase acknowledged that the machine "is not responsible, of course, for all our environmental lesions, but for enough to serve as a symbol. It also looks the part of a devouring monster from the Silurian age." A decade later, Richter's work gave visual form to Chase's critique.[3]

Science fiction offered additional depictions of construction equipment gone bad. Douglas Adams's *The Hitchhiker's Guide to the Galaxy* (1979) opens with alien-operated bulldozers poised to obliterate the earth to make way for a new interstellar highway. "The Thing Called . . . Killdozer," a 1974 Marvel comic, offers an even broader bulldozer-based narrative. The comic book

The cover of the May 1, 1971, issue of the *New Yorker* depicts a valiant white knight preparing to do battle with a sharp-toothed power shovel. Although reminiscent of the visual iconography of Virginia Lee Burton's "Mary Anne," this piece of construction equipment is depicted by the artist as much more vicious and dangerous. (Mischa Richter/The New Yorker/© Condé Nast)

is based on Theodore Sturgeon's 1944 short story "Killdozer," about a bulldozer that preys upon human victims. In Sturgeon's original version, eight construction workers disturb an ancient ruin while building an airfield on an unidentified Pacific Island. Their encounter unleashes a force that takes over the bulldozer and kills them all. Since it was written during World War II, this narrative's critical meditation on the power of machines over men is unusual. The story was not widely popularized, however, until it found a more receptive cultural climate in 1974, when both a comic book and a made-for-TV movie updated the tale for contemporary audiences. The comic relocates the action to "a day not far from tomorrow," when its bulldozer protagonist, a Caterpillar D7, wreaks havoc. Despite the comic's vivid cover, which depicts a crazed red bulldozer with fang-toothed blade attacking a slew of helpless victims, its Killdozer ultimately produces no human casualties. Other science fiction works from this era would fill in that gap.[4]

The 1973 film *Soylent Green* brings the bulldozer's violence closer to home when an all-powerful state coopts the machine for man-eating ends. The film is set in New York City in the year 2022, when government-issued foodstuffs provide the population with its only nourishment. When the citizens riot in response to a lack of food, the state unleashes tanks armored with bulldozer-like shovels. These violent marshals scoop up the demonstrators, cart them away, and deposit their bodies where they can be used as ingredients in the mysterious foodstuff called Soylent Green. Although the bulldozer serves as an agent of the state, the machine itself is also implicated in the violence. The bulldozer is so central, in fact, that a shovel appears prominently—elevated high and spilling over with terrorized victims—in the movie's promotional poster.[5]

In the novel, the comic book, and the film, construction equipment takes on ominous guises. As varied forces appropriate the instruments' power, the narratives warn that seemingly productive and progressive tools can also serve nefarious ends. They also remind audiences that real-world communities and landscapes are falling victim to these machines. These more critical cultural representations, along with burgeoning opposition to bulldozers on actual construction sites, changed the image of the machine. Once a symbol of progress, it had now become a symbol of destruction. This transformation

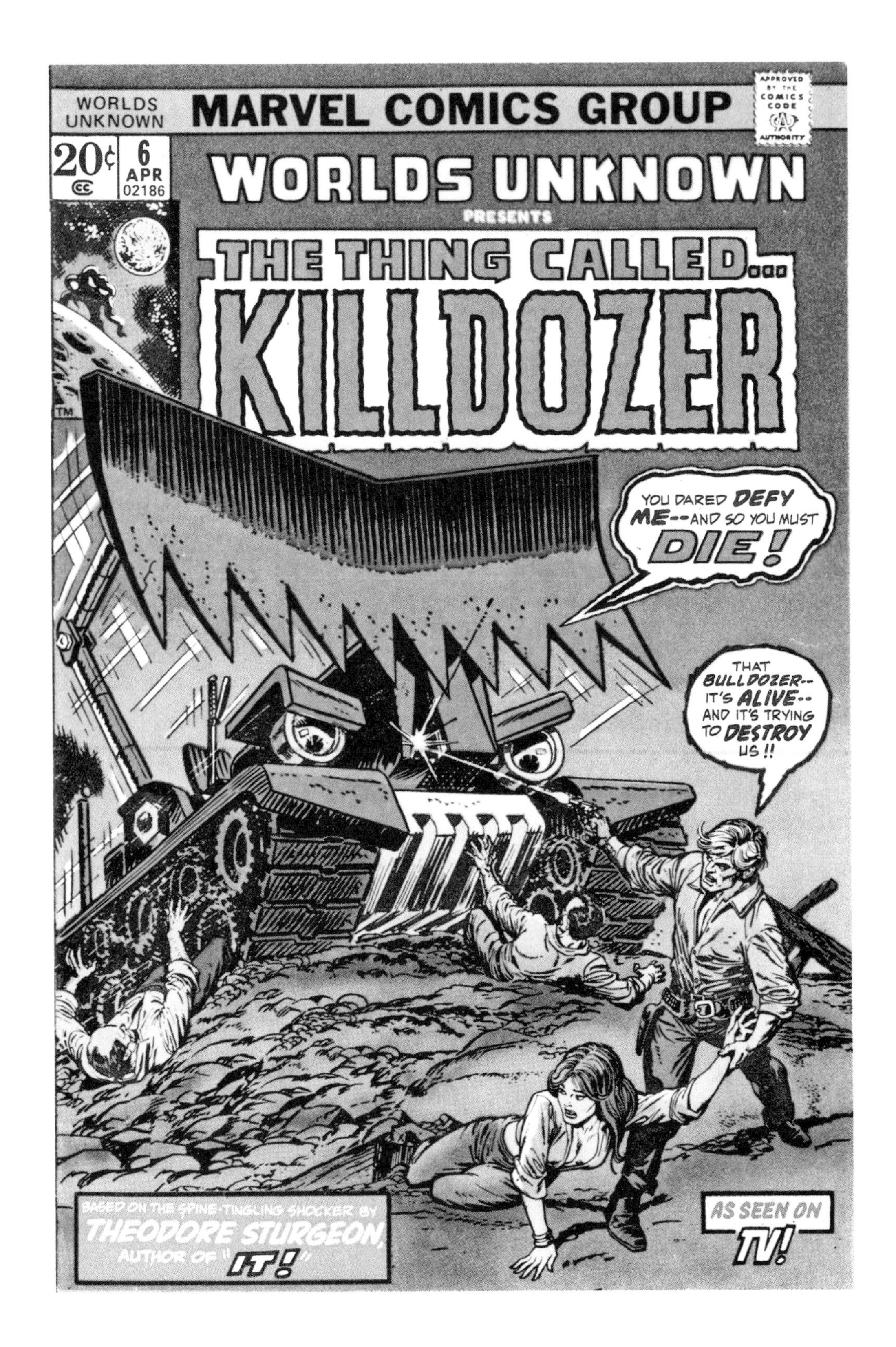
WORLDS UNKNOWN
MARVEL COMICS GROUP
APPROVED BY THE COMICS CODE AUTHORITY
20¢
6
APR
02186
WORLDS UNKNOWN
PRESENTS
THE THING CALLED...
KILLDOZER
TM
YOU DARED DEFY ME--AND SO YOU MUST DIE!
THAT BULLDOZER-- IT'S ALIVE-- AND IT'S TRYING TO DESTROY US!!
BASED ON THE SPINE-TINGLING SHOCKER BY THEODORE STURGEON, AUTHOR OF "IT!"
AS SEEN ON TV!

began during the 1960s, and it was fairly complete by the mid-1970s. By that time, the era of the culture of clearance had come to a close.

The Bulldozer as a Symbolic Object of Protest

The environmental journalist Harold Gilliam summarized the growing anti-clearance movement in a 1967 *Saturday Review* article aptly titled "Beating Back the Bulldozers." Focusing on activities in and around San Francisco, Gilliam described the rising tide of criticism and protest against clearance-based growth: "Every night of the week, someplace in California there is at least one indignation meeting where outraged residents are organizing committees of vigilance to turn back the bulldozers. The roots of the resentment are deep. Everyone who has lived in California more than a few months remembers a pleasant orchard that has been uprooted for a factory or shopping center, a favorite picnic spot that has been graded for tract houses . . . a rustic country lane that has been converted into a roaring highway. Thousands of residents have been displaced by freeways, and tens of thousands live in neighborhoods now dominated by their stench and racket. Native Californians watch the destruction of hallowed landmarks, man-made and natural, one by one." As Gilliam pointed out, years of ground-level clearance had yielded physical, social, and environmental damage. It had displaced residents, destroyed natural and built landscapes, and polluted the senses. By the mid-1960s, affected populations were mobilizing to alter this course of events, and the bulldozer became the object of their rallying cry. At times, groups of activists even stood valiantly in front of actual bulldozers in order to make their case.[6]

While federal funding spurred much clearance work, the resistance emerged largely at the local level. Freeway revolts were some of the first movements to gain force, but opposition to both urban renewal and suburban development soon followed. From African American neighborhoods in New Haven to more affluent southern California suburbs, residents turned out

---

(*Opposite*) The cover of Marvel's "The Thing Called . . . Killdozer" comic book, from 1974, depicts a possessed, fang-toothed bulldozer attacking human victims on "a day not far from tomorrow." (Reproduced courtesy of Marvel Entertainment)

African American protesters try to physically stop the urban renewal bulldozer at work in San Francisco's Fillmore neighborhood, ca. 1970, suggesting a battle between City Hall and Black Power.

into the streets and at public hearings to voice their opposition to the perpetuation of clearance practices they had recently observed cause much damage. While many of the earliest of these resistance efforts failed to stop demolition and development projects, they introduced a language of opposition and instigated an informal movement dedicated to protecting the landscape. Over time, these efforts helped transform, and even halt, certain individual projects.[7]

The high stakes of these resistance movements revealed themselves traumatically on a construction site in Cleveland, Ohio, in 1964. A group of residents had assembled to protest a new school planned for an African American neighborhood. Charging that the building would perpetuate de facto segregation, the residents tried to stop the construction. On April 6, several demonstrators lay down inside a trench being dug by a backhoe. The machine operator responded by dumping a load of dirt on top of them, and the police later hauled the protestors away. Photographs of this scene recalled World War II, when construction machinery was used to bury both enemy combatants mid-fight and American casualties after the fighting was through. In this case, however, the victims were American citizens battling the machines at home. While many of the protestors were African American students, the group was led by Bruce Klunder, a twenty-six-year-old white Presbyterian minister. On April 7, an equipment operator accidentally ran his bulldozer over Klunder as he lay prostrate beside the machine. He died shortly thereafter. The minister's widely reported death powerfully linked the symbolic machine with horrific physical and racial violence.[8]

Klunder's death also presaged the more recent ways in which political forces around the world have deployed bulldozers for violent uses. In 2003, an Israeli bulldozer fatally crushed the American activist Rachel Corrie as she stood in front of a Palestinian home facing demolition. Due in part to incidents like this one, the machine has become a potent symbol of Israeli efforts to dismantle Palestinian settlements in contested areas of Gaza and the West Bank. Similarly, government officials from Beijing to Cairo have used bulldozers to control popular protests waged in the name of democracy. From South Africa in the 1970s to Lima and Manila today, governments have also unleashed bulldozers on squatter settlements of their landless poor. Unlike the postwar United States, these countries have overtly deployed clearance as an end in itself, rather than as the first step in a construction process. Yet consistent with earlier American experiences, the machine has stood as a powerful symbol of violent order imposed from above.[9]

Earth First!, a group of environmentalists inspired by the fictionalized antics of Edward Abbey's *Monkey Wrench Gang,* incorporated violence into its radical anti-bulldozer activism. The national Earth First! movement formed in the early 1980s, with the explicit goal of militant ecodefense. Echoing Abbey, Earth First!ers based in the Los Angeles area poured sand into the crankcases of earthmoving equipment that was developing California's San Penasquitos Canyon. On other occasions, group members spiked northern California trees to deter lumber companies from harvesting them. Earth First!ers have also sawed down billboards, vandalized vehicles found driving off marked park paths, and at one point faced accusations of building bombs and attempting to sabotage nuclear facilities. The 1985 publication of *Ecodefense: A Field Guide to Monkeywrenching,* by Earth First! founder Dave

In 1964, a bulldozer accidentally ran over the Presbyterian minister Bruce Klunder after he lay down beside the machine to protest the racial injustices of postwar urban development on a Cleveland, Ohio, construction site. (The Plain Dealer/Landov)

Foreman, provided detailed instructions on how to "monkeywrench" successfully to create the greatest impact.[10]

Earth First!'s strategy included more than just the protests it staged. It also worked to mobilize popular media outlets to publicize its stunts. When one faction planned to dump cow manure on the desk of a bureaucrat with the San Diego Department of Fish and Game, the group alerted newspapers and television stations in advance. As expected, photographers were on hand to capture the moment. Journalistic photographs have also documented other Earth First! actions, including tree-ins staged to block logging projects and protestors chained to bulldozers to oppose new roadway construction. In imitation of Dick Lee's orchestration of photo-worthy demolition scenes for *Life* in the late 1950s, Earth First! harnessed the power of photography for its own ends. Whether in support of, or in opposition to, clearance work, the production of the representational image has proven to be as important as—or even more important than—the act itself.[11]

The End of the Culture of Clearance

In the course of a few decades, the machines of demolition and clearance were remade in the popular imagination. Their image changed from celebrated, patriotic icons to maligned symbols of destruction. Just as the ideology, technology, policy, and practice of large-scale destruction had created the culture of clearance, so did the gradual undoing of each of these pillars signal its decline. While criticism and protest had existed alongside the celebrated implementation of clearance from the start, by the late 1960s and early 1970s, a broader ideological transformation had occurred. New voices questioned the definition of progress, the costs at which it was achieved, the appropriate limits of power, and the authority afforded technology and its expert overseers. The ensuing legislative changes brought to a close many of the policies that had legally legitimized and economically incentivized large-scale clearance. At the same time, the practices these policies supported evolved as well. Due to increased regulation and practical knowledge that had been gained on the ground, it no longer seemed so easy, inexpensive, or progressive to tear down in order to build up. And although technology—the other key factor supporting the postwar culture of clearance—did not regress during this period, it alone was insufficient to propel the continued pace and scale of clearance. While the physical capability for large-scale destruction remained, the political, economic, and social will behind the practice had begun to wane.

Although no clear moment marks the demise of the culture of clearance, America's withdrawal from Vietnam in 1973 marks a suitable endpoint. From World War II to Vietnam, military conflicts bookend a period during which heavy construction equipment effectively waged war on international and domestic space. World War II established a critical context out of which the culture of clearance took shape and then flourished. The war's unprecedented mobilization of construction workers, equipment, and methods stimulated the postwar building trades. The war produced advanced technologies, skilled laborers, and improved clearance techniques. These practical gains also facilitated new ways of seeing the landscape and the actors and equipment that could reshape it. The implementation and representation of clearance during the war elevated the bulldozer into a powerful American hero. Its operator became a patriotic idol, with the scale of his destruction defining the measure of collective progress. When postwar Americans set their sights on their own blighted cities and undeveloped open space, they recognized opportunities for applying this war-honed vision of clearance to the home front.

Just as the earthmoving Seabees helped glamorize the bulldozer in the 1940s, military clearance in Vietnam helped tarnish it a few decades later. Not only did the U.S. military use bombs, chemical defoliants, and napalm to eviscerate the enemy landscape, it also bulldozed about 2 percent of South Vietnamese territory, typically with Rome Plows (armored tractors equipped with a blade at the front). The environmental damages were brutal, ranging from erosion—particularly significant in hilly regions with heavy rainfall—to the long-term depletion of vegetation and wildlife. Outside of large infrastructure like dams, the monster equipment that made this clearance possible had limited applications back home in the United States. Whereas treedozers had helped fell vegetation for the building of domestic roads and homes following World War II, for example, much larger tree crushers were necessary for clearing the jungles of Vietnam. With their huge tires and blade-studded steel rollers, these machines could push over trees and smash them in place. The *Hartford Courant* called the tree crusher "one of the most destructive machines ever to leave an engineer's drawing board." The Vietnam War, then, was less "useful" to advancing domestic clearance technology and practice than World War II had been.[12]

Vietnam also proved detrimental to the celebratory character of clearance. World War II had united a nation; Vietnam tore it apart. World War II had concluded decisively with an American victory; Vietnam ended with withdrawal, disenchantment, and defeat. The sizable antiwar movement, combined with the larger phenomenon of 1960s- and 1970s-era protests—yielded a quite different cultural landscape from that of the post–World War II era. The image of construction workers was also changing during this period from widely respected and skilled laborers to increasingly conservative, racist, and sexist muscle men. As "hardhats" in New York City attacked long-haired antiwar demonstrators in 1970, they helped further this changed popular perception. Even more broadly, popular faith in government, business, science, and technology gave way to suspicion of institutions of power and greater respect for the multilayered—and sometimes minority—viewpoints of ordinary people. The Vietnam War gave military and construction associations a substantially altered, and much less positive, cultural resonance.[13]

As thoughtful critiques and local protests started to undermine the domestic culture of clearance, lessons learned on construction sites exposed the challenges of clearance on the ground. These lessons included the inequity of forced property acquisition, the slow pace of implementation, the collateral damages of execution, and the visual and environmental hazards of abandonment that sometimes followed uncompleted projects. The end results could be equally unsatisfying, demonstrating that progress often came at too great a cost. On urban renewal sites, demolition ruptured the social fabric of communities, physically damaged neighboring properties, and left demoralizing, debris-filled lots in its wake. Suburban and rural land clearance depleted farmland, instigated landslides, and promoted sprawl. Finally, on contested highway terrain, implementing the seemingly simple logic of the engineering mind proved more complicated than its supporters had first acknowledged. While wartime experiences may have made clearance seem like a swift feat of technological conquest, domestic applications proved more complex. These challenges characterized both city and country, demonstrating the extent to which large-scale clearance was not an isolated urban, suburban, or rural phenomenon. It was a nationwide postwar process that traversed natural and built landscapes.

After decades of positive popular representations of construction work in various media—ranging from journalistic photographs to popular films and children's literature—other authors and artists began channeling the growing anti-bulldozer sentiment into their own works. The criticism emerged visibly in children's bulldozer books, where the laudatory depictions of the machines and their jobs started giving way to more thoughtful narratives that looked past the excitement of operating construction equipment to consider some of its more deleterious consequences. New categories of conceptual art emerged to increase the visibility of the material impacts of these machines. Even more explicit than these artistic criticisms was the outpouring of nonfiction writing that articulated the oppositional views. Two notable examples are Jane Jacobs's *Death and Life of Great American Cities* (1961) and Rachel Carson's *Silent Spring* (1962). Jacobs argued against postwar planners' "bulldozer approach" to destroying the multi-use, walking-scale built environments that she felt defined cities at their best. Exhibiting a corollary conservation ethic toward the natural environments of suburban and rural lands, Carson used the case of pesticides to advance an environmental critique of the postwar belief in science-based progress. While Carson did not target land clearance specifically, her ecological viewpoint informed the growing opposition to the damaging effects of technology and development on the landscape.[14]

On-site lessons learned in practice, and their dissemination in critical media, spurred policy changes that instituted new development practices and safeguards. These acts broadly coincided with the Vietnam War. Urban renewal ended with the Housing and Community Development Act of 1974, which replaced the large hand of the federal government with local control. The act also deemphasized the act of demolition, placing it among a range of revitalization practices. The National Historic Preservation Act of 1966 provided mechanisms for protecting individual buildings and neighborhoods from the wrecking ball. The Federal-Aid Highway Act of 1973 allowed for the reorientation of planned interstate routes and also expanded funding beyond highways to include multimodal transportation infrastructure. The National Environmental Policy Act of 1969 introduced review processes that, like historic preservation policy, helped ensure that subsequent landscape clearance—for whatever purpose—would occur only after a minimum level of analysis and impact assessment had occurred. These policies were markedly different from the early pieces of postwar legislation that had not only sanctioned large-scale clearance but had also tied substantial federal economic incentives to it.

The technology of demolition was continuing to progress during this period, but technology alone was insufficient to perpetuate the culture of clearance. Heavy equipment increased in size and power, but it was impractical to deploy machines of larger sizes or efficiencies on commercial construction sites. Instead, technological advances in implosion spurred a rise in dramatic spectacles of singular demolition, rather than the more mundane destructions of urban renewal. The built products of urban renewal construction often served as the objects of these implosions—from the Pruitt-Igoe public housing complex in Saint Louis in 1972 to the New Haven Coliseum in 2007. Along with the buildings, these implosions laid waste to the very idea of modernist, urban renewal–style redevelopment.[15]

By the 1970s, the technology behind demolition was no longer identified exclusively with American engineering and manufacturing dominance. Caterpillar had been the leading worldwide producer of construction equipment at the start of the postwar period. In 1971, however, *Business Week* announced, "Japanese Bulldozers Invade U.S. Market." By the 1980s, the new Japanese rival, Komatsu, was the second-largest manufacturer of earthmoving equipment worldwide. Ironically, one instigator of Komatsu's rise to prominence had been Japanese observations of American bulldozer prowess during World War II. After recovering a U.S. machine left on the battlefront, the company reverse-engineered the technology to develop its own product. Komatsu also benefited from low production costs, steadily improving quality, favorable exchange rates, and a growing foreign dealer organization. While Komatsu never usurped Caterpillar's industry dominance, the equipment manufacturer—like its Japanese counterparts in other industries—established a foothold in the American domestic market that it still maintains today. The foreign bulldozer became the enemy of American equipment manufacturers at the same time that domestic applications of the bulldozer were coming under increasingly critical assault.[16]

Over time, Caterpillar worked to protect its brand from both manufacturing competition and attacks on its image. The use of Caterpillar equipment by the Israeli Defense Fund (IDF) in the mass clearing of Palestinian settlements attracted negative attention. In 2004, Human Rights Watch issued a report urging the company to cease sale of its D9 bulldozers to the IDF. Caterpillar denied responsibility for the applications of its machines, and company officials noted that it was the IDF—and not the manufacturer—that militarized the equipment with armor and other modifications. One year earlier, the company had taken a more proactive role in its brand management. It brought several charges against Disney for including its trademark on bulldozers appearing in the movie *George of the Jungle 2*. In the film's most contentious scene, George and his animal friends battle against the Caterpillar-bulldozer-mounted henchmen of a jungle-destroying industrialist. In stark opposition to the celebration of clearance in early postwar children's media, voiceovers in the film describe the vehicles as "deleterious dozers" and "maniacal machines." The court dismissed Caterpillar's request for a temporary restraining order and issued a preliminary rejection of the company's trademark dilution claim. As the film clearly placed the blame for the destruction on the bulldozer operators rather than the machines, the court argued that Disney had not tarnished the image of Caterpillar's products. After the two parties settled out of court, Disney released the film. Despite its legal loss, Caterpillar achieved a subtle success: the court's written decision highlighted the frequently overlooked distinction between bulldozer technology and the acts of clearance to which various actors applied it.[17]

Looking Forward

Today, international landscapes offer the most fertile ground for clearance work. China is the exemplar, with current redevelopment plans proposing clearance of over a thousand urban villages—informal settlements that often become engulfed by modern cities—across the country. Their destruction will simultaneously displace millions of poor residents. Present-day demolition in China is as thoroughgoing, if not more so, than it was in the postwar United States. It also occurs on a greater scale, given the country's larger population, high rate of growth, and the stronger hand of the central government. Over a four-year period at the turn of the twenty-first century, for example, Shanghai demolished 20 percent of its residential area. At least 40 percent of Beijing's Old City fell between 1990 and 2002. In 2011, clearance nearly eradicated the entire village of Dachong, leveling over 10 million square feet that had housed more than seventy thousand residents. While resistance exists, its impact has been relatively limited. Wreckers continue to repeat the Dachong experience across the country. The rapid pace of construction in the instant cities of the Middle East offers comparable opportunities for contemporary land clearance. In advance of the armies of laborers who erect skyscrapers, airports, malls, and housing in places like Dubai, earthmoving machines first work to ready the land. Although desert landscapes vary substantially from the terrain that characterized typical American clearance projects, the mechanics and material objectives are comparable.[18]

Although foreign geographies may have become the dominant sites of contemporary clearance, some of the practices also continue in the United States. In 2011, American cities grew more rapidly than suburbs for the first time in almost a century. This population shift suggests that the oversupplies of housing in many cities were being reoccupied. It also reflects the densification of development in places like Anaheim and Los Angeles, which Smart Growth America, an anti-sprawl advocacy group, recently ranked, respectively, as the fourth- and seventh-most compact and connected large metropolitan areas in the country. Yet between 2000 and 2010, exurban areas—the scattered development located beyond the suburbs—experienced the most rapid population growth of all (a 60 percent growth rate versus the 10 percent average for that decade). This growth in exurban areas portends continued land clearance on the urban fringe, even as the popularity of city living resurges.[19]

The urban resurgence has not affected all cities equally, leading some municipalities to pursue demolition-based strategies to combat long-standing housing vacancies. The housing market collapse of 2007 and subsequent recession exacerbated patterns of vacancy that have plagued the deindustrializing Rust Belt since the 1970s. Buffalo announced plans in 2007 to demolish five thousand vacant buildings by 2012, accelerating the pace of demolition already in progress there. More ambitiously, in 2014, the Detroit Blight Removal Task Force recommended the clearance of roughly forty thousand of the city's buildings, at a price tag of $850 million. Financial difficulties have led some cities to sue mortgage holders for predatory lending practices that, they argue, spurred a rise in foreclosures and vacant properties. While many of these suits have been unsuccessful, in 2011 Bank of America announced plans to donate one hundred foreclosed homes to the city of Cleveland and pay for their demolition. Rather than continue to fund these properties through taxes and maintenance costs, the bank opted to tear down its financial obligations. In 2009, in southern California's Victorville suburb, another bank demolished sixteen new and partially completed homes it had acquired through foreclosure. In this case, demolition was cheaper than completing and selling the properties in the depressed housing market. It also assuaged the lender's fears about squatters and vandals.[20]

Although the revived language of blight as an urban cancer echoes sentiments from the urban renewal era, stark differences distinguish these two moments. Large-scale clearance is no longer the dominant development practice of the typical American city. Instead, it applies selectively to locales suffering from extreme economic maladies. In the absence of urban renewal legislation, challenging property acquisition processes and a dearth of funding impede the realization of most large-scale clearance plans. Further, whereas the urban renewal bulldozer largely displaced communities of occupied housing, these more recent examples of demolition often target vacant sites. There are meaningful differences in the social and cultural consequences of the two approaches as well. Rather than portending a return of the culture of clearance, these recent examples suggest relatively isolated practices.

Given these circumstances, we can see that although the United States no longer embraces a culture of clearance, it has not yet moved on to a culture of conservation. Celebratory bulldozer books like Sherri Duskey Rinker's *Goodnight, Goodnight Construction Site* (2011) and Candace Fleming's more recent *Bulldozer's Big Day* (2015) continue to capture young readers' attention. Online video games like *Demolition City* engage players with the laws of physics as they apply dynamite charges to topple an animated building. In 2000, in Kent, England, the chairman of a major European construction equipment supplier opened a theme park called Diggerland, where entire families can try their hand at operating earthmoving machinery. In 2014, the franchise expanded across the Atlantic, with Diggerland USA opening just outside Philadelphia. As these examples attest, clearance work and its equipment still retain widespread appeal.[21]

Yet the magnitude of domestic destruction has never returned to its postwar high. Whereas one out of every fifteen dwelling units fell during the 1960s, subsequent decades have seen losses on the order of one out of every thirty to sixty dwelling units. Total land clearance is less clearly quantified, but the reduction of farmland offers one window onto that metric. During the 1960s, nationwide farmland declined at a peak rate of 6.3 million acres per year. Today, that number is less than half: 2.6 million acres per year. Large-scale clearance practices continue, but at a slower pace and smaller scale compared to the height of the postwar period. The legacy of the culture of clearance still shapes our present moment.[22]

During the heyday of America's culture of clearance, neither city nor country, neither natural nor built environment, was immune to its influence. Its processes and consequences transformed the ideology, policy, and practice of large-scale landscape destruction, as well as the cultural perception of the technology that physically enabled it. Among these technologies, the bulldozer took the starring role. It remains a material and symbolic force today. Yet the machine's reputation is now tarnished by the scars that endure from the buildings, people, and land it cleared in the name of progress. The cheers that welcomed the machine's arrival—on the page, the movie screen, and the construction site—now compete with cries of protest, heralding the passing of a dramatic era in the remaking of the American landscape.

Notes

Abbreviations

AASHTO IHRP	American Association of State Highway and Transportation Officials' Interstate Highway Research Project, AASHTO Archives, AASHTO Headquarters, Washington, D.C.
IHC	International Harvester Company Corporate Archives Central File, 1819–1998 (McCormick Mss 6Z), McCormick-International Harvester Collection, Wisconsin Historical Society
NHRA Records	New Haven Redevelopment Agency Records (MS 1814), Manuscripts and Archives, Yale University Library. Because many folder names in the NHRA Records are long and inconsistent, I have provided standardized abbreviations in the notes below.
OHDNH	Oral Histories Documenting New Haven, Connecticut (RU 1055). Manuscripts and Archives, Yale University Library
RCL	Richard Charles Lee Papers (MS 318). Manuscripts and Archives, Yale University Library
VHP	Veterans History Project Collection, American Folklife Center, Library of Congress

Introduction

1. Stephen W. Meader, *Bulldozer* (New York: Harcourt, Brace, 1951), 21.

2. "Review: Bulldozer," *Kirkus Reviews,* September 20, 1951; Meader, *Bulldozer,* 38.

3. "Reading While Famous," *Literary Cavalcade* 57, no. 8 (May 2005): 30; Stephen King, *It* (rpt.; New York: Signet, 1987), 523. On Meader, see Chesley Howard Looney, "Stephen W. Meader: His Contributions to American Children's Literature" (Ph.D. diss., University of Maryland, College Park, 2005). Southern Skies recently republished Meader's work.

4. Meader, *Bulldozer,* 5, 239.

5. On "creative destruction," or the productive disposability of the land, see Joseph Schumpeter, *Capitalism, Socialism, and Democracy* (New York: Harper and Brothers, 1942). On the long history of land clearance in the United States, see David E. Nye, *America as Second Creation: Technology and Narratives of New Beginnings* (Cambridge: MIT Press, 2003); Walter Muir Whitehill, *Boston: A Topographical History* (Cambridge: Belknap Press of Harvard University Press, 1959); Michael Rawson, *Eden on the Charles: The Making of Boston* (Cambridge: Harvard University Press, 2010); Matthew W. Klingle, *Emerald City: An Environmental History of Seattle* (New Haven: Yale University Press, 2007).

6. Nationwide demolitions calculated based on U.S. Bureau of the Census, *U.S. Census of Housing: 1960,* vol. 4: *Components of Inventory Change* (Washington, D.C.: U.S. GPO, 1962); U.S. Bureau of the Census, *1970 Census of Housing, Components of Inventory Change* (Washington, D.C.: U.S. GPO, 1973). Total displaced urban renewal households from Herbert J. Gans, *The Urban Villagers: Group and Class in the Life of Italian-Americans,* updated and expanded ed. (New York: Free Press, 1982), 384–385. Highway displacement figures from 90th Congress, First Session, Subcommittee on Roads, Committee on Public Works, House, *Highway Relocation Assistance Study: A Study Transmitted by the Secretary of the Department of Transportation, as Required by the Federal-Aid Highway Act of 1966* (Washington, D.C., July 1967), 5. For another estimate of postwar demolitions, see Emily Talen, "Housing Demolition during Urban Renewal," *City and Community* 13, no. 3 (September 2014): 235, 238. Talen estimates that 10 percent of census tracts experienced dramatic loss (defined as 1,000 or more units, or at least 50 percent of their housing stock, in one decade) between 1940 and 1970.

7. The name "Seabee" derives from the phonetic pronunciation of the abbreviation "CB," for Construction Battalion. The group's name was also abbreviated "NCB," for its naval affiliation; "Seabee History: Formation of the Seabees and World War II," *Naval History and Heritage Command* Home Page, http://www.history.navy.mil/research/library/online-reading-room/title-list-alphabetically/s/seabee-history0/world-war-ii.html (accessed 5/9/15).

8. See Rob Nixon, *Slow Violence and the Environmentalism of the Poor* (Cambridge: Harvard University Press, 2011).

9. Richard Gordon Lillard, *Eden in Jeopardy* (New York: Knopf, 1966), 82; "The McCoy Outfit Tames the Hills," *Earth,* December 1973–January 1974, 31–32, Box 1365, Folder 16140, IHC.

10. See Albert M. Cole, "Address of Albert M. Cole, U.S. Housing Administrator, at Fourth National Construction Industry Conference, Hotel Sherman, Chicago, Ill., December 11, 1958, 1:00 PM," December 11, 1958, Box 166,

Folder: URA Speeches, NHRA Records; Gil Wyner Company, Inc., "Statement of Bidder's Qualifications," May 12, 1960, Box 123, Folder: Dem. Contr. 6A & 6B, Bid Documents, NHRA Records.

11. See Alan A. Altshuler, *The City Planning Process: A Political Analysis* (Ithaca: Cornell University Press, 1965), 339; Peter M. Wolf, *Land in America: Its Value, Use, and Control* (New York: Pantheon, 1981), 224–225.

12. For examples from the 1960s, see Martin Anderson, *The Federal Bulldozer: A Critical Analysis of Urban Renewal, 1949–1962* (Cambridge: MIT Press, 1964); Gans, *Urban Villagers,* especially "From the Bulldozer to Homelessness," 384–395 (first ed. published in 1962); Jane Jacobs, *The Death and Life of Great American Cities* (New York: Random House, 1961); Wolf von Eckhardt, *Bulldozers and Bureaucrats: Cities and Urban Renewal* (Washington, D.C.: New Republic, 1963). Recent scholarship on postwar urban renewal, suburbs, and highways is vast. Among these, Samuel Zipp explores bulldozer-based urban renewal in Cold War New York City, but primarily examines "city rebuilding"—rather than "city destroying"—and the protest that rose up against it; Adam Rome focuses less on the activities of the suburban bulldozer than on its consequences in spurring postwar environmentalism; Samuel Zipp, *Manhattan Projects: The Rise and Fall of Urban Renewal in Cold War New York* (New York: Oxford University Press, 2010); Adam Rome, *The Bulldozer in the Countryside: Suburban Sprawl and the Rise of American Environmentalism* (Cambridge: Cambridge University Press, 2001). For a valuable introductory survey of the construction equipment industry and earthmoving, see William R. Haycraft, *Yellow Steel: The Story of the Earthmoving Equipment Industry* (Urbana: University of Illinois Press, 2000). On demolition, see Max Page, *The Creative Destruction of Manhattan, 1900–1940* (Chicago: University of Chicago Press, 1999), esp. 69–110; Keller Easterling, *Enduring Innocence: Global Architecture and Its Political Masquerades* (Cambridge: MIT Press, 2005), 161–184; Jeff Byles, *Rubble: Unearthing the History of Demolition* (New York: Harmony Books, 2005); Bernard L. Jim, "Ephemeral Containers: A Cultural and Technological History of Building Demolition, 1893–1993" (Ph.D. diss., Case Western Reserve University, 2006).

13. "Construction Equipment Goes to War," *Engineering News-Record,* February 24, 1944, 132.

14. Historians have defined militarization as "the process by which war and national security became consuming anxieties and provided the memories, models, and metaphors that shaped broad areas of national life," and the "contradictory and tense social process in which civil society organizes itself for the production of violence": see, respectively, Michael S. Sherry, *In the Shadow of War: The United States Since the 1930s* (New Haven: Yale University Press, 1995), ix, and Michael Geyer, "The Militarization of Europe,

1914–1945," in *The Militarization of the Western World*, ed. John R. Gillis (New Brunswick: Rutgers University Press, 1989), 79. On militarization and U.S. cities, see Laura McEnaney, *Civil Defense Begins at Home: Militarization Meets Everyday Life in the Fifties* (Princeton: Princeton University Press, 2000); Roger W. Lotchin, *Fortress California, 1910–1961: From Warfare to Welfare* (New York: Oxford University Press, 1992).

On war and the U.S. built environment, see Sarah Jo Peterson, *Planning the Home Front: Building Bombers and Communities at Willow Run* (Chicago: University of Chicago Press, 2013); Jean-Louis Cohen, *Architecture in Uniform: Designing and Building for the Second World War* (Montreal: Canadian Centre for Architecture, Yale University Press, 2011); Donald Albrecht, ed., *World War II and the American Dream* (Washington, D.C.: National Building Museum, 1995); Andrew Michael Shanken, *194X: Architecture, Planning, and Consumer Culture on the American Home Front* (Minneapolis: University of Minnesota Press, 2009).

On World War II's postwar domestic impacts, see Kevin M. Kruse and Stephen Tuck, eds., *Fog of War: The Second World War and the Civil Rights Movement* (New York: Oxford University Press, 2012); John Morton Blum, *V Was for Victory: Politics and American Culture During World War II* (New York: Harcourt Brace Jovanovich, 1976); Michael C. Adams, *The Best War Ever: America and World War II* (Baltimore: Johns Hopkins University Press, 1994); James T. Sparrow, *Warfare State: World War II Americans and the Age of Big Government* (New York: Oxford University Press, 2011); Beatriz Colomina, *Domesticity at War* (Cambridge: MIT Press, 2007). On technology and war, see Paul S. Boyer, *By the Bomb's Early Light: American Thought and Culture at the Dawn of the Atomic Age* (New York: Pantheon, 1985); Elaine Tyler May, *Homeward Bound: American Families in the Cold War Era* (New York: Basic Books, 1988); Edmund Russell, *War and Nature: Fighting Humans and Insects with Chemicals from World War I to "Silent Spring"* (Cambridge: Cambridge University Press, 2001); Jennifer S. Light, *From Warfare to Welfare: Defense Intellectuals and Urban Problems in Cold War America* (Baltimore: Johns Hopkins University Press, 2003); Michael S. Sherry, *The Rise of American Air Power: The Creation of Armageddon* (New Haven: Yale University Press, 1987).

15. On the historical relationships between business and built space, see Alison Isenberg, *Downtown America: A History of the Place and the People Who Made It* (Chicago: University of Chicago Press, 2004); Jeffrey M. Hornstein, *A Nation of Realtors: A Cultural History of the Twentieth-Century American Middle Class* (Durham, N.C.: Duke University Press, 2005); Richard Harris, *Building a Market: The Rise of the Home Improvement Industry, 1914–1960*

(Chicago: University of Chicago Press, 2012). On the business of real estate development, see Dolores Hayden, *Building Suburbia: Green Fields and Urban Growth, 1820–2000* (New York: Pantheon, 2003); Marc A. Weiss, *The Rise of the Community Builders: The American Real Estate Industry and Urban Land Planning* (New York: Columbia University Press, 1987); Nicholas Dagen Bloom, *Merchant of Illusion: James Rouse, America's Salesman of the Businessman's Utopia* (Columbus: Ohio State University Press, 2004); David M. Freund, *Colored Property: State Policy and White Racial Politics in Suburban America* (Chicago: University of Chicago Press, 2007); Elihu Rubin, *Insuring the City: The Prudential Center and the Postwar Urban Landscape* (New Haven: Yale University Press, 2012); N. D. B. Connolly, *A World More Concrete: Real Estate and the Remaking of Jim Crow South Florida* (Chicago: University of Chicago Press, 2014); Sara Kathryn Stevens, "Developing Expertise: The Architecture of Real Estate, 1908–1965" (Ph.D. diss., Princeton University, 2012).

16. Through 1961, non-whites made up 66 percent of families living in and relocated from areas slated for urban renewal (among those for whom "color" was reported); United States Housing and Home Finance Agency, *Relocation from Urban Renewal Project Areas: Through December, 1961* (Washington, D.C.: U.S. GPO, 1962), 6. Among those relocated for urban renewal and neighborhood development projects by June 30, 1970, 40 percent of families who reported their race were white; U.S. Department of Housing and Urban Development, *1970 HUD Statistical Yearbook* (Washington, D.C.: U.S. GPO, 1971), 73. See also Larry Keating, *Atlanta: Race, Class and Urban Expansion* (Philadelphia: Temple University Press, 2001), 93; Kevin Michael Kruse, *White Flight: Atlanta and the Making of Modern Conservatism* (Princeton: Princeton University Press, 2005), 234; California Newsreel, Independent Television Service, and Public Broadcasting Service, *Race: The Power of an Illusion* (California Newsreel, 2003). African American ownership of new suburban homes calculated based upon new owner-occupied housing units created through new construction outside central cities, 1950–1970, can be found in U.S. Bureau of the Census, *U.S. Census of Housing: 1960,* vol. 4, *Components of Inventory Change,* Table 1; U.S. Bureau of the Census, *Census of Housing: 1970, Components of Inventory Change,* Table 2.

17. See Ben H. Bagdikian, "The Rape of the Land," *Saturday Evening Post,* June 18, 1966, 25–29, 86–94; William Worthy, *The Rape of Our Neighborhoods: And How Communities Are Resisting Take-Overs by Colleges, Hospitals, Churches, Businesses, and Public Agencies* (New York: Morrow, 1976). Urban renewal protestors in Boston's West End also circulated the poem "Smash Her Down . . . Watch Her Fall," which ends, "And after all is done,

with mouths agape,/Cry, "Dear Lord, I only watched the rape!": poem by Joseph C. Saltmarsh, in Committee to Save the West End, flyer, ca. 1956–1959, Box 23, Folder 9, Joseph Lee Papers, Boston Athenaeum.

18. On Chavez Ravine, see Dana Cuff, *The Provisional City: Los Angeles Stories of Architecture and Urbanism* (Cambridge: MIT Press, 2001). Other urban histories that seriously examine cultural forms as sources include: Eric Avila, *Popular Culture in the Age of White Flight: Fear and Fantasy in Suburban Los Angeles* (Berkeley: University of California Press, 2004); Max Page, *The City's End: Two Centuries of Fantasies, Fears, and Premonitions of New York's Destruction* (New Haven: Yale University Press, 2008). See also Isenberg, *Downtown America,* esp. 42–77; Carlo Rotella, *October Cities: The Redevelopment of Urban Literature* (Berkeley: University of California Press, 1998). Cultural landscape scholars have long viewed space as a category of analysis: John Brinckerhoff Jackson, *Landscape in Sight: Looking at America,* ed. Helen Lefkowitz Horowitz (New Haven: Yale University Press, 1997).

19. Bull-dose | bull-doze, *n.* and *v., Oxford English Dictionary,* 2nd ed., 1989; online version September 2011, http://www.oed.com/view/Entry/24526, earlier version first published in New English Dictionary, 1888. Bulldozer, *n., Oxford English Dictionary,* 2nd ed., 1989; online version September 2011, http://www.oed.com/view/Entry/24527, first published in *A Supplement to the OED I,* 1972.

20. "The Bulldozer: Ancestry of War's Surprise Weapon Traced to Primitive Planting Stick and Treadmill," *Automotive War Production,* July 1945, 8; Caterpillar Tractor Company, *Fifty Years on Tracks* (Peoria, Ill.: Caterpillar Tractor Company, 1954), 12.

21. Caterpillar Tractor Company, *Fifty Years on Tracks,* 14–18, 28; "Big Dirt Diggers," *Fortune,* January 1945, 133–134; Gilbert C. Nolde and Caterpillar Inc., *All in a Day's Work: Seventy-Five Years of Caterpillar* (New York: Forbes Custom Publishing, 2000), 7, 23.

22. Keith Haddock, *The Earthmover Encyclopedia: The Complete Guide to Heavy Equipment of the World* (St. Paul, Minn.: Motorbooks, 2007), 115–116; Haycraft, *Yellow Steel,* 70–73; Herbert L. Nichols, *Moving the Earth: The Workbook of Excavation* (Greenwich, Conn.: North Castle Books, 1955), 16-1.

Chapter 1. "A Dirt Moving War"

1. K. S. Andersson, "The Bulldozer—An Appreciation," *The Military Engineer,* October 1944, 339; John A. Menaugh, "Bulldozing the Enemy!" *Chicago Daily Tribune,* January 23, 1944, F2; Reybold on "victory," quoted in "Tractor Parade," *Time,* February 7, 1944, 87; International Harvester Company advertisement, "We're in a Dirt Moving War!" *Construction Methods,*

March 1944, 161; Reybold on "war's end," quoted in H. E. Foreman, Foreword to Van Rensselaer Sill, *American Miracle: The Story of War Construction Around the World* (New York: Odyssey, 1947), vi.

2. "Notes and Comment," *Southwest Builder and Contractor,* August 14, 1942, 15; Eugene Reybold, "General MacArthur Told Me: 'This Is Distinctly an Engineer's War,'" *Popular Science,* May 1945, 90; "Salute to Seabees," *Los Angeles Times,* December 28, 1950, A4; U.S. Bureau of Naval Personnel, *The Construction Battalion* (Washington, D.C.: U.S. GPO, 1956), 9; William Bradford Huie, "The Navy's Seabees: They Build the Roads to Victory," *Life,* October 9, 1944, 48; Andrew Hamilton, "We're Back in Business—the Seabees," *Popular Mechanics,* May 1951, 111.

3. "Seabee History: Formation of the Seabees and World War II," *Naval History and Heritage Command* Home Page, http://www.history.navy.mil/research/library/online-reading-room/title-list-alphabetically/s/seabee-history0/world-war-ii.html (accessed 5/9/15).

4. See Blanche D. Coll, *The Corps of Engineers: Troops and Equipment* (Washington, D.C.: Office of the Chief of Military History, Department of the Army, 1958), 5–12, 29–34; Jes Wilhelm Schlaikjer, *Corps of Engineers, United States Army, We Clear the Way,* Poster, 1942, Catalog ID Number: 663, George Marshall Foundation Library. See also U.S. Army Corps of Engineers, *The U.S. Army Corps of Engineers: A History* (Washington, D.C.: U.S. GPO, 2007).

5. See David Oakes Woodbury, *Builders for Battle: How the Pacific Naval Air Bases Were Constructed* (New York: Dutton, 1946), 398.

6. For a first-person account of Wake Island, see Walter L. J. Bayler, *Last Man off Wake Island* (Indianapolis: Bobbs-Merrill, 1943). For Hollywood's version, see *Wake Island,* directed by John Farrow, Paramount Pictures, 1942. See also "The Earth Movers I," *Fortune,* August 1943, 102; James Brooke, "Civilian Prisoners of War Recall Their Ordeals," *New York Times,* September 9, 1995, 6.

7. The precedent was set during World War I for a naval construction force through the formation of a lone regiment that worked on domestic projects in the Great Lakes area. See U.S. Bureau of Naval Personnel, *Construction Battalion,* 2; U.S. NCB, 8th, *Pieces of Eight . . .* (Allentown, Pa.: Schlechter's, 1946); Nathan A. Bowers, "Pacific Ocean Areas," in Waldo G. Bowman et al., *Bulldozers Come First: The Story of U.S. War Construction in Foreign Lands* (New York: McGraw-Hill, 1944), 163 (this volume is a compilation of articles originally published in the journal *Engineering News-Record*); U.S. Bureau of Yards and Docks, *Building the Navy's Bases in World War II: History of the Bureau of Yards and Docks and the Civil Engineer Corps, 1940–1946,* vol. 1 (Washington, D.C.: U.S. GPO, 1947), 148. For period histories of the Seabees, see William Bradford Huie, *Can Do! The Story of the Seabees*

(New York: Dutton, 1944); Hugh B. Cave, *We Build, We Fight! The Story of the Seabees* (New York: Harper and Brothers, 1944); Edmund Castillo, *The Seabees of World War II* (New York: Random House, 1963).

8. U.S. NCB, 107th, *The Log, 1943–1945: A Story of a Seabee Battalion Conceived in War—Dedicated to Peace!* (Baton Rouge: Army and Navy Pictorial Publishers, 1946), 100; Hamilton, "We're Back in Business—the Seabees," 112; U.S. NCB, 27th, *Danger; Fighting Men at Work: A Work-a-Day Tale of How the Job Was Actually Done by the 27th Seabees,* ed. Willard G. Triest (Baton Rouge: Army and Navy Pictorial Publishers, 1945), 14.

9. By late 1943, more than twelve thousand members of the International Union of Operating Engineers, which included heavy equipment operators, were serving in the military; International Union of Operating Engineers, "I'm Still a Member of My Union," *International Engineer,* November 1943, 145; Huie, "Navy's Seabees," 54; Lee Inman Thompson, interview by Stanley Schrader, DVD, January 20, 2010, Lee Inman Thompson Collection (AFC/2001/001/75848), VHP; Adolph Pisani, interview by Wilma Langley, DVD, February 19, 2004, Adolph Pisani Collection (AFC/2001/001/15326), VHP; United States Navy, Seabees, *Build and Fight with the Seabees and Follow Your Trade in the Navy* (U.S. Navy, 1943), 7, 9, 14–15; Hamilton, "We're Back in Business—the Seabees," 112; Huie, *Can Do!,* 85; ("I was always"): Albert Peter Johnson, interview by Andrew Davis and Heather Strangeby, transcript, February 2003, Albert Peter Johnson Collection (AFC/2001/001/8407), VHP.

10. Donald J. Smith, interview by Mary Krejchi, transcript, October 1, 2008, Donald J. Smith Collection (AFC/2001/001/65074), VHP; Keith Robert Parady, interview by Debra Parady Codiga and Mark S. Codiga, transcript, August 5, 2007, Keith Robert Parady Collection (AFC/2001/001/65892), VHP; William J. Schlumpf, interview by Maria Schlumpf, transcript, May 18, 2003, William J. Schlumpf Collection (AFC/2001/001/10373), VHP.

11. On the formal training of one typical Seabee, see Cave, *We Build, We Fight!,* 7–41. U.S. Bureau of Naval Personnel, *Construction Battalion,* 3–5; ("mastery of the bulldozer"): "The Seabees Can Do It," *Popular Science,* January 1944, 55.

12. U.S. NCB, 11th, *Southern Cross Duty, 1942–1944* (Baton Rouge: Army and Navy Pictorial Publishers, 1945), 79; U.S. NCB, 87th, *The Earthmover: A Chronicle of the 87th Seabee Battalion in World War II* (Baton Rouge: Army and Navy Pictorial Publishers, 1946), 260; U.S. NCB, 4th, *Lil' Short-Runner Presents the Fourth U.S. Naval Construction Battalion Penguin, 1944–45* (Baton Rouge: Army and Navy Pictorial Publishers, 1945), 36; U.S. NCB, 27th, *Danger; Fighting Men at Work,* 30, 96.

13. "Guam: U.S. Makes Little Island into Mighty Base," *Life,* July 2,

1945, 63–64; Samuel Eliot Morison, *The Struggle for Guadalcanal: August 1942–February 1943* (Boston: Little, Brown, 1950), 76.

14. U.S. NCB, 46th, *Following Invasions: Being an Account of the Forty-Sixth Seabee Battalion in the Southwest Pacific,* ed. Jones Gladston Emery (Oklahoma City: Harlow, 1947), 120; U.S. NCB, 27th, *Danger; Fighting Men at Work,* 63; Pisani, interview; Bowers, "Pacific Ocean Areas," 175–179, 181–182; U.S. NCB, 87th, *Earthmover,* 63; Susan D. Lanier-Graham, *The Ecology of War: Environmental Impacts of Weaponry and Warfare* (New York: Walker, 1993), 27; U.S. NCB, 88th, *In Review: A History and Pictorial Account of the 88th Naval Construction Battalion* (Jacksonville, Fla.: M. G. Lewis Printing, 1946), n.p.; Sill, *American Miracle,* 115–116.

15. See Philip S. Heisler, "Bulldozers Disinter Land of Mu Theory," *The Sun* (Baltimore), February 18, 1945, 25; Thomas M. O'Neill, "Raids Uncover Prolific Deposit of Roman Relics," *The Sun,* February 11, 1945, 1, 7; J. William Thompson and Kim Sorvig, *Sustainable Landscape Construction: A Guide to Green Building Outdoors* (Washington, D.C.: Island, 2007), 90; Waldo G. Bowman, "Britain, North Africa and the Middle East," in *Bulldozers Come First,* 17–18.

16. "Nature Stalemated by Contractors Putting up Grim Battle on Alaska Highway," *Southwest Builder and Contractor,* July 3, 1942, 38. See Harold W. Richardson, "Alaska and the Aleutians," in *Bulldozers Come First,* 127–129; Tracy Heather Strain, *Building the Alaska Highway,* Transcript (PBS Home Video, 2005), available at http://www.pbs.org/wgbh/americanexperience/features/transcript/alaska-transcript/.

17. See William E. Griggs, *The World War II Black Regiment That Built the Alaska Military Highway: A Photographic History* (Jackson: University Press of Mississippi, 2002), 67; Richardson, "Alaska and the Aleutians," 126–128, 134.

18. See Roxanne Willis, *Alaska's Place in the West: From the Last Frontier to the Last Great Wilderness* (Lawrence: University Press of Kansas, 2010), 16, 86–89.

19. See Heath Twichell, *Northwest Epic: The Building of the Alaska Highway* (New York: St. Martin's, 1992), 97–98; Charles Hendricks, "Race Relations and the Contributions of Minority Troops in Alaska: A Challenge to the Status Quo?" in *Alaska at War, 1941–1945: The Forgotten War Remembered,* ed. Fern Chandonnet (Fairbanks: University of Alaska Press, 2007), 279; Kenneth Coates, *North to Alaska* (Toronto: McClelland and Stewart, 1992), 104–105, 130–131; Griggs, *World War II Black Regiment That Built the Alaska Military Highway,* 51. African Americans also served in the Seabees, making up roughly 6 percent of the total by the end of 1944; see Dennis Denmark Nelson, *The Integration of the Negro into the United States Navy, 1776–1947*

([Washington, D.C.]: n.p., 1948), 69. On the circulation of the photographer's story, see, for example, Norman Rosten, *The Big Road: A Narrative Poem* (New York: Rinehart and Company, 1946); "Open Passage," *Time,* November 30, 1942.

20. Stephen Tuck, "'You Can Sing and Punch . . . but You Can't Be a Soldier or a Man': African American Struggles for a New Place in Popular Culture," in *Fog of War: The Second World War and the Civil Rights Movement,* ed. Kevin M. Kruse and Stephen Tuck (New York: Oxford University Press, 2012), 103–125; Dan Burley, "Combat Glamour Without Thanks and Headlines: Burma Saga Tells of Men of Supply—Heroes All; but Without Publicity," *New York Amsterdam News,* August 11, 1945; Dan Burley, "Whites Given Credit Where Negro GI's Did Hard Work," *New York Amsterdam News,* July 28, 1945; Nelson Grant Tayman, "Stilwell Road—Land Route to China," *National Geographic,* June 1945, 681–698; Dan Burley, "Burley Reports on Our Boys in Burma: An Intimate Dispatch of 'Well-Conked' GI's Spearheading One of Greatest Construction Jobs; Many Names Are Listed," *New York Amsterdam News,* June 9, 1945.

21. Woodbury, *Builders for Battle,* 180, 189; U.S. NCB, 4th, *Lil' Short-Runner Presents the Fourth U.S. Naval Construction Battalion Penguin,* 6–7, 12.

22. See Judith A. Bennett, *Natives and Exotics: World War II and Environment in the Southern Pacific* (Honolulu: University of Hawai'i Press, 2009), 158, 234. U.S. NCB, 27th, *Danger; Fighting Men at Work,* 54; U.S. NCB, 87th, *Earthmover,* 260. See Lamont Lindstrom, *Island Encounters: Black and White Memories of the Pacific War* (Washington, D.C.: Smithsonian Institution Press, 1990), 61.

23. Bennett, *Natives and Exotics,* 157–178; U.S. NCB, 87th, *Earthmover,* 260.

24. Further contradictory evidence to this "loyal islander" trope includes the cases of Pacific natives who organized strikes and walkouts to demand higher wages. Lindstrom, *Island Encounters,* 4, and, for "dig up my coconuts," 61; U.S. NCB, 46th, *Following Invasions,* 117–118; "Old Bermuda: Honeymoon Isles Become U.S. Defense Bastion," *Life,* August 18, 1941, 61–71.

25. Bruce J. Schulman, *From Cotton Belt to Sunbelt: Federal Policy, Economic Development, and the Transformation of the South, 1938–1980* (New York: Oxford University Press, 1991), 95–97.

26. "Building Bombs on the Plains," NebraskaStudies.org Home Page, http://www.nebraskastudies.org/; Don Greery, video and transcript, Wessels Living History Farm Home Page, http://www.livinghistoryfarm.org/farminginthe40s/movies/geery_life_09.html (accessed 4/7/11); NET Television, *Statewide: Cornhusker Ordnance Plant,* 2003, available at http://www.nebraskastudies.org/0800/stories/0801_0602cornhusker.html.

27. Bucyrus-Erie Company advertisement, "Fighting Today . . . Training for Tomorrow," *Excavating Engineer,* February 1945, 55; "Battling Bulldozers," *Excavating Engineer,* January 1945, 16–17.

28. Aurelio Tassone, "How the Tank Dozer Was Born," *Saturday Evening Post,* December 15, 1945, 6; U.S. NCB, 87th, *Earthmover,* 59.

29. "Aurelio Tassone," *Military Times Hall of Valor,* http://militarytimes.com/citations-medals-awards/recipient.php?recipientid=56533. Seabees also earned five Navy Crosses; see "Seabee History: Formation of the Seabees and World War II"; Tassone, "How the Tank Dozer Was Born," 6; "Construction Communiques," *Engineering News-Record,* July 13, 1944, 73; Sill, *American Miracle,* 179; "Engineer News and News Items," *The Military Engineer,* October 1944, 340–341.

30. ("Death traps"): Robert Sherrod, "Tarawa Today: The Historic Battlefield Is a Neglected Island," *Life,* August 5, 1946, 20; U.S. Marines, "Marines Use International TD-9 Tractor to Search for the Enemy," Photograph, 1944, McCormick-International Harvester Collection, Wisconsin Historical Society.

31. "The Abbott and Costello Program for Camel Cigarettes," Script (NBC, December 23, 1943), 22, Old Time Radio Research Group Archive, http://www.otrr.org/FILES/Scripts_pdf/Abbott_And_Costello/Abbott_And_Costello_43-12-23.pdf. "Seabee Who Killed Dozen with Bulldozer Identified," *New York Times,* December 13, 1943, 3; "Resistance Buried," *Time,* December 20, 1943, 72; "They Move the Earth," *Popular Mechanics,* April 1945, 20; "Construction Communiques," 73; H. V. Pehrson, "Baby Your Bulldozer," *Southern Lumberman,* September 15, 1944, 64; "The Bulldozer: Pacific Highways Are Pushed Across Trackless Island Wastes, Straight Towards Tokyo's Heart," *Automotive War Production,* June 1945, 5. Other publications that mentioned Tassone's story include *Engineering and Mining Journal, The Timberman,* the AFL-CIO's *American Federationist, Architectural Forum, Roads and Streets, The American Legion Magazine,* and *Carpenter.*

32. Buckeye Traction Ditcher Company advertisement, "You Should Have Seen Those Little Rats Run!" *Construction Methods,* March 1944, 18; Henry B. Lent, *Seabee: Bill Scott Builds and Fights for the Navy* (New York: Macmillan, 1944), jacket, 7, 173.

33. Harry Yeide, *Weapons of the Tankers: American Armor in World War II* (Saint Paul, Minn.: Zenith Imprint, 2006), 79; Tassone, "How the Tank Dozer Was Born," 6; Patrick Wright, *Tank: The Progress of a Monstrous War Machine* (London: Faber, 2000).

34. See Sill, *American Miracle,* 107; Fred W. Crismon, *U.S. Military Tracked Vehicles* (Osceola, Wisc.: Motorbooks International, 1992), 306; ("geysers" and "fighting partner"): "Engineer Wins DSC on 'D' Day," *International Engineer,* February 1945, 15. "William J. Shoemaker," *Military Times*

Hall of Valor, http://militarytimes.com/citations-medals-awards/recipient.php?recipientid=32786; "Vinton Walsh Dove," *Military Times Hall of Valor,* http://militarytimes.com/citations-medals-awards/recipient.php?recipientid=30794; Martin K. Gordon, "The Tank Dozer," in *Builders and Fighters: U.S. Army Engineers in World War II,* ed. Barry W. Fowle (Fort Belvoir, Va.: Office of History, U.S. Army Corps of Engineers, 1992), 176–177. The "Hellcat" was the nickname for the M-18 Tank Destroyer.

35. Lester R. King, *Hellcats Don't Leak Oil, They Mark Their Territory!* (Victoria, B.C.: Trafford, 2004), 121–123.

36. Gordon, "Tank Dozer," 175; "Can Do, Will Do—Did," *Time,* January 3, 1944, 59; "An American Bulldozer Does Its Stuff in Italy," *Engineering News-Record,* January 13, 1944, 53; U.S. NCB, 130th, *One Hundred and Thirtieth United States Naval Construction Battalion* (Baton Rouge: Army and Navy Pictorial Publishers, 1945), 89.

37. Example causes of death taken from handwritten notes in the Google Books version of U.S. NCB, 105th, *105 Naval Construction Battalion,* ed. Lester Colin (San Francisco: Crocker Union, 1945). Woodbury, *Builders for Battle,* 300–301; Peter Ference, interview by Ruth Osgood, transcript, July 4, 2004, Peter Ference Collection (AFC/2001/001/43117), VHP; "Seabee History: Formation of the Seabees and World War II; Strain, *Building the Alaska Highway.*

38. *The Fighting Seabees,* directed by Edward Ludwig (Republic Pictures, 1944). This is one of only two war movies (the other is *Sands of Iwo Jima*) in which Wayne dies in combat; Lawrence H. Suid, *Sailing on the Silver Screen: Hollywood and the U.S. Navy* (Annapolis: Naval Institute Press, 1996), 70. On Hollywood and World War II, see Thomas Patrick Doherty, *Projections of War: Hollywood, American Culture, and World War II* (New York: Columbia University Press, 1999); Clayton R. Koppes, *Hollywood Goes to War: How Politics, Profits, and Propaganda Shaped World War II Movies* (New York: Free Press, 1987).

39. Quoted in Sill, *American Miracle,* 3. On cowboys and American masculinity, see David Savran, *Communists, Cowboys, and Queers: The Politics of Masculinity in the Work of Arthur Miller and Tennessee Williams* (Minneapolis: University of Minnesota Press, 1992). On nose art, see Jeffrey L. Ethell, *World War II Nose Art in Color* (Osceola, Wisc.: Motorbooks International, 1993); Larry Davis and Don Greer, *Planes, Names and Dames* (Carrollton, Tex.: Squadron/Signal Publications, 1990).

40. The navy permitted Republic Pictures to film some scenes of wartime construction on Seabee bases, with members of the 105th NCB reliving boot camp, making beachheads, and disembarking from trucks in battle attire, all for the camera; see U.S. NCB, 105th, *105 Naval Construction Battalion;* "Ro-

mance of the Seven Seas," *New York Times,* March 18, 1944, 18; Edward E. Bloom, interview by Zachary Fox, transcript, February 9, 2003, Edward E. Bloom Collection (AFC/2001/001/6432), VHP.

41. "Guam: U.S. Makes Little Island into Mighty Base," 63–75; Bruce Poynter, interview by Darrell Pederson, DVD, n.d., Bruce Poynter Collection (AFC/2001/001/77603), VHP. The clarity of the image contrasts with the blurriness of the photographer Robert Capa's famous combat images (also published in *Life*) of the beach landings at Normandy. While Capa's work suggests the frenetic speed of combat, the crispness of Selby's portrait helps remove him conceptually from danger. See Patricia Vettel-Becker, *Shooting from the Hip: Photography, Masculinity, and Postwar America* (Minneapolis: University of Minnesota Press, 2005), 39–40.

42. "Waves Get Dumped," *Life,* September 10, 1945, 142–144. While the Coke advertisement offers a racist depiction of the islanders, it is of a piece with many of the stereotyped representations of foreigners in American wartime visual culture; Coca-Cola Company advertisement, "Now You're Talking . . . Have a Coca-Cola," *Life,* October 15, 1945, back cover. Other manufacturers, including Kohler and Studebaker, included Seabees in their wartime advertisements.

43. James A. Michener, *Tales of the South Pacific* (New York: Macmillan, 1947), 282–302.

44. Ibid., 26–49.

45. Chris Pearson, "'The Age of Wood': Fuel and Fighting in French Forests, 1940–1944," *Environmental History* 11, no. 4 (October 2006): 793.

46. See Jeffry M. Diefendorf, "Wartime Destruction and the Postwar Cityscape," in *War and the Environment: Military Destruction in the Modern Age,* ed. Charles E. Closmann (College Station: Texas A&M University Press, 2009), 185–186; Jeffry M. Diefendorf, "War and Reconstruction in Germany and Japan," in *Rebuilding Urban Japan After 1945,* ed. Carola Hein, Jeffry M. Diefendorf, and Yorifusa Ishida (New York: Palgrave Macmillan, 2003), 211–214; Douglas L. Oliver, *The Pacific Islands,* 3rd ed. (Honolulu: University of Hawai'i Press, 1989), 258.

47. Jörg Friedrich, *The Fire: The Bombing of Germany, 1940–1945,* trans. Allison Brown (New York: Columbia University Press, 2006), 462. For images of LeTourneau and LaPlant-Choate bulldozers performing such tasks in Italy and France, see "Battling Bulldozers," 46; "The Ruins of Cassino," *Life,* June 5, 1944, 27. Wickwire Spencer Steel Company advertisement, "Friend of the 'Cleaner Uppers,'" *Southern Lumberman,* August 15, 1944, 57.

48. Wartime photography in architectural periodicals similarly "refrained from demoralizing visual accounts of the destruction of monuments and cities," according to Jean-Louis Cohen, *Architecture in Uniform: Designing*

and Building for the Second World War (Montreal: Canadian Centre for Architecture, Yale University Press, 2011), 343, 383–385. On Blitz photography, see George Rodger, *The Blitz: The Photography of George Rodger* (London: Penguin, 1990); "In the Ruins," *Life,* September 27, 1943, 41–44.

49. *A Foreign Affair,* directed by Billy Wilder (Paramount Pictures, 1948). The film is one of several postwar "rubble films" that deployed the visual imagery of destroyed German cities as background setting and metaphor for the Germans' psychological state; see Robert R. Shandley, *Rubble Films: German Cinema in the Shadow of the Third Reich* (Philadelphia: Temple University Press, 2001), 2. See also Gerd Gemünden, "In the Ruins of Berlin: *A Foreign Affair* (1948)," in *A Foreign Affair: Billy Wilder's American Films* (New York: Berghahn, 2008), 54–75. Edward Steichen, *The Family of Man: The Photographic Exhibition Created by Edward Steichen for the Museum of Modern Art* (New York: Museum of Modern Art, 1955), 127.

50. On modernist planning and architecture, see Sigfried Giedion, *Space, Time and Architecture: The Growth of a New Tradition* (Cambridge: Harvard University Press, 1941); Peter Geoffrey Hall, *Cities of Tomorrow: An Intellectual History of Urban Planning and Design in the Twentieth Century* (Oxford: Blackwell, 1988).

51. As Sebald wrote, "There was a tacit agreement, equally binding on everyone, that the true state of material and moral ruin in which the country found itself was not to be described": Winfried Georg Sebald, *On the Natural History of Destruction,* trans. Anthea Bell (London: Hamish Hamilton, 2003), 10. Wright quoted in Cohen, *Architecture in Uniform,* 160–161 (and see also pp. 374–375); MacCornack quoted in "Plan to Mobilize Public Opinion for Vast Post-War Conrtruction [*sic*] Drive," *Southwest Builder and Contractor,* April 1942, 26. See Andrew Michael Shanken, *194X: Architecture, Planning, and Consumer Culture on the American Home Front* (Minneapolis: University of Minnesota Press, 2009), 29–39.

52. See Diefendorf, "Wartime Destruction and the Postwar Cityscape," 183; Cohen, *Architecture in Uniform,* 111–112; National Association of Housing and Redevelopment Officials, *Housing for the United States After the War,* 2nd ed. (Chicago: National Association of Housing Officials, 1944), 51. On militarized and medicalized language regarding slum clearance, see, for example, John H. Britton, "How King Wages War on Slums in Chicago—Showdown Looms: Moral Power," *Jet,* February 10, 1966, 14–20; "Slum Surgery in St. Louis," *Architectural Forum,* April 1951, 128–136; Harrison E. Salisbury, "New Weapons Against Slums," *Think,* 1959, 71.

53. Sherrod, "Tarawa Today," 19–20. Limited attention to environmental scars was true of period journalistic coverage, as well as later military histories. Environmental historians have recently turned increasing attention to

this subject; see Chris Pearson, Peter Coates, and Tim Cole, eds., *Militarized Landscapes: From Gettysburg to Salisbury Plain* (London: Continuum, 2010); Chris Pearson, *Scarred Landscapes: War and Nature in Vichy France* (New York: Palgrave Macmillan, 2008); Richard P. Tucker and Edmund Russell, eds., *Natural Enemy, Natural Ally: Toward an Environmental History of Warfare* (Corvallis: Oregon State University Press, 2004); Closmann, *War and the Environment;* Arthur H. Westing, *Arthur H. Westing: Pioneer on the Environmental Impact of War* (Heidelberg: Springer, 2013).

54. See Bennett, *Natives and Exotics,* 169, 191, 198–218; "They Move the Earth," 24–25.

55. See Lanier-Graham, *Ecology of War,* 25; Pearson, "'Age of Wood'"; J. R. McNeill, "Woods and Warfare in World History," *Environmental History* 9, no. 3 (July 2004): 399; Greg Bankoff, "Wood for War: The Legacy of Human Conflict on the Forests of the Philippines, 1565–1946," in *War and the Environment,* 42.

56. See Bennett, *Natives and Exotics,* 198, 302–303. On the challenge of making slow violence visible, see Rob Nixon, *Slow Violence and the Environmentalism of the Poor* (Cambridge: Harvard University Press, 2011). During World War II, conservationist goals forestalled the construction of the Wutoch Gorge dam, despite the structure's potential for valuable energy generation; see Frank Uekötter, "Total War? Administering Germany's Environment in Two World Wars," in *War and the Environment,* 92–111. Thomas M. Clement et al., *Engineering a Victory for Our Environment: A Citizens' Guide to the U.S. Army Corps of Engineers* (Washington, D.C.: Institute for the Study of Health and Society, July 7, 1971).

57. "The Earth Mover," *Time,* May 3, 1954.

Chapter 2. Prime Movers

1. Caterpillar Tractor Company, *Annual Report for the Year 1944* (Peoria, Ill.: Caterpillar Tractor Company, 1945), cover, 21.

2. See Donald Marr Nelson, *Arsenal of Democracy: The Story of American War Production* (New York: Harcourt Brace Jovanovich, 1946). See also Marilynn S. Johnson, *The Second Gold Rush: Oakland and the East Bay in World War II* (Berkeley: University of California Press, 1993); Sarah Jo Peterson, *Planning the Home Front: Building Bombers and Communities at Willow Run* (Chicago: University of Chicago Press, 2013). On wartime technological advances, see Alan S. Milward, *War, Economy and Society, 1939–1945* (Berkeley: University of California Press, 1977), 177–180, 193–202; Paul S. Boyer, *By the Bomb's Early Light: American Thought and Culture at the Dawn of the Atomic Age* (New York: Pantheon, 1985); Michael S. Sherry, *The Rise*

of *American Air Power: The Creation of Armageddon* (New Haven: Yale University Press, 1987); Edmund Russell, *War and Nature: Fighting Humans and Insects with Chemicals from World War I to "Silent Spring"* (Cambridge: Cambridge University Press, 2001). Franklin Roosevelt's commitment to the role of the United States as the "arsenal of democracy" behind the production of a "crushing superiority of equipment" facilitated a strategy focused on fighting a "war of machines," rather than men: David M. Kennedy, *Freedom from Fear: The American People in Depression and War, 1929–1945* (New York: Oxford University Press, 1999), 615–668.

3. Accounting for inflation, sales nearly tripled during this period: "Construction Equipment Goes to War," *Engineering News-Record,* February 24, 1944, 131–132; "Construction Equipment Enters Buyers' Market," *Business Week,* January 29, 1949, 25.

4. See Fred W. Crismon, *U.S. Military Tracked Vehicles* (Osceola, Wisc.: Motorbooks International, 1992), 7; International Harvester Company advertisement, "Power and More Power," *Life,* September 8, 1941, 39.

5. In late 1942, for example, WPB limits reduced Harvester's production to 14 percent of 1940 levels, although vigorous protest from farmers and manufacturers raised those quotas less than a year later. See Fowler McCormick, "Both Jobs Have Been Done: A Statement Concerning International Harvester Company's Farm Equipment Production," May 1944, Box 103, Folder 06164, IHC; International Harvester Company, *1939 Annual Report* (Chicago: The Company, 1940), 3; Dale Cox to Albert Haring, June 28, 1945, 1–2, Box 1360, Folder 486, IHC; "Memorandum of the Company's Principal War Activities," April 12, 1943, Box 1360, Folder 482, IHC; International Harvester Company, "Harvester Called to the Colors," 1942, 5, Box 1360, Folder 489, IHC. See also Karl B. Mickey, *Food in War and Peace* (Chicago: International Harvester Company, 1944).

6. International Harvester Company, *Annual Report,* for the years 1941–1945 (Chicago: The Company); "Memorandum of the Company's Principal War Activities," 1, 5; International Harvester Company, "International Harvester Company, Summary, War Orders, January 1, 1940 to December 31, 1943," January 1944, Table II, Box 102, Folder 1672, IHC; International Harvester Company, *1940 Annual Report* (Chicago: The Company, n.d.), 19, 23; C. Alfred Campbell, "Trucks for War Purposes," *Military Engineer,* February 1943, 63; "Notes and Comment," *Southwest Builder and Contractor,* July 3, 1942, 15; "[International Harvester Ads]," 1943, Box 108, Folder 18178, IHC; Fred W. Crismon, *International Trucks* (Osceola, Wisc.: Motorbooks International, 1995), 173; Crismon, *U.S. Military Tracked Vehicles,* 8–10. Original quotation regarding the M-5 tractor from *Field Artillery Journal,* March

1945, reproduced in "Harvester War Production Report," May 1945, 3–4, Box 107, Folder 13881, IHC.

7. Caterpillar Tractor Company, *Annual Report for the Year 1942* (Peoria, Ill.: Caterpillar Tractor Company, 1943), 18–19; "Fifty Years of Crawler Tractors," *Southern Lumberman,* May 15, 1954, 40, Box 1082, Folder 12228, IHC.

8. Caterpillar Tractor Company, *Fifty Years on Tracks* (Peoria, Ill.: Caterpillar Tractor Company, 1954), 54; Robert Haralson Selby, "Earthmovers in World War II: R. G. LeTourneau and His Machines" (Ph.D. diss., Case Western Reserve University, 1970), 30, 43; Robert G. LeTourneau, "Scraper," U.S. Patent 1530779, filed January 12, 1924, and issued March 24, 1925; Robert G. LeTourneau, "Telescoping Scraper," U.S. Patent 1598864, filed December 28, 1925, and issued September 7, 1926; Robert Gilmour LeTourneau, "Carrying Scraper," U.S. Patent 2127223, filed December 13, 1937, and issued August 16, 1938. See also Robert Gilmour LeTourneau, *Mover of Men and Mountains: The Autobiography of R. G. LeTourneau* (Englewood Cliffs, N.J.: Prentice-Hall, 1960).

9. International Harvester Company, *Profits in Wartime* (Chicago: International Harvester Company, 1944), 4–5; Caterpillar Tractor Company, *Annual Report for the Year 1944,* 4.

10. Caterpillar Tractor Company, *Annual Report for the Year 1942,* 11.

11. David G. Hammond, "'Recent Developments' in Military Equipment for Highway and Airport Construction," *Construction Equipment News,* April 1951, 7; ("the crucible"): Crismon, *U.S. Military Tracked Vehicles,* 9. On the "misleading" impression of "lightning technological breakthroughs effected under the tremendous pressure of war," see Milward, *War, Economy and Society, 1939–1945,* 169–207.

12. Crismon, *International Trucks,* 173; "Engineer News and News Items," *Military Engineer,* October 1944, 340–341; Crismon, *U.S. Military Tracked Vehicles,* 312; ("ceased to be"): Caterpillar Tractor Company, *Fifty Years on Tracks,* 28; William R. Haycraft, *Yellow Steel: The Story of the Earthmoving Equipment Industry* (Urbana: University of Illinois Press, 2000), 101–102; Caterpillar Tractor Company, *Annual Report for the Year 1943* (Peoria, Ill.: Caterpillar Tractor Company, 1944), 11.

13. International Harvester Company, *1944 Annual Report* (Chicago: The Company, 1945), 12; Caterpillar Tractor Company, *Annual Report for the Year 1942,* 11; Eric C. Orlemann, *R. G. LeTourneau Heavy Equipment: The Mechanical Drive Era, 1921–1953* (Hudson, Wisc.: Iconografix, 2008), 81–85, 131; "They Move the Earth," *Popular Mechanics,* April 1945, 21–22; Hammond, "'Recent Developments' in Military Equipment for Highway and Airport Construction," 8; "A Bulldozer Takes a Ride," *Engineering News-Record,*

April 1, 1943, 2; "Small Construction Machines, No Longer Farm Tractors with Gadgets," *Roads and Streets,* February 1973, 29; "Big Dirt Diggers," *Fortune,* January 1945, 136; Hickman Powell, "Weld It, Says R. G. LeTourneau," *Popular Science,* January 1942, 57; LeTourneau, "Tournapulls Save Freight and Time Between Projects," *Military Engineer,* January 1943, 13.

14. Crismon, *U.S. Military Tracked Vehicles,* 306; Lida Mayo, *The Ordnance Department: On Beachhead and Battlefront* (Washington, D.C.: Center of Military History, U.S. Army, U.S. GPO, 1968), 196–199; Eisenhower quoted in Martin K. Gordon, "The Tank Dozer," in *Builders and Fighters: U.S. Army Engineers in World War II,* ed. Barry W. Fowle (Fort Belvoir, VA: Office of History, U.S. Army Corps of Engineers, 1992), 171–174.

15. Harry Yeide, *Weapons of the Tankers: American Armor in World War II* (Saint Paul, Minn.: Zenith, 2006), 79; International Harvester Company, *1943 Annual Report* (Chicago: The Company, 1944), 13; "They Move the Earth," 21; International Harvester Company, *1944 Annual Report,* 13; International Harvester Company, *1943 Annual Report,* 13.

16. Carl E. Kramer, *The Brandeis Century: Constant Values in Changing Times* (Jeffersonville, Ind.: Sunnyside, 2008), 72; "The Earthmover: Man and Machines," *Progressive Architecture,* April 1967, 128; Nathan A. Bowers, "Pacific Ocean Areas," in Waldo G. Bowman et al., *Bulldozers Come First: The Story of U.S. War Construction in Foreign Lands* (New York: McGraw-Hill, 1944), 172; "G.I. Bulldozer," *Business Week,* August 26, 1944, 73; Eric C. Orlemann, *R. G. LeTourneau Heavy Equipment: The Electric Drive Era, 1953–1970* (Hudson, Wisc.: Iconografix, 2009), 14; Orlemann, *R. G. LeTourneau Heavy Equipment: The Mechanical Drive Era,* 131.

17. Hal W. Hunt, "Notable Practices on War Construction," *Engineering News-Record,* February 11, 1943, 122; Crismon, *U.S. Military Tracked Vehicles,* 325; ("emerged from the war"): Fred W. Crismon, *U.S. Military Wheeled Vehicles* (Sarasota, Fla.: Crestline, 1983), 427.

18. International Harvester Company, *1939 Annual Report,* 1; International Harvester Company, *1945 Annual Report* (Chicago: The Company, 1946), 15; Caterpillar Tractor Company, *Annual Report for the Year 1945* (Peoria, Ill.: Caterpillar Tractor Company, 1946), 13; Caterpillar Tractor Company, *Annual Report for the Year 1939* (Peoria, Ill.: Caterpillar Tractor Company, 1940), 10; Caterpillar Tractor Company, *Annual Report for the Year 1944,* 9; Caterpillar Tractor Company, *Annual Report for the Year 1943,* 14. Unusually high employee turnover also characterized other industries during this period; see Gerald D. Nash, *The American West Transformed: The Impact of the Second World War* (Bloomington: Indiana University Press, 1985), 66.

19. Caterpillar Tractor Company, *Annual Report for the Year 1944,* 10;

Winning Against Odds (Caterpillar Tractor Company, 1940s), http://www.archive.org/details/winning_against_odds; International Harvester Company, *1942 Annual Report* (Chicago: The Company, n.d.), 26; Caterpillar Tractor Company, *Annual Report for the Year 1942,* 12.

20. Caterpillar Tractor Company, *Annual Report for the Year 1942,* 16, 18–19; McCormick, "Both Jobs Have Been Done," 5; "Harvester War Production Report," 6.

21. "Harvester's Own: Tank Maintenance Battalion of 859 Men Recruited from Company's Rolls," *Business Week,* July 25, 1942, 90; International Harvester Company, "News Release," July 8, 1942, Box 1360, Folder 489, IHC; International Harvester Company, "Harvester Called to the Colors"; Caterpillar Tractor Company, *Annual Report for the Year 1942,* 16; Gilbert C. Nolde and Caterpillar Inc., *All in a Day's Work: Seventy-Five Years of Caterpillar* (New York: Forbes Custom Publishing, 2000), 196.

22. Bowers, "Pacific Ocean Areas," 164; International Harvester Company, *1945 Annual Report,* 12, 15.

23. "Notes and Comment," July 3, 1942, 15; Harold W. Richardson, "Alaska and the Aleutians," in *Bulldozers Come First,* 89–90.

24. William Bradford Huie, "The Navy's Seabees: They Build the Roads to Victory," *Life,* October 9, 1944, 50; Caterpillar Tractor Company, *Annual Report for the Year 1943,* 5; Bowers, "Pacific Ocean Areas," 163.

25. Mary Alice Sentman and Patrick S. Washburn, "How Excess Profits Tax Brought Ads to Black Newspaper in World War II," *Journalism Quarterly* 64, no. 4 (Winter 1987): 769–774, 863; Ed Adams, "Combatting Advertising Decline in Magazines During WWII: Image Ads Promoting Wartime Themes and the War Loan Drives," *Web Journal of Mass Communication Research* 1, no. 1 (December 1997), at http://www.scripps.ohiou.edu/wjmcr/vol01/1-1a.HTM.

26. Gerald E. Stedman, "Are You Writing Your Company's Wartime History?" *Industrial Marketing,* January 1943, 42, 46, Box 1360, Folder 486, IHC; Dale Cox to Fowler McCormick, August 4, 1944, Box 1360, Folder 486, IHC; "Our Authors," *Military Engineer,* April 1943; Cynthia Lee Henthorn, *From Submarines to Suburbs: Selling a Better America, 1939–1959* (Athens: Ohio University Press, 2006), 6; T. J. Jackson Lears, *Fables of Abundance: A Cultural History of Advertising in America* (New York: Basic, 1994), 233–249.

27. See, for example, Caterpillar Tractor Company advertisement, "Let Freedom Ring," *Saturday Evening Post,* 1944; International Harvester Company advertisement, "To Keep the Battle Machines Slugging," *Life,* October 26, 1942; "Statement Concerning Forthcoming Advertising Program," *International Engineer,* January 1943, 23.

28. "G.I. Bulldozer," 73; Caterpillar Tractor Company advertisement, "'Caterpillar' Diesels Are Worth Waiting For!" *Construction Methods,* January 1944, 119.

29. Thew-Lorain advertisement, "The Whole World Is Their Proving Ground," *Construction Methods,* January 1944, 30; Caterpillar Tractor Company advertisement, "What's Ahead in Earthmoving?" *Excavating Engineer,* January 1945, 25; Gar Wood Industries, Inc., advertisement, ". . . In Peacetime and Wartime—Powerful Trailbuilders and Bulldozers," *Construction Methods,* April 1944, 40; International Harvester advertisement, "Nothing Changed but the Paint," *Construction Methods,* January 1944, 115.

30. Baker Manufacturing Company advertisement, "Direct Down-Pressure on the Axis!" *Engineering News-Record,* April 22, 1943, 131; Boris Artzybasheff and Wickwire Spencer Steel Company, *Axis in Agony! Presenting a Series of Caricatures from the Brush of Boris Artzybasheff* (New York: Wickwire Spencer Steel Company, 1944); Caterpillar Tractor Company advertisement, "What the Little Jap General Forgot," *Military Engineer,* April 1943, n.p.; Caterpillar Tractor Company advertisement, "Coming at You, Schicklgruber!" *Military Engineer,* June 1943, 12.

31. Bucyrus-Erie Company, *In War and Peace Progress Starts with Excavation . . .* (South Milwaukee: Bucyrus-Erie Company, 1945); Revere Copper and Brass advertisement, "Let YOURS Be These Helping Hands," *Saturday Evening Post,* October 14, 1944, 70; Andrew Michael Shanken, *194X: Architecture, Planning, and Consumer Culture on the American Home Front* (Minneapolis: University of Minnesota Press, 2009), 133; Rodgers Hydraulic Inc., advertisement, "Mother Earth Is Going to Have Her Face Lifted!" *Military Engineer,* May 1944, inside cover.

32. Caterpillar Tractor Company, *Annual Report for the Year 1942,* 19–20; "Don't Bring It Back," *Engineering News-Record,* May 6, 1943, 94; E. J. Amberg, "Letter to the Editor: Postwar Thinking," *Engineering News-Record,* June 3, 1943, 93.

33. L. H. Houck, "U.S. Engineers Redistribute Used Material," *Excavating Engineer,* April 1945, 194; Harold W. Richardson, "What Happens at a Surplus Equipment Sale," *Engineering News-Record,* March 22, 1945, 95–97.

34. "Construction Equipment Goes to War," 133; "Surplus War Equipment Disposal Plan," *Excavating Engineer,* May 1945, 249–250; Richard Thruelson, "What You Can Buy from the Army," *Saturday Evening Post,* December 1, 1945, 128–129; "Surplus War Equipment," *Engineering News-Record,* February 10, 1944, 58.

35. "Construction Dealers Weigh Marshall Plan," *Washington Post,* February 15, 1948, R3; William Clark, "Red Tape Ties Vast Surplus in Pacific Isles: Costs of Salvage Held Prohibitive," *Chicago Daily Tribune,* March 15, 1946, 1;

Thruelson, "What You Can Buy from the Army," 129–130; Vernon Lee LaMoreaux, interview by Joanne Cargill, DVD, April 17, 2008, Vernon Lee LaMoreaux Collection (AFC/2001/001/71506), VHP; Marvin Miles, "Seabees Reclaiming Stranded Machinery," *Los Angeles Times,* March 15, 1951, 2; Bruce Adams, *Rust in Peace: South Pacific Battlegrounds Revisited* (Sydney: Antipodean Publishers, 1975).

36. James Joseph, "Out of Jungles Come Millions," *Nation's Business,* October 1951, 76–77.

37. Thruelson, "What You Can Buy from the Army," 129–130; U.S. Congress, Senate, Special Subcommittee on Surplus Property, Committee to Study Problems of American Small Business, *Problems of Surplus Property Disposal; Preliminary Report,* July 21, 1944, LexisNexis Congressional Research Digital Collection; ("We didn't buy"): Kramer, *Brandeis Century,* 80.

38. "Earthmoving: Everyone Gets in the Act," *Business Week,* August 15, 1953, 58.

39. International Harvester Company, *1945 Annual Report,* 6; International Harvester Company, *1943 Annual Report,* 10.

40. International Harvester Company, *1945 Annual Report,* 11; Caterpillar Tractor Company, *Annual Report for the Year 1946* (Peoria, Ill.: Caterpillar Tractor Company, 1947), 6; Haycraft, *Yellow Steel,* 118–119, 123, 125.

41. E. A. Braker, "International Crawler Tractor History," March 30, 1955, Box 249, Folder 11157, IHC; International Harvester Company, "History of the Construction Equipment Division," [1950s], 1, Box 183, Folder 682, IHC.

42. "Big Dirt Diggers," 135; Caterpillar Tractor Company, *Annual Report for the Year 1944,* 18; Allis-Chalmers Manufacturing Company, *Annual Report 1952* (West Allis, Wisc.: Allis-Chalmers Manufacturing Company, 1953), 22; Allis-Chalmers Manufacturing Company, *Annual Report 1953* (West Allis, Wisc.: Allis-Chalmers Manufacturing Company, 1954), 13; Allis-Chalmers Manufacturing Company, *Annual Report 1955* (West Allis, Wisc.: Allis-Chalmers Manufacturing Company, 1956), 19; Haycraft, *Yellow Steel,* 125.

43. "Earthmoving: Everyone Gets in the Act," 56; "Clark Muscles into Earth Moving," *Business Week,* November 30, 1957, 74–76, 78.

44. On the history of the Bobcat Company, see Marty Padgett, *Bobcat: Fifty Years of Opportunity, 1958–2008* (Saint Paul: MBI, 2007). Melroe Company advertisement, "The Melroe BOBCAT . . . Lovely Little Homewrecker," *Wrecking and Salvage Journal,* July 1969, 12.

45. Caterpillar Tractor Company, *Annual Report for the Year 1946,* 18–21; Caterpillar Tractor Company, *Annual Report 1949* (Peoria, Ill.: Caterpillar Tractor Company, 1950), 2–3; "Caterpillar Tractor Earnings Early in 1946 to Run Below Year Ago: Only Price Increases Will Turn Tide—Sales Are About

Double Best Pre-War Levels," *Wall Street Journal,* January 15, 1946, 9; Haycraft, *Yellow Steel,* 108; Caterpillar Tractor Company, *Annual Report 1950* (Peoria, Ill.: Caterpillar Tractor Company, 1951), 6–7.

46. George E. Sadler, "Construction Machinery: Unit Man-Hour Trends, 1945–47," *Monthly Labor Review* 68 (January 1949): 24–25; Caterpillar Tractor Company, *Annual Report 1956* (Peoria, Ill.: Caterpillar Tractor Company, 1957), 20–21. Caterpillar profit growth calculated based on annual reports.

Chapter 3. Grading Groves and Moving Mountains

1. The information on Levitt is based on David Halberstam's interview with Bill Levitt, summarized in David Halberstam, *The Fifties* (New York: Villard, 1993), 132–133.

2. James Dunn, "Hugh: The Biography of Hugh Codding—Chapter 3," *Sonoma Business,* September 1993, 43–49; Dunn, "Hugh: The Biography of Hugh Codding—Chapter 5," *Sonoma Business,* March 1994, 40; Hugh Codding, interview by Gaye LeBaron, transcript, January 1995, Gaye LeBaron Collection Oral Histories, Sonoma State University Library; Gaye LeBaron, "Hugh Codding Dies at 92," *Press Democrat,* April 3, 2010, available at http://www.pressdemocrat.com/article/20100404/ARTICLES/100409836.

3. "Design for Lifting," *International Trail,* November 1953, 7, Box 15, Folder: "McCormick-IHC Publications, International Trail, 1953," IHC. On Halpin, see "Gerald T. Halpin," Shenandoah University Home Page, http://www.su.edu/business/about/gerald-t-halpin/ (accessed 5/19/15).

4. Gerald D. Nash, *The American West Transformed: The Impact of the Second World War* (Bloomington: Indiana University Press, 1985), 63.

5. Glenn C. Altschuler and Stuart M. Blumin, *The GI Bill: A New Deal for Veterans* (New York: Oxford University Press, 2009), 182.

6. Dolores Hayden, *Building Suburbia: Green Fields and Urban Growth, 1820–2000* (New York: Pantheon Books, 2003), 162–164.

7. "California Has a New Boom," *Life,* June 10, 1946, 31–37; George B. Leonard, "California: A Promised Land for Millions of Migrating Americans," *Look,* September 25, 1962, 30–31; "400 New Angels Every Day," *Life,* July 13, 1953, 23–29; "Orange County Spells Progress," *Los Angeles Times,* September 20, 1964, W22. Large counties include those with population of a hundred thousand or more in 1950; U.S. Bureau of the Census, *Census of Population,* 1950, 1960, 1970.

8. Helen Johnson, "Men Move Mountains to Help City of Orange Grow," *Los Angeles Times,* June 21, 1959, OC1.

9. On postwar American suburbs, see, for example, Hayden, *Building Suburbia;* Kenneth T. Jackson, *Crabgrass Frontier: The Suburbanization of the*

United States (New York: Oxford University Press, 1985); Robert Fishman, *Bourgeois Utopias: The Rise and Fall of Suburbia* (New York: Basic, 1987); Kevin Michael Kruse and Thomas J. Sugrue, eds., *The New Suburban History* (Chicago: University of Chicago Press, 2006). On southern California suburbs, see Eric Avila, *Popular Culture in the Age of White Flight: Fear and Fantasy in Suburban Los Angeles* (Berkeley: University of California Press, 2004); Lisa McGirr, *Suburban Warriors: The Origins of the New American Right* (Princeton: Princeton University Press, 2001); Rob Kling, Spencer C. Olin, and Mark Poster, eds., *Postsuburban California: The Transformation of Orange County Since World War II* (Berkeley: University of California Press, 1991). On first and second nature, see William Cronon, *Nature's Metropolis: Chicago and the Great West* (New York: Norton, 1991).

10. Walter Muhonen, "Ebb and Flow," *Los Angeles Times,* April 10, 1960, OC1; Richard Gordon Lillard, *Eden in Jeopardy* (New York: Knopf, 1966), 82; Raymond Fredric Dasmann, *The Destruction of California* (New York: Macmillan, 1965), 132–133; J. Herbert Snyder, "A New Program for Agricultural Land Use Stabilization: The California Land Conservation Act of 1965," *Land Economics* 42, no. 1 (February 1966): 30; Conservation Foundation, *State of the Environment: A View Toward the Nineties* (Washington, D.C.: Conservation Foundation, 1987), 208, 210.

11. Dewey Linze, "Sylmar Olive Groves Give Way to Progress," *Los Angeles Times,* January 1, 1962, B1; "Subdivision Blooms in Orange Grove," *Los Angeles Times,* August 30, 1970, I1. In contrast to the bevy of statistics on demographics and construction, the *Progress Report* relegated its orchard census to just a few pages, suggesting the groves' limited contribution to "progress": Board of Supervisors, *Orange County Progress Report,* for the years 1954–1973 (Santa Ana, Calif.).

12. See George J. Sánchez, *Becoming Mexican American: Ethnicity, Culture, and Identity in Chicano Los Angeles, 1900–1945* (New York: Oxford University Press, 1993), 88–90; Leo Marx, *The Machine in the Garden: Technology and the Pastoral Ideal in America* (New York: Oxford University Press, 1964). On this earlier epoch in southern California's orchard history, see Douglas Cazaux Sackman, *Orange Empire: California and the Fruits of Eden* (Berkeley: University of California Press, 2005); Matt García, *A World of Its Own: Race, Labor, and Citrus in the Making of Greater Los Angeles, 1900–1970* (Chapel Hill: University of North Carolina Press, 2001).

13. U.S. Bureau of the Census, *Census of Agriculture, 1940* (Washington, D.C.: U.S. GPO, 1942), 3, 84; H. M Butterfield, *A History of Subtropical Fruits and Nuts in California* (Berkeley: University of California, Division of Agricultural Sciences, Agricultural Extension Service, Agricultural Experiment Station, 1963), 32; U.S. Department of Agriculture, National Agricultural

Statistics Service, *2010 California Citrus Acreage Report,* July 15, 2010; California Department of Food and Agriculture, California Agricultural Statistics, *2014 California Citrus Acreage Report* (Sacramento, August 4, 2014), 2; Board of Supervisors, *Orange County Progress Report,* for the years 1954–1973; John O'Dell, "Money Is in Land, Not Crops," *Los Angeles Times,* June 2, 1981, OC-CC.

14. See Board of Supervisors, *Orange County Progress Report,* for the years 1954–1973.

15. Ibid.; James P. Degnan, "Santa Clara: The Bulldozer Crop," *Nation,* March 8, 1965, 243; "Land, Say the Homebuilders, Is Our Most Critical Problem," *House and Home,* August 1960, 100.

16. During the mid-1950s, while California growers paid taxes averaging thirty-five dollars per acre (up to forty-five in Orange County), their Florida counterparts paid only nine dollars: Alvin David Sokolow and California Department of Conservation, *The Williamson Act: 25 Years of Land Conservation* ([Sacramento]: California Department of Conservation; [Davis]: University of California Agricultural Issues Center, 1990), 9; "Need Reappraisal of Taxes," *California Citrograph,* March 1952, 177; O'Dell, "Money Is in Land, Not Crops," OC-CC; "Land, Say the Homebuilders, Is Our Most Critical Problem," 102; Paul F. Griffin and Ronald L. Chatham, "Population: A Challenge to California's Changing Citrus Industry," *Economic Geography* 34, no. 3 (July 1958): 273; "Costs Hit New High in Citrus Industry," *Los Angeles Times,* July 2, 1956, 36.

17. Lloyd N. Brown, "Westside Dust Test Plots," *California Agriculture,* October 1951, 8; Lillard, *Eden in Jeopardy,* 78–79; Jones & Stokes Associates, California Office of Land Conservation, and California Department of Conservation, *The Impacts of Farmland Conversion in California* (Sacramento: California Department of Conservation, Office of Land Conservation, January 24, 1991), 3-7; Griffin and Chatham, "Population," 273; Pierre Laszlo, *Citrus: A History* (Chicago: University of Chicago Press, 2007), 51–52; Howard A. Miller, "Farm and City Relations," *California Citrograph,* April 1948, 242; John T. Middleton, "Air Pollution Effect on Citrus," *California Citrograph,* July 1955, 330, 352–353; Harry W. Lawton and Lewis G. Weathers, "The Origins of Citrus Research in California," in *Citrus Industry: Crop Protection, Postharvest Technology and Early History of Citrus Research in California,* ed. Walter Reuther (Berkeley: University of California, Division of Agricultural Sciences, 1989), 323; C. N. Roistacher, "Diagnosis and Management of Virus and Virus Like Diseases of Citrus," in *Diseases of Fruits and Vegetables: Diagnosis and Management,* ed. S. A. M. H. Naqvi (Norwell, Mass.: Kluwer, 2004), 113. ("That morning"): "marlcal," comment on "Standpipe in

Orange Grove in 1940's," http://www.flickr.com/photos/30399736@N03/2996050935/in/set-72157617659558203/ (accessed 6/8/15).

18. See "Land, Say the Homebuilders, Is Our Most Critical Problem," 115; "Dead Orange Trees Held Fire Hazard," *Los Angeles Times,* April 30, 1961, OC6; Jones & Stokes Associates, California Office of Land Conservation, and California Department of Conservation, *Impacts of Farmland Conversion in California,* 4-4.

19. Muhonen, "Ebb and Flow," OC1; Marvin P. Miller and Wallace Sullivan, "What's the Cost to Change an Orchard?" *California Citrograph,* February 1955, 148. Garnett's much-discussed images appeared in multiple media, including *Life* magazine and Peter Blake, *God's Own Junkyard: The Planned Deterioration of America's Landscape* (New York: Holt, Rinehart and Winston, 1964).

20. Miller and Sullivan, "What's the Cost to Change an Orchard?" 147; Eric C. Orlemann, *LeTourneau Earthmovers* (Saint Paul, Minn.: Motorbooks International, 2001), 54–56. On treedozers, see Herbert L. Nichols, *Moving the Earth: The Workbook of Excavation,* 1st ed. (Greenwich, Conn.: North Castle Books, 1955), 15-53.

21. Weldon Field, "An Interview with Weldon Field," interview by Anne Riley, 1978, Oral History Series Program, Fullerton College.

22. Ibid.; "Builders and Developers Race Against Fire Ban," *Los Angeles Times,* December 28, 1967, OC1.

23. "Builders and Developers Race Against Fire Ban," OC1; Bruce Poynter, interview by Darrell Pederson, DVD, n.d., Bruce Poynter Collection (AFC/2001/001/77603), VHP.

24. See "Holden Homes, San Mateo California," *NAHB Correlator,* May 1951, 110; Lillard, *Eden in Jeopardy,* 83; Barbara M. Kelly, "The Houses of Levittown in the Context of Postwar American Culture," in *Preserving the Recent Past,* ed. Deborah Slaton and Rebecca A. Shiffer (Washington, D.C.: Historic Preservation Education Foundation, 1995), II-147–II-156, available at http://www.nps.gov/history/nr/publications/bulletins/suburbs/Kelly.pdf; Jackson, *Crabgrass Frontier,* 236.

25. "The Economics of Trees," *House and Home,* April 1953, 131; William Garnett, quoted in "Trenching Lakewood, California," J. Paul Getty Museum Education Department, http://www.getty.edu/education/teachers/classroom_resources/curricula/historical_witness/downloads/garnett_lakewood.pdf (accessed 4/14/15).

26. Christopher Sellers, "Suburban Nature and Environmentalism in Levittown," in *Second Suburb: Levittown, Pennsylvania,* ed. Dianne Harris (Pittsburgh: University of Pittsburgh Press, 2010), 294; M. H. Kimball, "Are

Your Dream Trees Dying?" *Los Angeles Times,* October 10, 1948, H17; Jones & Stokes Associates, California Office of Land Conservation, and California Department of Conservation, *Impacts of Farmland Conversion in California,* 4-2.

27. The highest grade soils included Classes I and II, which required the least conservation and had the fewest limits on their suitability to plants. See Dewey Linze, "Citrus Industry Juicier After Housing Squeeze," *Los Angeles Times,* January 22, 1962, C10; Nicolai V. Kuminoff, Alvin David Sokolow, and Daniel A. Sumner, *Farmland Conversion: Perceptions and Realities,* AIC Issues Brief ([Sacramento]: University of California Agricultural Issues Center, May 2001), 3; Degnan, "Santa Clara," 243; Sokolow and California Department of Conservation, *Williamson Act;* Jones & Stokes Associates, California Office of Land Conservation, and California Department of Conservation, *Impacts of Farmland Conversion in California,* 3-15–3-19; Ed Ainsworth, "Famous Oranges Still Persisting," *Los Angeles Times,* January 26, 1962, A9; "Farms on Hills of California Prove Success: Terraces Are Often More Than Mile in Length—Trees Save Slides," *China Press* (Shanghai), March 7, 1931, 10.

28. See "Don't Oversimplify Land Encroachment," *California Farmer,* November 26, 1955, 2; "Major Orange Packing House Closes Doors," *Los Angeles Times,* May 24, 1965, OC8.

29. "Patrick," comment on Orange County Archives, http://www.flickr.com/photos/ocarchives/2851366182/ (accessed 6/8/15); "Boy, 3½, Hurt Critically in Scoop by Bulldozer," *Los Angeles Times,* March 1, 1950, A1.

30. Roy Hitchcock, "Put Houses, Not Farms, on Hilltops," *California Farmer,* March 7, 1953, 259; Laura R. Barraclough, *Making the San Fernando Valley: Rural Landscapes, Urban Development, and White Privilege* (Athens: University of Georgia Press, 2011).

31. Kelly Brown and Stanford Environmental Law Society, Williamson Act Study Group, *The Property Tax and Preservation of Open Space Land in California: A Study of the Williamson Act* (Stanford, Calif.: Stanford Environmental Law Society, 1974), 1–9; Sokolow and California Department of Conservation, *Williamson Act,* 10, 12, 21, 31, 40; Marlow Vesterby and Ralph E. Heimlich, "Land Use and Demographic Change: Results from Fast-Growth Counties," *Land Economics* 67, no. 3 (1991): 279; Rebecca Conard, "Green Gold: 1950s Greenbelt Planning in Santa Clara County, California," *Environmental Review* 9, no. 1 (April 1, 1985): 13–16.

32. Brown and Stanford Environmental Law Society, Williamson Act Study Group, *Property Tax and Preservation of Open Space Land in California,* 9–37.

33. Ibid.

34. Hitchcock, "Put Houses, Not Farms, on Hilltops," 259; "Cut and Fill Variations," *Progressive Architecture,* April 1967, 154; National Association of

Home Builders (hereafter NAHB), *Home Builders Manual for Land Development,* rev. ed. (Washington, D.C.: NAHB, 1953), 225; "Look How You Can Cut Land Costs If You Know How to Profit from Today's Earthmoving Machinery," *House and Home,* August 1960, 160; "Look How You Can Now Build on Land That Others Wouldn't Buy," *House and Home,* August 1960, 151; NAHB, *Land Development Manual* (Washington, D.C.: NAHB, 1969), 92.

35. Anaheim Hills, Inc., advertisement, "Trees Bring New Life to the Land of Anaheim Hills," *Los Angeles Times,* January 28, 1973, K13.

36. See Hitchcock, "Put Houses, Not Farms, on Hilltops," 259; Reyner Banham, *Los Angeles: The Architecture of Four Ecologies* (New York: Harper and Row, 1971), 88–90; Horst J. Schor, *Landforming: An Environmental Approach to Hillside Development, Mine Reclamation and Watershed Restoration* (Hoboken, N.J.: Wiley, 2007), 5–6.

37. See Jack Boettner, "Council's Action Expected to Speed Canyon Development," *Los Angeles Times,* September 13, 1972, OC-A1; "Plans Unveiled for 'Hills' Development," *Los Angeles Times,* February 23, 1971, D1; W. B. Rood, "Controversy Perches on the Hillsides," *Los Angeles Times,* April 23, 1972, OC1.

38. Anaheim Hills, Inc., advertisement, "Trees Bring New Life to the Land of Anaheim Hills," K13; Hardy quoted in "The Earth: Discussing the Basic Issues," *Progressive Architecture,* April 1967, 178.

39. On counterarguments to the dangers of sprawl, see Sam Staley and Reason Public Policy Institute, *The Sprawling of America: In Defense of the Dynamic City,* Policy Study 251 ([Los Angeles]: Reason Public Policy Institute, 1999); Robert Bruegmann, *Sprawl: A Compact History* (Chicago: University of Chicago Press, 2005).

40. Crook Company advertisement, "'Our Le Tourneau—Westinghouse C Fullpaks Are Great' Says W. E. McKnight," *Southwest Builder and Contractor,* December 12, 1958, cover; "The Earthmover: Man and Machines," *Progressive Architecture,* April 1967, 129–130; Eric C. Orlemann, *R. G. LeTourneau Heavy Equipment: The Mechanical Drive Era, 1921–1953* (Hudson, Wisc.: Iconografix, 2008), 76–77; "Look How You Can Cut Land Costs If You Know How to Profit from Today's Earthmoving Machinery," 163.

41. Herbert L. Nichols, *Moving the Earth: The Workbook of Excavation,* 3rd ed. (New York: McGraw-Hill, 1976), 15-1, 15-2, 15-62, 16-1.

42. See George W. Eger, "Material: The Common Denominator," September 16, 1968, Box 771, Folder 9217, IHC; "The Earthmover: Man and Machines," 128, 134; "Look How You Can Cut Land Costs If You Know How to Profit from Today's Earthmoving Machinery," 162.

43. Caterpillar Tractor Company, "Let's Talk Bulldozers . . . Built by Caterpillar" (Peoria: Caterpillar Tractor Company, 1951), trade catalogs, Na-

tional Museum of American History Library; Caterpillar Tractor Company, "Bulldozers, Straight and Angling Blade: Comparative Specifications" (Peoria: Caterpillar Tractor Company, 1952), trade catalogs, National Museum of American History Library.

44. See Eric C. Orlemann, *Caterpillar Chronicle: The History of the World's Greatest Earthmovers* (Saint Paul, Minn.: Motorbooks International, 2000), 67; Nichols, *Moving the Earth,* 3rd ed., 15-1.

45. See "Look How You Can Cut Land Costs If You Know How to Profit from Today's Earthmoving Machinery," 161; Richard Campbell, "The Caterpillar 666," *Contractor,* June 2007; Orlemann, *Caterpillar Chronicle,* 67–70; "The McCoy Outfit Tames the Hills," *Earth,* December 1973–January 1974, 32–33, Box 1365, Folder 16140, IHC.

46. Richard A. Miller, "Bulldozer Architecture," *Architectural Forum,* August 1960, 88–90; ("with greater comfort"): E. A. Braker to John W. Vance, February 4, 1965, Box 486, Folder 12797, IHC; "Land, Say the Homebuilders, Is Our Most Critical Problem," 106.

47. See "Cut and Fill Variations," 154; Braker, "New Developments in Construction Power and Equipment," ca. 1959, 2, Box 249, Folder 12785, IHC; Johnson quoted in "The Earth: Discussing the Basic Issues," 176.

48. Hubert Kelley, "Earth Movers Shape the Future," *Nation's Business,* October 1954, 50; Ruth Jaffe, "Following the Trail of the 'Cat,'" *Landscape Architecture,* Winter 1959, 90.

49. "Notes and Comment," *Southwest Builder and Contractor,* July 3, 1942, 15; Miller, "Bulldozer Architecture," 90.

50. Kelley, "Earth Movers Shape the Future," 29; "Technology—1957," *Architectural Forum,* January 1957, 93; "A New Approach to Landshaping," *Architectural Forum,* January 1957, 97; "Cut and Fill Variations," 154.

51. Frank Rowsome Jr., "'Bulldozer U' Students Push Campus Around," *Popular Science,* August 1956, 72–76, 220, 222.

52. "School for Earthmoving," *Business Week,* December 24, 1955, 96–98; "A Cut Up Campus," *Life,* April 22, 1957, 80; "Training Men for a Better Life," *Ebony,* December 1973, 94–100.

53. U.S. Bureau of the Census, *Census of Construction Industries, 1972,* vol. 1 (Washington, D.C.: U.S. GPO, 1976), sect. 23.

54. "McCoy Outfit Tames the Hills," 30–33.

55. Urban Land Institute and NAHB, *Home Builders Manual for Land Development* (Washington, D.C.: NAHB, 1950), 5–6; NAHB, *Home Builders Manual for Land Development* [1953], 224–226; NAHB, *Land Development Manual* [1969], 92, 239–259.

56. NAHB, *Home Builders Manual for Land Development* [1953], 52; NAHB, *Land Development Manual* [1969], 110–117, 246–247.

57. See Nichols, *Moving the Earth,* 1st ed., 15-46; Nichols, *Moving the Earth,* 3rd ed., 15-60; "Look How You Can Cut Land Costs If You Know How to Profit from Today's Earthmoving Machinery," 163–164; "McCoy Outfit Tames the Hills," 31–32.

58. See NAHB, *Home Builders Manual for Land Development* [1953], 225; Schor, *Landforming,* 179–185.

59. Horst Schor, "Landform Grading: Building Nature's Slopes," *Pacific Coast Builder,* June 1980, 80–83; John O'Dell, "Grading on the Curve: Developer Goes for Natural Look in Sculpting Hills for Talega Project," *Los Angeles Times,* July 12, 1991, OC-D6.

60. See Schor, *Landforming,* 150; NAHB, *Land Development Manual* [1969], 248–250.

61. J. William Thompson and Kim Sorvig, *Sustainable Landscape Construction: A Guide to Green Building Outdoors* (Washington, D.C.: Island Press, 2007), 45; NAHB, *Land Development Manual* [1969], 254–259; Jaffe, "Following the Trail of the 'Cat,'" 90.

62. See Tracy H. Abell, "'View Lots' Replace Natural Mountain Landscapes," *Landscape Architecture,* Winter 1959–1960, 94–95; John Gregory, "Anaheim Hills: 'Trial Balloon' Draws Heavy Fire," *Los Angeles Times,* September 30, 1973, OC-A1; Scott Moore, "School District Moves to Halt Housing Project," *Los Angeles Times,* October 20, 1973, OC1; Robert M. Gettemy, "2 Groups Sue to Halt Development in Anaheim Hills," *Los Angeles Times,* December 1, 1973, OC1.

63. Gregory, "Anaheim Hills," OC-A1; "The Menace in Santa Ana Canyon," *Los Angeles Times,* October 3, 1973, OC-A8.

64. See Jack Boettner, "Plans Submitted for Partial Restoring of Robbers' Roost," *Los Angeles Times,* April 28, 1973, 10A; Thomas Fortune, "Robbers' Roost to Be Restored, but Won't Look the Same," *Los Angeles Times,* December 18, 1973, OC1; Dorothy L. Pillsbury, "Machine Churns Arizona Desert, but an Archaeologist Walks Ahead," *Christian Science Monitor,* October 27, 1950, 13; Gene Bylinsky, "Probing the Past," *Wall Street Journal,* April 9, 1959, 1; Bynum Shaw, "Roadbuilders and Archaeologists Following the Bulldozers," *The Sun* (Baltimore), July 28, 1962, 10; Charles Hillinger, "Archaeologists Beat the Freeway Bulldozers," *Los Angeles Times,* May 2, 1966, A1; "State Archeologist Curbing Bulldozers," *New York Times,* April 28, 1974, 91; Lou Desser, "Concern for Ecology Shapes New Projects," *Los Angeles Times,* October 24, 1971, 12; "Scientists to Precede Bulldozers," *Los Angeles Times,* July 18, 1971, O26.

65. Jaffe, "Following the Trail of the 'Cat,'" 91; Rood, "Controversy Perches on the Hillsides," OC1; "The Earth: Discussing the Basic Issues," 179. See William Bronson, *How to Kill a Golden State* (Garden City, N.Y.:

Doubleday, 1968); Blake, *God's Own Junkyard;* Banham, *Los Angeles,* 95–109.

66. See William H. Whyte, Jr., "A Plan to Save Vanishing U.S. Countryside," *Life,* August 17, 1959; Robert E. Cubbedge, *The Destroyers of America* (New York: Macfadden-Bartell, 1964); Margo Tupper, *No Place to Play* (Philadelphia: Chilton Books, 1966); Dasmann, *Destruction of California;* Lillard, *Eden in Jeopardy;* Adam Rome, *The Bulldozer in the Countryside: Suburban Sprawl and the Rise of American Environmentalism* (Cambridge: Cambridge University Press, 2001); Christopher C. Sellers, *Crabgrass Crucible: Suburban Nature and the Rise of Environmentalism in Twentieth-Century America* (Chapel Hill: University of North Carolina Press, 2012).

67. See Stephen J. Pyne, *Fire in America: A Cultural History of Wildland and Rural Fire* (Princeton: Princeton University Press, 1982), 405–409; Mike Davis, *Ecology of Fear: Los Angeles and the Imagination of Disaster* (New York: Metropolitan Books/Henry Holt, 1998), 99–112; Marcida Dodson, "Dispute Involves High-Priced Homes in Anaheim Hills," *Los Angeles Times,* July 8, 1981, OC-A1; Don Smith and Richard O'Reilly, "San Clemente Fire Disaster," *Los Angeles Times,* January 22, 1976, 1; "Firebreaks Built at Community," *Los Angeles Times,* June 25, 1978, O10.

68. "Landslides Wreck Another Two Los Angeles Homes," *House and Home,* October 1959, 55; "$4 Million Judgment for Slides," *Los Angeles Times,* March 6, 1963, 1; Lillard, *Eden in Jeopardy,* 111–112, 114; Leslie Berkman, "Landslides: Strict Laws, Watchdogs, Advice Could Help," *Los Angeles Times,* December 3, 1978, OC-A1; Anne Whiston Spirn, *The Granite Garden: Urban Nature and Human Design* (New York: Basic, 1984), 116–117; Rome, *Bulldozer in the Countryside,* 166–173; Art Seidenbaum, "The Decline and Fall of the Los Angeles Hills," *Los Angeles Times,* August 11, 1968, A26.

69. See Jack Boettner, "Anaheim Council Delays Action on Controversial Grading Law," *Los Angeles Times,* September 3, 1974, C8; Berkman, "Landslides," OC-A1; Davis, *Ecology of Fear,* 79.

70. See Christine M. Rodrigue, "Impact of Internet Media in Risk Debates: The Controversies over the Cassini-Huygens Mission and the Anaheim Hills, California, Landslide," *Australian Journal of Emergency Management* 16, no. 1 (Autumn 2001): 58.

71. See Smith and O'Reilly, "San Clemente Fire Disaster," 1; "Firebreaks Built at Community," O10.

Chapter 4. "Armies of Bulldozers Smashing Down Acres of Slums"

1. Frances Glennon to Richard C. Lee, February 18, 1958, Box 6, Folder 406, RCL; "City Clean-up Champion," *Life,* February 17, 1958, 87–88, 90.

2. Outtakes from Box 154, RCL.

3. Peter Myers, "[Daily Demolition Progress Report]," January 21, 1957, Box 120, Folder: 8-1 Dem. Contr. #1 Corresp., NHRA Records.

4. ("For propaganda purposes"): Michael DePalma to Frank O'Brion, July 1, 1960, Box 637, Folder: R-1, Dem., Gen. Corresp., NHRA Records; Jeanne R. Lowe, "Lee of New Haven and His Political Jackpot," *Harper's,* October 1957, 41–42; Joe Alex Morris, "He Is Saving a 'Dead' City," *Saturday Evening Post,* April 19, 1958, 115; ("Some mayors"): "City Clean-up Champion," 88. Lee also welcomed local families to visit demolition sites for Sunday afternoon entertainment: Jeanne R. Lowe, *Cities in a Race with Time: Progress and Poverty in America's Renewing Cities* (New York: Random House, 1967), 426.

5. Richard C. Lee to Frances Glennon, February 12, 1958, Box 16, Folder 406, RCL; Richard C. Lee to Pete Villano, February 13, 1958, Box 6, Folder 406, RCL; L. S. Rowe, "Letter to the Editors: City Clean-Up Champion," *Life,* March 10, 1958, 14; Pete Villano to Richard C. Lee, "Re: Life," February 14, 1958, Box 6, Folder 406, RCL.

6. The Housing Act of 1949 focused on "redevelopment," or slum clearance and reconstruction, versus "urban renewal" (officially launched with the Housing Act of 1954), with its incorporation of rehabilitation alongside redevelopment. In this book I follow the convention of many historians and journalists and use the blanket term "urban renewal" for both. See William L. Slayton, "Operation and Achievements of the Urban Renewal Program," in *Urban Renewal: The Record and the Controversy,* ed. James Q. Wilson (Cambridge: MIT Press, 1966), 203–208. On implosions, see Keller Easterling, *Enduring Innocence: Global Architecture and Its Political Masquerades* (Cambridge: MIT Press, 2005), 161–184; Jeff Byles, *Rubble: Unearthing the History of Demolition* (New York: Harmony, 2005); Bernard L. Jim, "Ephemeral Containers: A Cultural and Technological History of Building Demolition, 1893–1993" (Ph.D. diss., Case Western Reserve University, 2006).

7. On urban renewal, see Jon C. Teaford, *The Rough Road to Renaissance: Urban Revitalization in America, 1940–1985* (Baltimore: Johns Hopkins University Press, 1990); Hilary Ballon and Kenneth T. Jackson, eds., *Robert Moses and the Modern City: The Transformation of New York* (New York: Norton, 2007); Samuel Zipp, *Manhattan Projects: The Rise and Fall of Urban Renewal in Cold War New York* (New York: Oxford University Press, 2010); Christopher Klemek, *The Transatlantic Collapse of Urban Renewal: Postwar Urbanism from New York to Berlin* (Chicago: University of Chicago Press, 2011). On Senator Sparkman, see "Senator Sparkman Sets Home Fire Burning as Warmup to Congress' Anti-Slum Action," *Journal of Housing,* February 1959, 43.

8. See Capital Grant Reservations by Cities as of December 31, 1965, from *Urban Renewal Directory,* December 31, 1965, and *U.S. Decennial Census 1960,* as quoted in Allan R. Talbot, *The Mayor's Game: Richard Lee of New Haven and the Politics of Change* (New York: Harper and Row, 1967), 160; Lowe, *Cities in a Race with Time,* 405; Douglas W. Rae, *City: Urbanism and Its End* (New Haven: Yale University Press, 2003), 28; New Haven Redevelopment Agency (hereafter NHRA), "Demolition Report, March 1, 1970," Box 330, Folder: Monthly Demolition Reports, NHRA Records.

9. On the influence of business in other aspects of urban renewal, including promoting the program and influencing the shape of future projects, see G. William Domhoff, *Who Really Rules? New Haven and Community Power Reexamined* (New Brunswick, N.J.: Transaction, 1978); Elihu Rubin, *Insuring the City: The Prudential Center and the Postwar Urban Landscape* (New Haven: Yale University Press, 2012). For before and after photographs, see "Urban Planning and Urban Revolt: A Case Study," *Progressive Architecture,* January 1968, 134–156.

10. Albert M. Cole, "Address of Albert M. Cole, U.S. Housing Administrator, at Fourth National Construction Industry Conference, Hotel Sherman, Chicago, Ill., December 11, 1958, 1:00 PM," December 11, 1958, Box 166, Folder: URA Speeches, NHRA Records.

11. See Mabel L. Walker, *Urban Blight and Slums: Economic and Legal Factors in Their Origin, Reclamation, and Prevention* (Cambridge: Harvard University Press, 1938), 132–144; Lawrence M. Friedman, *Government and Slum Housing: A Century of Frustration* (Chicago: Rand McNally, 1968), 68–72; National Association of Housing and Redevelopment Officials, *Demolition of Substandard Housing: An Outline of Current Principles and Practice* (Chicago: National Association of Housing Officials, 1938).

12. U.S. Bureau of the Census, *Census of Construction Industries, 1972, volume I, Industry and Special Statistics* (Washington, D.C.: U.S. GPO, 1976), 24-4.

13. Gil Wyner Company, Inc., "Statement of Bidder's Qualifications," May 12, 1960, Box 123, Folder: Dem. Contr. 6A & 6B, Bid Documents, NHRA Records.

14. Gibraltar Wrecking & Supply Company, "Bid for Demolition and Site Clearance [Contract #6]," May 12, 1960, Box 124, Folder: Dem. Contr. 6A & 6B, Bid Documents, NHRA Records; Caterpillar Tractor Company advertisement, "He's Renewing a City's Pride," *U.S. News and World Report,* January 1, 1959, 91; Martin E. Avroch to Louis J. Ferony, "Re: Contract No. 5B, Wooster Square Renewal Area," December 16, 1960, Box 122, Folder: R-1 Dem. Contr. #5B Corresp., NHRA Records.

15. Jim, "Ephemeral Containers," 24–30; Saul Greenberg to NHRA,

"Re: Podell's, 229–235 Commerce St, New Haven, Conn.," May 16, 1957, Box 120, Folder: 8-1 Dem. Contr. #1 Corresp., NHRA Records; Commercial Contractors Corporation to NHRA, "Re: Fire Damage—Peat & Voigt Building, Block A Parcel 1 Section B, Crown and Park Streets," November 23, 1957, Box 120, Folder: 8-1 Demo. Contr. #2 Corresp., NHRA Records; Saul Greenberg to NHRA, July 10, 1958, Box 120, Folder: R-2 Dem. Contr. 3 Corresp., NHRA Records; NHRA, "Church Street Redevelopment and Renewal Area, Project U-R Conn. R-2, Bid Opening, August 8, 1961," August 8, 1961, Box 125, Folder: R-2 #8 DeFonce Construction Bids, NHRA Records; Nancy Love, "Wrecking the Joint," *Philadelphia Magazine,* January 1966, 143–144; Herbert T. Duane, telephone interview with author, July 2009.

16. Lawrence Gochberg to Melvin J. Adams, August 28, 1964, Box 127, Folder: R-2 Dem. Contr. #14, Sargent Property, Seymour Berkowitz, Inc., Contractor, Gen. Corresp., NHRA Records.; Joel Cogen to Lawrence Gochberg, September 3, 1964, Box 127, Folder: R-2 Dem. Contr. #14, Sargent Property, Seymour Berkowitz, Inc., Contractor, Gen. Corresp., NHRA Records.

17. Commercial Contractors, Bid Documents, Contracts #1, #2, #3, and #5B, NHRA Records; "Invitation for Bids," n.d., Box 120, Folder: 8-1 Dem. Contr. #1 Corresp., NHRA Records; Armand Aloi to Byrne Stoddard, July 19, 1967, Box 330, Folder: Dem., Gen. Corresp., NHRA Records; Byrne Stoddard to Project Directors, July 25, 1967, Box 330, Folder: Dem., Gen. Corresp., NHRA Records; "An Estimation of Demolition Costs Based on Size and Type of Building," January 10, 1973, Box 330, Folder: Dem., Gen. Corresp., NHRA Records.

18. Commercial Contractors Corporation, "Weekly Payroll: Contract No. 1," January 31, 1957–April 23, 1958, Box 120, Folder: 8-1 Dem. Contr. #1 Corresp., NHRA Records; U.S. Department of Labor, Wage and Hour Division, "History of Federal Minimum Wage Rates Under the Fair Labor Standards Act, 1938–2009," http://www.dol.gov/whd/minwage/chart.htm (accessed 5/20/15); Avroch to Ferony, "Re: Contract No. 5B, Wooster Square Renewal Area"; U.S. Department of Labor, [Schedule of Wages], March 7, 1975, contained in NHRA and Dunn Bros., Inc., "Agreement for Demolition and Site Clearance," May 15, 1975, Box 131, Folder: R-28, R-91, R-106, A-8-1, A-8-2, R-20, Dem. Contr. #34, Executed Contract May 15, 1975, Dunn Bros. Inc., NHRA Records; U.S. Department of Labor, Bureau of Labor Statistics, "CPI Detailed Report: Data for March 2015," http://www.bls.gov/cpi/cpid1503.pdf (accessed 5/20/15), Table 24.

19. "U.S. District Court, District of Connecticut, Gil Wyner Co., Inc., v. NHRA, Civil Action No. 9542, Continued Deposition of Charles Locke, v. IV, pp. 201–341," January 16, 1965, Box 121, NHRA Records. At preconstruction

conferences held in the 1970s, one rare company reported 50 percent minority representation in its skilled labor force; Jim O'Connor to Files, "Re: Preconstruction Conference, Manafort Brothers, Dem. Contr. #31, Held Thursday, June 28, 1973, at 10:00 a.m., in the City Plan Conference Room, 10th Floor, 157 Church St, New Haven, Conn.," June 29, 1973, Box 130, Folder: R-2, R-79, Dem. Contr. #31, Gen. Corresp., Manafort Bros., NHRA Records. See also Manafort Brothers and NHRA, "Agreement," January 17, 1977, Box 131, Folder: R-2, A-8-2, R-28, A-8-1, R-91, Airport, Dem. Contr. #37, Executed Contract, Manafort Bros., Inc., January 17, 1977, NHRA Records; Brian Frederick to Henry Fisher, "Re: Preconstruction Conference with Stamford House Wrecking Company, Inc. Concerning Demolition of Fire Training Station; Dem. Contr. #28," May 30, 1972, Box 130, Folder: LW, Dem. Contr. #28, Contract and Gen. Corresp., Stamford House Wrecking Co., Contractor (Fire Training Station), NHRA Records. On African American construction workers, urban renewal, and affirmative action, see David A. Goldberg and Trevor Griffey, eds., *Black Power at Work: Community Control, Affirmative Action, and the Construction Industry* (Ithaca: ILR Press/Cornell University Press, 2010); Jennifer Hock, "Political Designs: Architecture, Activism, and Urban Renewal in the Civil Rights Era" (Ph.D. diss., Harvard University, 2012).

20. See Lowe, *Cities in a Race with Time,* 405–551.

21. Peter Myers to Tom Appleby, "Re: Demolition Contracts Commercial Contractors," September 17, 1959, Box 120, Folder: 8-1 Dem. Contr. #1 Corresp., NHRA Records; Peter Myers to Tom Appleby, "Re: Demolition 8-1, Demolition Cost Estimates and Proposals Tabulated," 1958, Box 120, Folder: 8-1 Dem. Contr. #1 Corresp., NHRA Records.

22. Saul Greenberg to H. Ralph Taylor, March 26, 1957, Box 120, Folder: 8-1 Dem. Contr. #1 Corresp., NHRA Records; H. Ralph Taylor to Charles J. Horan, "Re: Proposed Change Order to Contract for Demolition and Site Clearance—UR Conn 8-1 and Conn R-2," August 14, 1958, Box 120, Folder: R-2 Dem. Contr. 3 Corresp., NHRA Records; W. Adam Johnson to L. Thomas Appleby, May 23, 1958, Box 330, Folder: Dem., Gen. Corresp., NHRA Records. Although the city had originally considered using fire as a demolition method, officials decided against it because of the anticipated increase in time, expense, and logistical challenges; George Mulligan to H. Ralph Taylor, "Re: Demolition," October 24, 1956, Box 120, Folder: 8-1 Dem. Contr. #1 Corresp., NHRA Records.

23. Suellen Wood to Margot Mitchell, "Re: Audit Findings," January 23, 1970, Box 330, Folder: Demolition, Gen. Corresp., NHRA Records; "Hamden Sets Six-Month Deadline for Ending City Renewal Dumping," May 7, 1967, Box 330, Folder: Removal of Demolition Debris, Contr. and Corresp., NHRA

Records; Margaret Leavy to Melvin J. Adams, "Re: Claims Made by Commercial Contractors, Inc.," June 28, 1962, Box 123, Folder: R-1 Dem. Contr. #6B Gen. Corresp. 1962, Commercial Contractors #2, NHRA Records.

24. In 1962, for example, an excavator preparing for new construction on a former Commercial site encountered 1,300 cubic yards of wood and other materials subject to rot: Melvin J. Adams to Saul Greenberg, November 16, 1962, Box 123, Folder: R-1 Dem. Contr. #6B Gen. Corresp. 1962, Commercial Contractors #2, NHRA Records. On disposal of debris, see also Ralph Brown to John Spain, "Re: Revised Demolition Contracts," January 11, 1965, Box 330, Folder: Dem. Contr. Corresp., NHRA Records; Joel Cogen to Mike Catania, March 5, 1968, Box 330, Folder: Removal of Demolition Debris, Contr. and Corresp., NHRA Records; ("digging up"): Joel Cogen to Lou Ferony, "Re: Commercial Contractor," October 6, 1960, Box 123, Folder: R-1 Dem. Contr. #6B Gen. Corresp. 1960–1961, Commercial Contractors #1, NHRA Records; ("left virtually"): L. Thomas Appleby to Richard C. Lee, September 4, 1962, Box 123, Folder: R-1 Dem. Contr. #6B Gen. Corresp. 1962, Commercial Contractors #2, NHRA Records.

25. Armand Aloi to Chuck Shannon, August 1, 1967, Box 128, Folder: R-1, R-2, R-20, R-28, Dem. Contr. #18, Gen. Corresp., Barnum Lumber Co., Inc., Contractor, NHRA Records. See also NHRA, "Summary of Problems with Barnum Lumber as of August 1968," August 1968, 2–3, Box 129, Folder: R-1, R-2, R-20, R-28, R-71, R-91, R-96, R-106, Dem. Contr. #19, Gen. Corresp., Barnum Lumber Co., Contractor, NHRA Records.

26. ("One sweep"): "U.S. District Court, District of Connecticut, Gil Wyner Co., Inc., v. NHRA, Civil Action No. 9542, Continued Deposition of Charles Locke, v. IV, pp. 201–341"; "U.S. District Court, District of Connecticut, Gil Wyner Co., Inc., v. NHRA, Civil Action No. 9542, Continued Deposition of Charles Locke, v. II, pp. 55–123," December 1, 1964, Box 121, NHRA Records; William R. Murphy to NHRA, "Re: NHRA, Wyner vs.," September 7, 1965, Box 121, Folder: R-2 Gil Wyner Dem. Contr. #4 Litigation, NHRA Records; "Resolution of the NHRA Authorizing Settlement of Gil Wyner Co., Inc. Litigation," n.d., Box 121, Folder: R-2 Gil Wyner Dem. Contr. #4 Litigation, NHRA Records; Gil Wyner Company, Inc., "Statement of Bidder's Qualifications," May 12, 1960, Box 123, Folder: Dem. Contr. 6A & 6B, Bid Documents, NHRA Records.

27. Joel Cogen to NHRA, "Re: Litigation; Wyner vs. Redevelopment Agency," January 25, 1966, Box 121, Folder: R-2 Gil Wyner Dem. Contr. #4 Litigation, NHRA Records; Bob Dolezal to Chuck Shannon, "Re: Manual Change Recommendations—First Submission," January 9, 1963, Box 330, Folder: Dem. Contr. Corresp., NHRA Records; Armand Aloi to Chuck Shannon, June 26, 1968, Box 128, Folder: R-1, R-2, R-20, R-28, Dem. Contr.

#18, Gen. Corresp., Barnum Lumber Co., Inc., Contractor, NHRA Records; Charles I. Shannon to Charles J. Horan, August 4, 1965, Box 126, Folder: R-1 Dem., Dem. Contr. #11A, Gen. Corresp., Dunn Bros.—Contractor, NHRA Records.

28. ("The cost of"): Ernest Laden to Mike Catania, June 14, 1965, Box 126, Folder: R-20 & R-1 Dem. Contr. #9 Corresp., NHRA Records; ("the time for"): Murray Goldblum to Armand Aloi, November 13, 1967, Box 128, Folder: R-2, R-20, R-71, R-7, Dem. Contr. #15, Gen. Corresp., The Stamford House Wrecking Co., Contractor, NHRA Records. See Charles I. Shannon to Charles J. Horan, "Re: Request for Concurrence in Change Orders, Dem. Contr. #16 (Conn. R-1, Conn. R-71 and Conn. R-79) to Dem. Contr. #17 (All Projects)," April 10, 1968, Box 128, Folder: R-1, R-71, R-79, R-91, Dem. Contr. #16, Gen. Corresp., Industrial Wrecking Co., Contractor, NHRA Records; Ralph I. Brown to Chuck Shannon, May 6, 1966, Box 330, Folder: Dem. Contr. Corresp., NHRA Records. On the widespread nature of these challenges on projects across the country, see United States, Real Estate Research Corporation and RTKL Associates, *Evaluating Local Urban Renewal Projects: A Simplified Manual* (Washington, D.C.: Department of Housing and Urban Development, Office of Community Planning and Development, Office of Evaluation, 1975), 55; United States, Real Estate Research Corporation and RTKL Associates, *Guidelines for Urban Renewal Land Disposition* (Washington, D.C.: Department of Housing and Urban Development, Office of Community Planning and Development, Office of Evaluation, 1975), 13.

29. According to one postwar general contractor, some low bidders on public contracts expected to make their profits through change orders, by submitting price increases for individual buildings after their overall low bid had been accepted; perhaps Commercial participated in this practice: Samuel C. Florman, *Good Guys, Wise Guys, and Putting Up Buildings: A Life in Construction* (New York: Thomas Dunne, 2012), 131–132. Joel Cogen to Chuck Shannon, August 13, 1965, Box 126, Folder: R-1 Dem., Dem. Contr. #11A, Gen. Corresp., Dunn Bros.—Contractor, NHRA Records; Appleby to Lee, September 4, 1962. On Cleveland Wrecking Company, see Love, "Wrecking the Joint," 141–145. Commercial's profitability is calculated based on weekly payroll accounts, contractor invoices, contemporary salvage estimates, and equipment rental rates found in NHRA Records.

30. Commercial Contractors Corporation, "Weekly Payroll: Contract No. 5B, 6A, 6B," August 31, 1960, Box 121, Folder: R-1 Payroll Affidavits, Dem. Contr. #5, NHRA Records; Thomas F. Lydon, Jr., "Fire Investigation of Four-Alarm Fire August 29, 1960 at 513 Chapel Street," September 1960, Box 122, Folder: R-1 Dem. Contr. #5B Corresp., NHRA Records.

31. Lydon, "Fire Investigation of Four-Alarm Fire August 29, 1960."

32. Ibid.; "4-Alarm Fire Hospitalizes 12 Firemen Here," *New Haven Journal-Courier,* August 30, 1960, 5; "Heroic Fire-Fighting," *New Haven Journal-Courier,* September 1, 1960, 10; "Human Reaction at Fire Scene Intensifies Dramatic Moments," *New Haven Journal-Courier,* August 30, 1960, 5.

33. "Negligence in Blaze Denied by Wreckers," *New Haven Journal-Courier,* August 30, 1960, 3; Scoppetta quoted in "Wooster Area Fires Speed Up Relocation," *New Haven Journal-Courier,* September 1, 1960, 2.

34. Richard C. Lee, "Remarks of Richard C. Lee, Mayor, City of New Haven; Occasion: Luncheon—Oak Street Demolition Ceremony; Place: Towne House; Time: 12:00 Noon, January 30, 1957," January 30, 1957, Box 120, Folder: Dem. #2 Contr. Corresp., NHRA Records; Richard C. Lee, "News Release: Redevelopment—End to Fire Hazards," October 8, 1959, Box 9, Album "1959 Mayoralty Campaign," Lee Accession 2001-M-010, Manuscripts and Archives, Yale University Library; Anthony V. Riccio, *The Italian American Experience in New Haven: Images and Oral Histories* (Albany: State University of New York Press, 2006), 320–322.

35. Thomas J. Collins to Mayor Lee, n.d., Box 120, Folder: Dem. #2 Contr. Corresp., NHRA Records. The article was "Destruction Hurtles Down," *Daily News* (New York), March 15, 1957, 1, 3, centerfold.

36. Lydon, "Fire Investigation of Four-Alarm Fire August 29, 1960"; "Chapel Street Fire Wreaks Huge Loss," *New Haven Journal-Courier,* August 13, 1960, 9.

37. Melvin J. Adams to Files, "Re: Demolition Contract No. 5B," September 2, 1960, Box 122, Folder: R-1 Dem. Contr. #5B Corresp., NHRA Records; "Demolition Even on Sunday," *New Haven Journal-Courier,* September 5, 1960, 1; "A Silent Sunday," *New Haven Journal-Courier,* September 5, 1960, 12; NHRA to Charles J. Horan, March 9, 1961, Box 122, Folder: R-1 Dem. Contr. #5B Corresp., NHRA Records.

38. "Small Chestnut St. Fire Spurs New Criticism," *New Haven Journal-Courier,* August 31, 1960; Harold Grabino to Charles J. Horan, "Re: Wooster Square Project, Conn. R-1, Demolition Contract 5B, Claim for Extra Payment and Deletion of Parcels," March 28, 1961, Box 122, Folder: R-1 Dem. Contr. #5B Corresp., NHRA Records; Martin E. Avroch to Louis J. Ferony, "Re: Contract No. 5B, Wooster Square Renewal Area—Superseded," December 16, 1960, Box 122, Folder: R-1 Dem. Contr. #5B Corresp., NHRA Records; L. Thomas Appleby to Richard C. Lee, "Re: Demolition of Vacant Structures for the July 4 Weekend," June 30, 1960, Box 122, Folder: R-1 Dem. Contr. #5B Corresp., NHRA Records; Armand Aloi to Francis J. Sweeney, June 29, 1966, Box 330, Folder: Vacant Buildings—Unsafe—Abandoned (to be Demolished) thru 1975, Gen. Corresp. and Reports, NHRA Records; Eugene N. Sosnoff to Donald W. Celotto, "Re: Industrial Wrecking Company," March 27, 1969,

Box 128, Folder: R-1, R-2, R-20, R-28, R-71 & R-91, Dem. Contr. #17, Gen. Corresp., Industrial Wrecking Co.—Contractor, NHRA Records; Charles I. Shannon to Stanley Goldstein, "Re: Reply to Audit Finding, Consolidated Audit #3," March 11, 1969, Box 128, Folder: R-1, R-71, R-79, R91, Dem. Contr. #16, Gen. Corresp., Industrial Wrecking Co., Contractor, NHRA Records.

39. Board of Aldermen of the City of New Haven, "Resolution of Redevelopment Agency Requiring All Demolition Contractors to Abide by City Ordinances," October 6, 1960, Box 330, Folder: Dem., Gen. Corresp., NHRA Records.

40. Alvin A. Mermin, "New Haven's Experience in Family Relocation (Oak Street)," n.d., 4, Box 582, Folder: 8-1 Residential Relocation Reports, NHRA Records; Alvin A. Mermin to Richard C. Lee, "Re: Status of Family Relocation, as Requested," July 19, 1957, Box 582, Folder: 8-1 Residential Relocation Reports, NHRA Records; Charles Eliot to Tom Appleby, "Re: Family Relocation," March 11, 1958, Box 582, Folder: 8-1 Residential Relocation Reports, NHRA Records; Rae, *City,* 338–340.

41. Through 1961, nonwhites made up 66 percent of families living in and relocated from areas slated for urban renewal (among those for whom "color" was reported); United States, Housing and Home Finance Agency, *Relocation from Urban Renewal Project Areas: Through December, 1961* (Washington, D.C.: U.S. GPO, 1962), 6. Through June 30, 1970, 40 percent of families relocated for urban renewal and neighborhood development projects were white (among those who reported their race); U.S. Department of Housing and Urban Development, *1970 HUD Statistical Yearbook* (Washington, D.C.: U.S. GPO, 1971), 73; Rae, *City,* 340–341; Mermin, "New Haven's Experience in Family Relocation (Oak Street)," 4. On the experience of San Francisco's homosexual roomer population, see Clayton C. Howard, "The Closet and the Cul de Sac: Sex, Politics, and Suburbanization in Postwar California" (Ph.D. diss., University of Michigan, 2010); Ira Nowinski, *No Vacancy: Urban Renewal and the Elderly* (San Francisco: C. Bean Associates, 1979).

42. Norma Barbieri and Theresa Argento, interview by Sarah Barca, February 24, 2005, transcript, 12, OHDNH; Theresa Argento, interview by Sarah Barca, March 9, 2004, 18, transcript, OHDNH.

43. Argento, interview, 4; Richard Abbatiello, interview by Doug London, July 15, 2004, transcript, 2, 13, OHDNH.

44. Studies of the psychosocial effects of urban renewal relocation in other cities substantiate the New Haven results: see Herbert J. Gans, *The Urban Villagers: Group and Class in the Life of Italian-Americans* (New York: Free Press, 1962); Daniel Thursz, *Where Are They Now? A Study of the Impact of Relocation on Former Residents of Southwest Washington, Who Were Served in an HWC Demonstration Project* (Washington, D.C.: Health and Wel-

fare Council of the National Capital Area, 1966); Marc Fried, "Grieving for a Lost Home," in *The Urban Condition: People and Policy in the Metropolis,* ed. Leonard J. Duhl (New York: Basic Books, 1963), 151–171; Mindy Thompson Fullilove, *Root Shock: How Tearing Up City Neighborhoods Hurts America, and What We Can Do About It* (New York: One World/Ballantine, 2004). Harry DeBenedet, interview by Katherine K. Smith, November 20, 2003, transcript, 11, OHDNH; Barbieri and Argento, interview, 23.

45. Bret Bissell, interview by Emily Johnson, February 6, 2005, transcript, 21–22, OHDNH. The city's first relocation officer resigned in 1956, complaining that racial prejudice made his job impossible: Rae, *City,* 338. See also David M. Freund, *Colored Property: State Policy and White Racial Politics in Suburban America* (Chicago: University of Chicago Press, 2007).

46. NHRA, "Special Conditions for Demolition and Clearance: Contract #2," n.d., Box 120, NHRA Records; Alvin A. Mermin to File, December 1957, Box 582, Folder: 8-1 Residential Relocation Reports, NHRA Records; *Price & Lee's New Haven (New Haven County, Conn.) City Directory, Including West Haven, East Haven, and Woodbridge* (New Haven: Price & Lee Company, n.d.).

47. United States, Real Estate Research Corporation and RTKL Associates, *Evaluating Local Urban Renewal Projects,* 70; "Wooster Area Fires Speed Up Relocation," 1; *Price & Lee's New Haven;* Alvin A. Mermin to H. Ralph Taylor, "Re: Status of Relocation from State Properties in Oak St. Area as of 1/31/57," February 1, 1957, Box 582, Folder: 8-1 Residential Relocation Status Rep., NHRA Records; ("extensive smoke"): Peter S. Myers to Saul Greenberg, "Re: Demolition Contract #3, Notice to Proceed #25," April 2, 1959, Box 120, Folder: R-2 Dem. Contr. #3 Proceed Orders 8-1 & R-2, NHRA Records.

48. David Berman to Industrial Wrecking Co., Inc., October 31, 1968, Box 128, Folder: R-1, R-2, R-20, R-28, R-71 & R-91, Dem. Contr. #17, Gen. Corresp., Industrial Wrecking Co.—Contractor, NHRA Records; Joan G. DeCarlo to William T. Donohue, March 17, 1978, Box 130, Folder: R-20, R-28, R-96, Dem. Contr. #33, Gen. Corresp., NHRA Records.

49. Mark R. Farfel et al., "A Study of Urban Housing Demolition as a Source of Lead in Ambient Dust on Sidewalks, Streets, and Alleys," *Environmental Research* 99, no. 2 (October 2005): 204–213; Mark R. Farfel et al., "A Study of Urban Housing Demolitions as Sources of Lead in Ambient Dust: Demolition Practices and Exterior Dust Fall," *Environmental Health Perspectives* 111, no. 9 (July 2003): 1228–1234; J. Bowie, "Community Experiences and Perceptions Related to Demolition and Gut Rehabilitation of Houses for Urban Redevelopment," *Journal of Urban Health: Bulletin of the New York Academy of Medicine* 82, no. 4 (October 2005): 532–542.

50. L. Thomas Appleby to Commercial Contractors Corporation, March 9, 1960, Box 122, Folder: R-1 Dem. Contr. #5B Corresp., NHRA Records; Louis J. Ferony to Commercial Contractors Corporation, "Re: Wooster Square Project Demolition Contract #5B," February 2, 1960, Box 122, Folder: R-1 Dem. Contr. #5B Corresp., NHRA Records; Armand Aloi to Chuck Shannon, "Re: Industrial Wrecking—Boarding up of 152–164 Brewery Street," March 7, 1966, Box 128, Folder: R-1, R-71, R-79, R-91, Dem. Contr. #16, Gen. Corresp., Industrial Wrecking Co., Contractor, NHRA Records; Lou Ferony to Tom Appleby, "Re: Demolition," September 1, 1960, Box 122, Folder: R-1 Dem. Contr. #5B Corresp., NHRA Records.

51. Francis V. McManus to Melvin J. Adams, April 30, 1965, Box 409, Folder: R-20, Dem., Gen. Corresp., NHRA Records. In late summer 1964, for example, rioters in Philadelphia threw bricks from demolition projects at firemen attempting to put out a blaze: "Rioters Hurl Rocks, Insults at Firemen Fighting Blaze," *Philadelphia Evening Bulletin,* April 29, 1964, Folder: Riots—Miscellaneous—1964 August, Temple University Libraries, Urban Archives, Philadelphia. Armand Aloi to Joan Cates, "Re: Demolished Properties," April 22, 1969, Box 330, Folder: Monthly Demolition Reports, NHRA Records. On the August 1967 riot and its aftermath, see Rae, *City,* 351–355; Mandi Isaacs Jackson, *Model City Blues: Urban Space and Organized Resistance in New Haven* (Philadelphia: Temple University Press, 2008), 138–152.

52. Vincent Villano to William T. Donohue, "Re: Ralph Valanzuolo, 109–111 Washington Ave, New Haven, CT," March 7, 1975, Box 511, Folder: R-96 and NDP A-8-2, Dem., Gen. Corresp., 1973–, NHRA Records.

53. NHRA, *New Haven Redevelopment Agency Annual Report 1971, Financial Information* (New Haven: NHRA, n.d.); Rae, *City,* 343–346; Dolezal to Shannon, "Re: Manual Change Recommendations—First Submission"; NHRA, "Resolution of the NHRA Establishing a Policy for the Payment of Moving Expenses and Direct Losses of Property to Businesses, Families and Individuals Displaced by the Church Street Project," April 18, 1958, Box 701, Folder: NHRA Minutes, 1958, NHRA Records; Murray Trachten, interview by Casey Miner, April 20, 2005, transcript, 10, OHDNH. Among the eighty-five surviving businesses, forty-three relocated to blocks close to their former sites, thirty-four relocated elsewhere within the city, and eight left New Haven. Half reported an improvement in business, while the rest split evenly in assessing business as either flat or down: Talbot, *Mayor's Game,* 68. Urban renewal put out of business roughly fourteen hundred New Haven neighborhood retailers: Rae, *City,* 345.

54. See Teaford, *The Rough Road to Renaissance,* 156; United States, Real Estate Research Corporation and RTKL Associates, *Guidelines for Urban Re-*

newal Land Disposition, 6, 13, 376, 379; ("hole city"): Lowe, *Cities in a Race with Time,* 408.

55. ("I always remember"): DeBenedet, interview, 10; Lowe, *Cities in a Race with Time,* 188; ("Okay, so we"): Samuel Teitelman, interview by Casey Miner, February 16, 2005, transcript, 10–11, OHDNH.

56. United States, Real Estate Research Corporation and RTKL Associates, *Guidelines for Urban Renewal Land Disposition,* 256, 392, 396–397. On interwar property owners' purposeful creation of parking lots for tax saving purposes, see Alison Isenberg, *Downtown America: A History of the Place and the People Who Made It* (Chicago: University of Chicago Press, 2004), 218, 241.

57. Argento, interview, 5; Trachten, interview, 8.

58. Robert Silverman, interview by Andy Horowitz, transcript, May 6, 2004, 29, OHDNH; Argento, interview, 18.

59. See Richard C. Lee to Marion Morra, December 9, 1966, Box 144, Folder 2489, RCL.

60. By 1961, the city was forecasting 815 demolished structures, versus 498 rehabilitated ones, in Wooster Square. Ten years later, rehabilitation indeed surpassed demolition: 1,045 to 843: NHRA, *1961 Annual Report of the Redevelopment Agency* (New Haven: NHRA, n.d.); NHRA, *New Haven Redevelopment Agency Annual Report 1972, Financial Information* (New Haven: NHRA, n.d.). For case studies of urban renewal rehabilitation in New Haven and other postwar U.S. cities, see Margaret Carroll, *Historic Preservation Through Urban Renewal: How Urban Renewal Works, Two Areas of Emphasis, Broad Requirements, Preservation and Renewal in Action* (Washington, D.C.: Urban Renewal Administration, Housing and Home Finance Agency, 1963).

61. On organized resistance to New Haven's urban renewal plans, see Jackson, *Model City Blues.*

62. Kenneth Lassen, interview by Andy Horowitz, June 17, 2005, partial transcript, available at New Haven Oral History Project Home Page, http://research.yale.edu/nhohp/content/the-collection/featured-interview/ken-lassen-on-rebuilding-louis-lunch (accessed 11/11/2009). See also Andy Horowitz, "Our Beef with Texas," *New York Times,* January 28, 2007, sec. 14CN, 15.

63. Kenneth C. Lassen to Melvin J. Adams, January 20, 1970, Box 621, Folder: Relocation—Louis' Lunch, 202 George Street, Relocation Corresp., 1963–72, NHRA Records; Nicholas Bros., Inc. and NHRA, "Agreement [to Move Louis' Lunch]," July 9, 1975, Box 621, Folder: Relocation—Louis' Lunch, 202 George St, Relocation Corresp., 1973–75, NHRA Records; Armand Aloi to George W. Simrell, "Re: Demolition Costs—Temple-George Project," February 2, 1968, Box 621, Folder: Relocation—Louis' Lunch, 202 George St, Relocation Corresp., 1963–72, NHRA Records; Donohue quoted in

"'Louis Lunch' Moved out of Path of Destruction," *Journal of Housing,* October 1975, 459.

64. "'Louis Lunch' Moved out of Path of Destruction," 459; Kenneth Lassen, interview.

65. Jane Jacobs, *The Death and Life of Great American Cities* (New York: Random House, 1961); Gans, *Urban Villagers;* Martin Anderson, *The Federal Bulldozer: A Critical Analysis of Urban Renewal, 1949–1962* (Cambridge: MIT Press, 1964).

66. See Stephanie R. Ryberg, "Historic Preservation's Urban Renewal Roots: Preservation and Planning in Midcentury Philadelphia," *Journal of Urban History* 39, no. 2 (March 1, 2013): 193–213.

67. Mary K. Nenno, "Overview: Lessons of Experience," and J. Robert Dumouchel, "Urban Revitalization: Evolution," in *Housing and Community Development: A 50-Year Perspective,* by National Association of Housing and Redevelopment Officials (Washington, D.C.: National Association of Housing and Redevelopment Officials, 1986), 8–9, 75.

68. Herbert J. Gans, "Gans Responds to Granovetter," *American Journal of Sociology* 80, no. 2 (1974): 529–530; Jackson, *Model City Blues.*

Chapter 5. "The Intricate Blending of Brains and Brawn"

1. "Builders of the Great Thruway," *New York Times Magazine,* December 17, 1950, SM10.

2. Ibid., SM10, SM52.

3. See American Association of State Highway and Transportation Officials [hereafter AASHTO], *The States and the Interstates: Research on the Planning, Design and Construction of the Interstate and Defense Highway System* (Washington, D.C.: AASHTO, 1991), 59; Roger Biles, "Expressways Before the Interstates: The Case of Detroit, 1945–1956," *Journal of Urban History* 40, no. 5 (September 2014): 843–854; Mark H. Rose and Bruce E. Seely, "Getting the Interstate System Built: Road Engineers and the Implementation of Public Policy, 1955–1985," *Journal of Policy History* 2, no. 1 (1990): 29. On the New York Thruway, see Michael R. Fein, *Paving the Way: New York Road Building and the American State, 1880–1956* (Lawrence: University Press of Kansas, 2008).

4. "Builders of the Great Thruway," SM10. The AASHTO contracted with the Public Works Historical Society to conduct oral history interviews with key officials in ten state highway departments. During 1987 and 1988, a team led by Mark Rose and Bruce Seely completed over one hundred such interviews. These testimonies offer an invaluable window into highway builders' mindsets both during and after the interstate highway era. See Rose

and Seely, “Getting the Interstate System Built,” 49; AASHTO, *States and the Interstates.* On “design politics,” see Lawrence J. Vale, *Purging the Poorest: Public Housing and the Design Politics of Twice-Cleared Communities* (Chicago: University of Chicago Press, 2013).

5. Judy Lehr, “Jacques and Eugene Yeager Are at the Intersection of Transportation and Higher Education,” University of California, Riverside, *Fiat Lux,* April 2000, at http://www.fiatlux.ucr.edu/cgi-bin/display.cgi?id=430 (accessed 5/15/15); Walter Hjelle (North Dakota State Highway Department) and Charles Gullicks (North Dakota Department of Transportation), interview by John T. Greenwood, transcript, March 2, 1988, 83–84, AASHTO IHRP. Hereafter “State Highway Department” will be abbreviated as “SHD,” and “Department of Transportation” as “DOT.”

6. Dwight D. Eisenhower, *At Ease: Stories I Tell to Friends* (Garden City, N.Y.: Doubleday, 1967), 157; U.S. Federal Highway Administration (hereafter FHWA), *America's Highways, 1776–1976: A History of the Federal-Aid Program* (Washington, D.C.: U.S. GPO, 1977), 142.

7. U.S. Public Roads Administration, *Toll Roads and Free Roads,* House Document, 76th Congress, 1st Session, House, no. 272 (Washington, D.C., 1939), vii, 1, 65.

8. See Warren Cremean (Ohio DOT), interview by Mark Rose and Bruce Seely, transcript, August 18, 1987, 2, AASHTO IHRP; U.S. FHWA, *America's Highways, 1776–1976,* 273–277.

9. U.S. FHWA, *America's Highways, 1776–1976,* 142; Henry Hanley (Illinois DOT), interview by Howard Rosen, transcript, ca. 1988, 8, AASHTO IHRP; Eisenhower, *At Ease,* 166–167.

10. Tom Lewis, *Divided Highways: Building the Interstate Highways, Transforming American Life* (London: Viking Penguin, 1997), 105–107; “Earthmoving: Everyone Gets in the Act,” *Business Week,* August 15, 1953, 56.

11. John F. Bauman, Roger Biles, and Kristin M. Szylvian, *The Ever-Changing American City: 1945–Present* (Lanham, Md.: Rowman and Littlefield, 2011), 43; Hjelle and Gullicks, interview, 54.

12. Edward Haase (Colorado SHD), interview by Mark Rose and Bruce Seely, transcript, August 24, 1988, 41, AASHTO IHRP; U.S. Department of Commerce, Bureau of Public Roads, “Highway History: The Size of the Job,” 1961, FHWA Home Page, http://www.fhwa.dot.gov/infrastructure/50size.cfm (accessed 2/26/13).

13. U.S. FHWA, *America's Highways, 1776–1976,* 467; “Road Builders' Biggest Year,” *Life,* December 12, 1949, 142; Caterpillar Tractor Company, *Annual Report 1956* (Peoria, Ill.: Caterpillar Tractor Company, 1957), 26–27; Charles W. Wixom, *A Pictorial History of Roadbuilding* (Washington, D.C.: American Road Builders' Association, 1975), 144.

14. Richard Zettel (consultant to the California legislature), interview by David Jones, transcript, February 28, 1988, 24–25, AASHTO IHRP; Special Subcommittee on Federal-Aid Highway Program, Committee on Public Works, House, *Defense Highway Needs,* February 9–10, 16–19, 1960, 75, 79.

15. D. K. Chacey testimony in *Defense Highway Needs,* 78; Frank Turner (BPR and Clay Committee), interview by John T. Greenwood, transcript, February 7, 1988, AASHTO IHRP, 64–65; "Missile Altering U.S. Highway Plan: Overpasses Are Too Low for Truck-Borne Rockets—Change May Cost Billion," *New York Times,* February 10, 1960, 4; "8,000 Road Bridges Are Called Too Low," *New York Times,* February 11, 1960, 29; U.S. DOT, FHWA, "Right of Passage: The Controversy over Vertical Clearance on the Interstate System," *Highway History,* June 30, 2014, http://www.fhwa.dot.gov/infrastructure/50vertical.cfm (accessed 7/3/14).

16. "Topics of the Times," *New York Times,* August 20, 1956, 20. On midcentury ideological conflicts between engineers and planners, see Louis Ward Kemp, "Aesthetes and Engineers: The Occupational Ideology of Highway Design," *Technology and Culture* 27, no. 4 (October 1986): 759–797.

17. On the engineering mentality and pre-interstate highways, see Bruce Edsall Seely, *Building the American Highway System: Engineers as Policy Makers* (Philadelphia: Temple University Press, 1987). Mumford quoted in Earl Swift, *The Big Roads: The Untold Story of the Engineers, Visionaries, and Trailblazers Who Created the American Superhighways* (Boston: Houghton Mifflin Harcourt, 2011), 238, 241; AASHTO, *States and the Interstates,* 5; Rose and Seely, "Getting the Interstate System Built," 39–40; Chuck Pivetti (California Division of Highways), interview by David Jones, transcript, February 19, 1988, 28, AASHTO IHRP.

18. Pivetti, interview, 28; Wayne J. Capron (Colorado SHD), interview by Mark Rose and Bruce Seely, transcript, August 23, 1987, 36, AASHTO IHRP.

19. Kevin Starr, *Golden Dreams: California in an Age of Abundance, 1950–1963* (Oxford: Oxford University Press, 2009), 251; Leo Trombatore (California DOT), interview by David Jones, Transcript, 1988, 1, AASHTO IHRP; Haase, interview, 16; James Siebels (California Department of Highways), interview by Mark Rose, transcript, August 24, 1988, 14, AASHTO IHRP; Scott Lathrop (California Division of Highways), interview by David Jones, transcript, April 12, 1989, 16, AASHTO IHRP; ("was the lowest offer"): W. M. Lackey (Kansas DOT), interview by Mark Rose, transcript, August 25, 1988, 19, AASHTO IHRP; ("most of them"): Walter G. Johnson and Richard L. Peyton (Kansas DOT), interview by Mark Rose, transcript, August 26, 1987, 27–29, AASHTO IHRP.

20. Leon O. Talbert et al. (Ohio SHD), interview by Mark Rose and Bruce

Seely, transcript, 1989, 31, AASHTO IHRP; Mike Lackey (Kansas DOT), interview by Sherry Schermer, transcript, February 5, 1988, 16–18.

21. Glenn Anschutz (Kansas DOT), interview by John T. Greenwood, transcript, November 3, 1986, 100–102, AASHTO IHRP.

22. Bertram D. Tallamy (New York Thruway Authority and FHWA), interview by Darwin Stolzenbach, transcript, September 21, 1987, 44, AASHTO IHRP; "New Vistas of the Road," *Life,* November 19, 1956, 84.

23. Haase, interview, 23, 37; Dale Dugan (Kansas DOT), interview by John T. Greenwood, transcript, November 3, 1987, 130–131, AASHTO IHRP.

24. ("go through," "didn't hesitate"): Johnson and Peyton, interview, 20, 21; Thomas D. Moreland (Georgia SHD), interview by Darwin Stolzenbach, transcript, February 28, 1989, 21, AASHTO IHRP. See also AASHTO, *The States and the Interstates,* 52; Siebels, interview, 35; Cremean, interview, 30–31; Jeremy Louis Korr, "Washington's Main Street: Consensus and Conflict on the Capital Beltway, 1952–2001" (Ph.D. diss., University of Maryland, College Park, 2002), 155–156.

25. Pivetti, interview ("I think a lot"): 33, ("well, it's not an interstate"): 6–7; Lathrop, interview, 39–40; Swift, *Big Roads,* 219–220.

26. Douglas Fugate (Virginia State Highway and Transportation Department), interview by John T. Greenwood, transcript, May 27, 1987, 57, AASHTO IHRP; Anschutz, interview, 7–12; U.S. FHWA, *America's Highways, 1776–1976,* 396; ("the most economical"): Austin Hunsberger (Virginia SHD), interview by John T. Greenwood, transcript, n.d., 4–5, AASHTO IHRP; ("to determine earthwork"): Haase, interview, 13–14; Jack Jones (Mississippi SHD), interview by Michael Robinson, transcript, March 1, 1989, 5, AASHTO IHRP.

27. Special Subcommittee on Federal-Aid Highway Program, Committee on Public Works, House, *Disposition of Right-of-Way Improvements in Florida,* March 1–3, 7–10, 1961, 3–4.

28. Dugan, interview, 10–11; U.S. FHWA, *America's Highways, 1776–1976,* 459.

29. Charles Shumate (Colorado SHD), interview by Erin Christensen and Ellis Armstrong, transcript, February 9, 1989, 1, AASHTO IHRP; ("A farmer"): Johnson and Peyton, interview, 20; Dugan, interview, 21.

30. Cremean, interview, 27; Swift, *Big Roads,* 220–221.

31. Jacob Dekema (California DOT), interview by David Jones, transcript, March 27, 1989, 13, AASHTO IHRP; ("I was proud"): Zettel, interview, 39; ("If I had my druthers"): Lathrop, interview, 82.

32. Dekema, interview, 29–31.

33. AASHTO, *States and the Interstates,* ix.

34. Hanley, interview, 4–5; *Disposition of Right-of-Way Improvements in Florida,* 286.

35. Peter Paravalos, *Moving a House with Preservation in Mind* (Oxford: Rowman Altamira, 2006), 1–14; Charles H. Faulkner, "Moved Buildings: A Hidden Factor in the Archaeology of the Built Environment," *Historical Archaeology* 38, no. 2 (January 1, 2004): 55–67; *Los Angeles Times* quoted in Todd Douglas Gish, "Building Los Angeles: Urban Housing in the Suburban Metropolis, 1900–1936" (Ph.D. diss., University of Southern California, 2007), 309–310; Susan Feyder, "Moving Your House? Beware of Expense, Legal Roadblocks," *Chicago Tribune,* March 26, 1978, S-B1; Robert Stloukal, "Real Estate: To Pete Friesen, Moving Means Not the Contents but the House," *Chicago Tribune,* February 3, 1984, SD28.

36. On the house-moving process, see Paravalos, *Moving a House with Preservation in Mind,* 21–35. *Disposition of Right-of-Way Improvements in Florida,* 91; "House-Moving Party," *Life,* September 13, 1948, 25, 73–76.

37. Victor Boesen, "Look, Ma! There Goes Our House," *Saturday Evening Post,* November 5, 1949, 57; *Disposition of Right-of-Way Improvements in Florida,* 92–94.

38. Boesen, "Look, Ma! There Goes Our House," 20–21; *Disposition of Right-of-Way Improvements in Florida,* 66–67, 93.

39. *Disposition of Right-of-Way Improvements in Florida,* 188.

40. Ibid., 29, 36, 46–47.

41. Ibid., 103.

42. Ibid., 57–58, 296.

43. Haase, interview, 25–26, 52.

44. Ibid., 33; Pivetti, interview, 5.

45. Haase, interview, 35; Fugate, interview, 85.

46. Grady Clay, "The Tiger Is Through the Gate," *Landscape Architecture,* Winter 1958–1959, 80–81; Talbert et al., interview, 36–37; AASHTO, *States and the Interstates,* 207.

47. Swift, *Big Roads,* 220; AASHTO, *States and the Interstates,* 63.

48. See Raymond A. Mohl, "Stop the Road," *Journal of Urban History* 30, no. 5 (July 2004): 674–706; Raymond A. Mohl, "The Interstates and the Cities: The U.S. Department of Transportation and the Freeway Revolt, 1966–1973," *Journal of Policy History* 20, no. 2 (April 2008): 193–226; Christopher Klemek, *The Transatlantic Collapse of Urban Renewal: Postwar Urbanism from New York to Berlin* (Chicago: University of Chicago Press, 2011); Karilyn Michelle Crockett, "'People Before Highways': Reconsidering Routes to and from the Boston Anti-Highway Movement" (Ph.D. diss., Yale University, 2013); Andrew M. Giguere, "'. . . and Never the Twain Shall Meet': Baltimore's East-West Expressway and the Construction of the 'High-

way to Nowhere'" (M.A. thesis, Ohio University, 2009); Eric Avila, *The Folklore of the Freeway: Race and Revolt in the Modernist City* (Minneapolis: University of Minnesota Press, 2014).

49. On gains justifying means, see Talbert et al., interview, 37, 69; ("for the most part"): AASHTO, *The States and the Interstates,* 169; Fugate, interview, 30; Bernard B. Hurst (Ohio DOT), interview by Mark Rose, transcript, August 19, 1987, 12, AASHTO IHRP.

50. Dugan, interview, 129. See Mark H. Rose, *Interstate: Express Highway Politics, 1939–1989,* rev. ed. (Knoxville: University of Tennessee Press, 1990), 107.

51. See Wayne F. Kennedy, "Blatnik to Now: A Modern History of U.S. Transportation Legislation," *Right of Way,* February 1997, 17. John Kemp (FHWA and Kansas DOT), interview by Sherry Sherer, transcript, February 23, 1988, 12–13, AASHTO IHRP. On the debate over whether to include relocation compensation in the 1956 act, see Raymond A. Mohl, *The Interstates and the Cities: Highways, Housing, and the Freeway Revolt* (Poverty and Race Research Action Council, 2002), 15–18, at http://www.prrac.org/pdf/mohl.pdf (accessed 1/15/15).

52. See U.S. FHWA, *America's Highways, 1776–1976,* 65; 90th Congress, 1st session, Subcommittee on Roads, Committee on Public Works, House, *Highway Relocation Assistance Study: A Study Transmitted by the Secretary of the Department of Transportation, as Required by the Federal-Aid Highway Act of 1966* (Washington, D.C., July 1967), 29–30, 45.

53. A. Q. Mowbray, *Road to Ruin* (Philadelphia: Lippincott, 1969); Helen Leavitt, *Superhighway—Superhoax* (Garden City, N.Y.: Doubleday, 1970); Albert Benjamin Kelley, *The Pavers and the Paved* (New York: Brown, 1971).

54. Fugate, interview, 62; Harold W. Monroney and Thomas R. Mracek (Illinois DOT), interview by Mark Rose and Howard Rosen, transcript, April 1988, 14, AASHTO IHRP; Stan Hyatt (North Carolina DOT), interview by Rob Amberg, transcript, November 30, 2000, 29, Southern Oral History Program Collection (#4007), Southern Historical Collection, the Wilson Library, University of North Carolina at Chapel Hill.

55. Hyatt, interview, 27; Haase, interview, 36.

56. Pivetti, interview, 18–19; Capron, interview, 37; Fugate, interview, 63.

57. Floyd Diemoz (Citizens Advisory Committee, Glenwood Canyon Project, I-70, Colorado), interview by Erin Christensen and Karen Waddell, transcript, October 20, 1987, 27, AASHTO IHRP; Anschutz, interview, 76; Shumate, interview, 18.

58. See Hyatt, interview, 29; Adam Rome, *The Bulldozer in the Countryside: Suburban Sprawl and the Rise of American Environmentalism* (Cambridge: Cambridge University Press, 2001), 221–253; Laura A. Watt, Leigh

Raymond, and Meryl L. Eschen, "On Preserving Ecological and Cultural Landscapes," *Environmental History* 9, no. 4 (October 2004): 620–647.

59. Talbert et al., interview ("in fact, you couldn't"): 38, ("created the greatest upheaval"): 49, ("up to that point" and "historic preservation people"): 50, 75; Tallamy, interview, 37–38; ("You know, we now go"): Hurst, interview, 39–40.

60. See Capron, interview, 75; AASHTO, *States and the Interstates,* 6, 64; P. J. Culhane, "NEPA's Impacts on Federal Agencies, Anticipated and Unanticipated," *Environmental Law* 20, no. 3 (1990): 681–702.

61. James C. Wright, "Highway Robbery," *Saturday Evening Post,* November 30, 1963, 20; U.S. FHWA, *America's Highways, 1776–1976,* 484.

62. In response to the 1973 act, twenty-one states withdrew a total of fifty segments, measuring 343 miles. These were predominantly urban routes. Passage of the Intermodal Surface Transportation Efficiency Act (ISTEA) in 1991 officially ended the interstate construction era: see "Technical Memorandum Task 1: The Interstate and National Highway System—A Brief History and Lessons Learned," June 13, 2006, 37, at http://www.dsmic.org/documentstore/Plans%20and%20Studies%20%28Corridor%29/EndionNeighborhoodTransportationPlan/Interstate%20System%201956.pdf (accessed 5/27/15).

Chapter 6. Unearthing "Benny the Bulldozer"

1. Edith Thacher Hurd, *Benny the Bulldozer,* illus. Clement Hurd (New York: Lothrop, Lee, and Shepard, 1947), n.p.

2. On cities of fact and feeling, see Carlo Rotella, *October Cities: The Redevelopment of Urban Literature* (Berkeley: University of California Press, 1998). The sociologist Milton Albrecht has identified three kinds of relationships between literature and society: reflection, expression of emerging themes, and social control: Milton C. Albrecht, "The Relationship of Literature and Society," *American Journal of Sociology* 59, no. 5 (March 1954): 425–427. See also David E. Nye, *America as Second Creation: Technology and Narratives of New Beginnings* (Cambridge: MIT Press, 2003), 1–7.

3. Among bulldozer books identified from the midcentury and postwar period, four titles remain in print and still occupy several thousand library shelves; another handful of texts circulate with about five hundred copies each; over a dozen endure at the level of about a hundred copies each; and a final half-dozen or so exist in smaller quantities. Figures based on a survey of WorldCat library holdings conducted July 2011. WorldCat includes academic and public libraries but excludes the tens of thousands of school libraries.

4. Lois R. Kuznets, *When Toys Come Alive: Narratives of Animation, Meta-*

morphosis, and Development (New Haven: Yale University Press, 1994), 7; Beverly Lyon Clark, *Kiddie Lit: The Cultural Construction of Children's Literature in America* (Baltimore: Johns Hopkins University Press, 2003), 2, 5; Julia L. Mickenberg, *Learning from the Left: Children's Literature, the Cold War, and Radical Politics in the United States* (New York: Oxford University Press, 2006), 15; Milton Meltzer, "Where Do All the Prizes Go? The Case for Nonfiction," in *Children and Their Literature: A Readings Book,* ed. Jill P. May (West Lafayette, Ind.: Children's Literature Association Publications, 1983), 92–97; James Cross Giblin, "More Than Just the Facts: A Hundred Years of Children's Nonfiction," *Horn Book Magazine,* August 2000, 413–424; Betty Bacon, "The Art of Nonfiction," *Children's Literature in Education* 12, no. 1 (March 1981): 7.

5. Louisa Della Rocca Oriente, "Images of City Life as Depicted in Contemporary Realistic Fiction for Children, Ages Eight to Twelve" (Ed.D. diss., Columbia University, 1976), 36–37; Paul Hazard, *Books, Children and Men,* trans. Marguerite Mitchell, 2nd ed. (Boston: Horn Book, 1960), 1; Burton quoted in Joe Goddard, "Virginia Lee Burton's *Little House* in Popular Consciousness: Fuelling Postwar Environmentalism and Antiurbanism?" *Journal of Urban History* 37, no. 4 (July 2011): 563; Ellen Handler Spitz, *Inside Picture Books* (New Haven: Yale University Press, 1999), 14, 208; R. Gordon Kelly and Lucy Rollin, "Children's Literature," in *The Greenwood Guide to American Popular Culture,* ed. M. Thomas Inge and Dennis Hall (Westport, Conn.: Greenwood Press, 2002), 214; Joyce A. Thomas, "Nonfiction Illustration: Some Considerations," in *Children and Their Literature,* 122.

6. Jean Poindexter Colby, *Tear Down to Build Up: The Story of Building Wrecking,* illus. Joshua Tolford (New York: Hastings House, 1960); Jean Poindexter Colby, *Building Wrecking: The How and Why of a Vital Industry,* illus. Corinthia Morss (New York: Hastings House, 1972); David C. Cooke, *How Superhighways Are Made* (New York: Dodd, Mead, 1958); Tonka Toys, Inc., *Annual Report, 1961* (Mound, Minn.: The Company, 1961), n.p.

7. Clark, *Kiddie Lit,* 72; Margery Fisher, *Matters of Fact: Aspects of Non-Fiction for Children* (New York: Crowell, 1972), 12, 20; Lucy Sprague Mitchell, *Here and Now Story Book, Two-to-Seven Year Olds; Experimental Stories Written for the Children of the City and Country School (formerly the Play School) and the Nursery School of the Bureau of Educational Experiments* (New York: Dutton, 1921), 8. *Here and Now Story Book* was developed in the same spirit as two New York City schools Sprague co-founded: the Play School, which educated young children through physical engagement with their environment, and Bank Street School, which taught educational instructors and children's-book writers like Margaret Wise Brown. Margaret Wise Brown's most famous work is *Goodnight Moon,* also illustrated by Clement Hurd. The

Play School eventually became the City and Country School, which continues today. For more on Mitchell, see Joyce Antler, *Lucy Sprague Mitchell: The Making of a Modern Woman* (New Haven: Yale University Press, 1987). For more on Mitchell's influence, see Mickenberg, *Learning from the Left,* 40–45; Dora V. Smith, *Fifty Years of Children's Books, 1910–1960: Trends, Backgrounds, Influences* (Champaign, Ill.: National Council of Teachers of English, 1963), 28–30.

8. Giblin, "More Than Just the Facts"; Cooke, *How Superhighways Are Made,* 5. For more on the history of the NDEA, see Barbara Barksdale Clowse, *Brainpower for the Cold War: The Sputnik Crisis and National Defense Education Act of 1958* (Westport, Conn.: Greenwood, 1981).

9. Colby, *Tear Down to Build Up,* 54, n.p.

10. Virginia Lee Burton, *Mike Mulligan and His Steam Shovel* (Boston: Houghton Mifflin, 1939); Norman Bate, *Who Built the Dam? A Picture Story* (New York: Scribner's, 1958), n.p. See also Norman Bate, *Who Built the Highway? A Picture Story* (New York: Scribner's, 1953).

11. James Browning, *The Busy Bulldozer,* illus. Dorothy Grider (Chicago: Rand McNally—Tip-Top Elf Books, 1952); Carla Greene, *I Want to Be a Road-Builder,* illus. Irma Wilde and George A. Wilde (Chicago: Children's Press, 1958).

12. Margo McWilliams and Patricia Reisdorf, *Let's Go to Build a Highway,* illus. Albert Micale (New York: Putnam's, 1971), 10; Colby, *Tear Down to Build Up,* 27. In the 1972 updated reissue of *Tear Down to Build Up*—which Colby titled *Building Wrecking: The How and Why of a Vital Industry*—the author conveys slightly more sensitivity. Toning down her language on page 44, she describes these incidents as "unusual," rather than "amusing." Cooke, *How Superhighways Are Made,* 14–15; Virginia Lee Burton, *The Little House* (Boston: Houghton Mifflin, 1942). On "escape syndromes," see Katherine M. Heylman, "The Little House Syndrome vs. Mike Mulligan and Mary Anne: A Mobilization of Juvenile Books on Ecology, Conservation, and Pollution," *Library Journal,* April 15, 1970, 1562. Max Page notes similarly, "Even young readers must be bothered by the easy ending" to this innocent story of urbanization: Max Page, *The Creative Destruction of Manhattan, 1900–1940* (Chicago: University of Chicago Press, 1999), 254–260. See also Goddard, "Virginia Lee Burton's *Little House* in Popular Consciousness."

13. Frank Tashlin, *The Bear That Wasn't* (New York: Dutton, 1946).

14. McWilliams and Reisdorf, *Let's Go to Build a Highway,* 48.

15. Herbert S. Zim and James R. Skelly, *Tractors,* illus. Lee J. Ames (New York: Morrow, 1972), dust jacket, 55–56; Constance Cappel and Raymond Montgomery, *Vermont Roadbuilder,* illus. Larry Barns (Waitsfield: Vermont

Crossroads Press, 1975); Jack McClellan and Millard Black, *What a Highway!* (Boston: Houghton Mifflin—Citizens All Series, 1967), 50–51, 61–62; Colby, *Building Wrecking,* 47–48; Herbert S. Zim and James R. Skelly, *Hoists, Cranes, and Derricks,* illus. Gary Alan Ruse (New York: Morrow, 1969), 63.

16. See K. A. Cuordileone, *Manhood and American Political Culture in the Cold War* (New York: Routledge, 2005); David K. Johnson, *The Lavender Scare: The Cold War Persecution of Gays and Lesbians in the Federal Government* (Chicago: University of Chicago Press, 2004).

17. Vera Edelstadt, *A Steam Shovel for Me!* (New York: Stockes, 1933) (Mickenberg also makes this observation about these illustrations; Mickenberg, *Learning from the Left,* 44–45); Lewis Wickes Hine, *Men at Work: Photographic Studies of Modern Men and Machines* (New York: Macmillan, 1932); Michael Bronski, "From Victorian Parlor to 'Physique Pictorial': The Male Nude and Homosexual Identity," in *Passing: Identity and Interpretation in Sexuality, Race, and Religion,* ed. María Carla Sánchez and Linda Schlossberg (New York: New York University Press, 2001), 135–159; Jonathan Weinberg, *Male Desire: The Homoerotic in American Art* (New York: Abrams, 2004).

18. Elisabeth MacIntyre, *The Affable, Amiable Bulldozer Man* (New York: Knopf, 1965); Cappel and Montgomery, *Vermont Roadbuilder,* 30–31; Zim and Skelly, *Tractors,* 60; Browning, *Busy Bulldozer,* n.p.

19. Burton, *Mike Mulligan and His Steam Shovel;* Cappel and Montgomery, *Vermont Roadbuilder,* 21–23.

20. Zim and Skelly, *Tractors,* 60; Colby, *Tear Down to Build Up,* 22, 29; Cooke, *How Superhighways Are Made,* 62.

21. Zim and Skelly, *Hoists, Cranes, and Derricks,* 63; McWilliams and Reisdorf, *Let's Go to Build a Highway,* 9; Greene, *I Want to Be a Road-Builder;* E. Joseph Dreany, *The Big Builders* (Racine, Wisc.: Whitman, 1961). See William H. Whyte, *The Organization Man* (New York: Simon and Schuster, 1956).

22. H. Joseph Schwarcz, "Machine Animism in Modern Children's Literature," in *A Critical Approach to Children's Literature: The Thirty-First Annual Conference of the Graduate Library School, August 1–3, 1966,* ed. Sara Innis Fenwick (Chicago: University of Chicago Press, 1967), 80, 92–93. On "sentient machines" in children's books of the first half of the twentieth century, see Nathalie Op de Beeck, *Suspended Animation: Children's Picture Books and the Fairy Tale of Modernity* (Minneapolis: University of Minnesota Press, 2010), 119–167.

23. Virginia Lee Burton, *Katy and the Big Snow* (Boston: Houghton Mifflin, 1943), 33; Catherine Danner, *Buster Bulldozer,* illus. Mary Alice Stoddard (Racine, Wisc.: Whitman, 1952), n.p.

24. Schwarcz, "Machine Animism in Modern Children's Literature," 95; Cooke, *How Superhighways Are Made,* 12.

25. Dick Ashbaugh, "Gangway! The Kids Are Rolling!" *Saturday Evening Post,* 1951, 36, 117–118.

26. Ibid., 36, 118–119. More recently, in an example of cultural rather than engineering borrowing, Ford introduced a pickup truck with its looks, model name, and durability modeled on Tonka toys: Dan McCosh, "Tonka Truck: A Pickup with Design Cues from the Toy Gains Fuel-Saving Tech," *Popular Science,* February 2002, 67–70.

27. Ashbaugh, "Gangway! The Kids Are Rolling!," 117; Melvin H. Williams, "Fortunes in Toy Ideas," *Popular Mechanics,* November 1953, 92, 270; Ray Funk, "Doepke 'Model Toys,'" in *Collecting Toy Cars and Trucks: A Collector's Identification and Value Guide,* ed. Richard O'Brien, vol. 1 (Florence, Ala.: Books Americana, 1994), 138.

28. Tonka Toys, Inc., *Annual Report, 1961;* Tonka Toys, Inc., *Annual Report, 1965* (Mound, Minn.: The Company, 1965).

29. Tonka Toys, Inc., *Annual Report,* for the years 1961–1965 (Mound, Minn.: The Company, 1961–1965).

30. Tonka Toys, Inc., *Annual Report, 1963* (Mound, Minn.: The Company, 1963); Ashbaugh, "Gangway! The Kids Are Rolling!," 118; "Banks on Bulldozer," *Southern Skies Publishing,* January 13, 2014, http://southernskies publishing.wordpress.com/.

31. Doepke Manufacturing Company advertisement, "Ever Watch a Bulldozer Work?" *Popular Science,* October 1952, back cover; Hubert Kelley, "Earth Movers Shape the Future," *Nation's Business,* October 1954, 50.

32. "Child-Size Bulldozer Looks and Works Like the Real Thing," *Popular Mechanics,* August 1950, 126; "PM's September News Briefs," *Popular Mechanics,* September 1966, 91.

33. David Nye refers to such counternarratives as alternative narratives "written by or addressed to groups that had been silenced in or absent from the original formulation": Nye, *America as Second Creation,* 12. Don Freeman, *Fly High, Fly Low* (New York: Viking, 1957).

34. Roald Dahl, *Fantastic Mr. Fox* (New York: Knopf, 1970); Catherine Woolley, "Butch the Bulldozer," in *The Animal Train, and Other Stories,* illus. Robb Beebe (New York: Morrow, 1953), 57–65.

35. Suzanne Hilton, *How Do They Get Rid of It?* (Philadelphia: Westminster, 1970), 39; Pearl Augusta Harwood, *Mr. Bumba and the Orange Grove* (Minneapolis: Lerner Publications, 1964), n.p.

36. McClellan and Black, *What a Highway!,* 1–3, 6–12, 88–93; Zim and Skelly, *Tractors,* 61–63. For the authors' earlier, more exclusively positive, perspectives, see Zim and Skelly, *Hoists, Cranes, and Derricks;* Herbert S.

Zim and James R. Skelly, *Machine Tools,* illus. Gary Alan Ruse (New York: Morrow, 1969).

37. Hila Colman, *Andy's Landmark House,* illus. Fermin Rocker (New York: Parents' Magazine Press, 1969), 45; Toby Talbot, *My House Is Your House,* illus. Robert Weaver (New York: Cowles Book Company, 1970), 1.

38. Helen Palmer, *Johnny's Machines,* illus. Cornelius DeWitt (New York: Simon and Schuster, 1949); P. D. Eastman, *Are You My Mother?* (New York: Random House—Beginner Books, 1960); Talbot, *My House Is Your House,* 33.

39. David Macaulay, *Unbuilding* (Boston: Houghton Mifflin, 1980). See also David Macaulay, *Cathedral: The Story of Its Construction* (Boston: Houghton Mifflin, 1973); David Macaulay, *Pyramid* (Boston: Houghton Mifflin, 1975); Jon C. Stott, "Architectural Structures and Social Values in the Non-Fiction of David Macaulay," *Children's Literature Association Quarterly* 8, no. 1 (Spring 1983): 15–17.

40. Macaulay, *Unbuilding,* 78.

41. Ibid., n.p.

Chapter 7. Bulldozers as Paintbrushes

1. Walter De Maria, "Art Yard," in *An Anthology of Chance Operations,* ed. La Monte Young (Bronx, N.Y.: La Monte Young and Jackson Mac Low, 1963), n.p.

2. Robert Smithson, "Toward the Development of an Air Terminal Site," *Artforum,* Summer 1967, 38.

3. See Jeffrey Kastner, "Preface," in Jeffrey Kastner, ed., *Land and Environmental Art,* abridged, rev., and updated ed. (New York: Phaidon, 2010), 12.

4. *Earth Works,* advertisement, *Artforum,* October 1968. Smithson derived the title for this show from a dystopian science fiction novel about a future world of environmental catastrophe and socioeconomic inequality, Brian W. Aldiss's *Earthworks* (New York: Doubleday, 1966).

5. See David Bourdon, *Designing the Earth: The Human Impulse to Shape Nature* (New York: Abrams, 1995).

6. Grace Glueck, "Moving Mother Earth," *New York Times,* October 6, 1968, D38; Howard Junker, "Getting Down to the Nitty Gritty," *Saturday Evening Post,* November 2, 1968, 44; Neil Jenney, quoted in "The Symposium," in *Earth Art,* by Andrew Dickson White Museum of Art (Ithaca: Office of University Publications, Cornell University, 1970), n.p.

7. Junker, "Getting Down to the Nitty Gritty," 46. See Suzaan Boettger, *Earthworks: Art and the Landscape of the Sixties* (Berkeley: University of California Press, 2002), 152. On entropy (disorder, decay, and dissipation over time), see Robert Smithson, "Entropy and the New Monuments," *Artforum,*

June 1966, 26–31; Robert Smithson, "Entropy Made Visible—Interview with Alison Sky (1973)," in *The Writings of Robert Smithson: Essays with Illustrations,* ed. Nancy Holt (New York: New York University Press, 1979).

8. See Grace Glueck, "An Artful Summer," *New York Times,* May 19, 1968, D35; Junker, "Getting Down to the Nitty Gritty," 44–45; "The Earth Movers," *Time,* October 11, 1968, 98; Dennis Oppenheim, *Dennis Oppenheim: Explorations,* ed. Germano Celant (Milan: Charta, 2001), 52; Boettger, *Earthworks,* 143.

9. Robert Smithson, interview by Paul Cummings, transcript, July 14, 1972, Archives of American Art, Smithsonian Institution; "The Symposium."

10. Robert Smithson, "The Sedimentation of the Mind: Earth Projects," *Artforum,* September 1968, 45.

11. Quoted from Menil Collection website in "Walter De Maria and Barnett Newman: Who's Afraid of Red, John Deere Yellow and Blue?" *Houston Chronicle,* September 23, 2011, available at http://blog.chron.com/29-95/2011/09/walter-de-maria-and-barnett-newman-whos-afraid-of-red-john-deere-yellow-and-blue/ (accessed 5/29/15). See also Walter De Maria, *Walter De Maria: Trilogies,* ed. Clare Elliott, Josef Helfenstein, and Neville Wakefield (New Haven: Yale University Press, 2011).

12. Virginia Dwan, interview by Charles Stuckey, Archives of American Art, Smithsonian Institution, May 1984, tape 7, page 24, quoted in Suzaan Boettger, "Patronage: Behind the Earth Movers," *Art in America,* April 2004, 58; Boettger, *Earthworks,* 116–117, 136. De Maria re-created this earth room piece at least two more times, including a 1977 showing in a SoHo gallery and a more permanent installation at New York's 141 Wooster Street gallery.

13. ("Might have affected"): "Discussions with Heizer, Oppenheim, Smithson," *Avalanche,* Fall 1970, 64; ("started making this stuff"): quoted in Douglas C. McGill, *Michael Heizer: Effigy Tumuli: The Reemergence of Ancient Mound Building* (New York: Abrams, 1990), 11. See Junker, "Getting Down to the Nitty Gritty," 42; Boettger, *Earthworks,* 194–195.

14. Calvin Tomkins, "Onward and Upward with the Arts: Quantum Leap," *New Yorker,* February 5, 1972, 44; David Bourdon, "What on Earth," *Life,* April 25, 1969, 86.

15. David Bourdon, "The Razed Sites of Carl Andre," *Artforum,* October 1966, 17; "Discussions with Heizer, Oppenheim, Smithson," 50; Smithson quoted in James Dickinson, "Journey into Space: Interpretations of Landscape in Contemporary Art," in *Technologies of Landscape: From Reaping to Recycling,* ed. David E. Nye (Amherst: University of Massachusetts Press, 1999), 55; ("I make something"): Heizer quoted in William Wilson, "Don't Know Trenches, but We Know What We Like," *Los Angeles Times,* July 27, 1969, sec. WEST Magazine, N17.

16. See Lizabeth Cohen, *A Consumers' Republic: The Politics of Mass Consumption in Postwar America* (New York: Knopf, 2003); Oppenheim, *Dennis Oppenheim: Explorations,* 64; Diane Waldman, "Holes Without History," *Art News,* May 1971, 48; Tomkins, "Onward and Upward with the Arts: Quantum Leap," 44.

17. "The Symposium"; "High Priest of Danger," *Time,* May 2, 1969, 56; Tomkins, "Onward and Upward with the Arts: Quantum Leap," 44; Bourdon, "What on Earth," 86. See also Jane Kramer, "Profiles: Man Who Is Happening Now," *New Yorker,* November 26, 1966, 64–120.

18. Lucy R. Lippard, *Six Years: The Dematerialization of the Art Object from 1966 to 1972: A Cross-Reference Book of Information on Some Esthetic Boundaries* (New York: Praeger, 1973), xv; "Discussions with Heizer, Oppenheim, Smithson," 54.

19. Andrew Dickson White Museum of Art, *Earth Art* (Ithaca: Office of University Publications, Cornell University, 1970).

20. Oppenheim quoted in Glueck, "An Artful Summer," D35; Tomkins, "Onward and Upward with the Arts: Quantum Leap," 43; "Discussions with Heizer, Oppenheim, Smithson," 50; Waldman, "Holes Without History," 47.

21. See John Beardsley, *Earthworks and Beyond: Contemporary Art in the Landscape,* 4th ed. (New York: Abbeville, 2006), 89; Dickinson, "Journey into Space: Interpretations of Landscape in Contemporary Art," 56–57.

22. Brian Wallis, "Survey," in *Land and Environmental Art,* 32; Robert Smithson, "Frederick Law Olmsted and the Dialectical Landscape," *Artforum,* February 1973, 65; Robert Smithson, *The Writings of Robert Smithson: Essays and Illustrations,* ed. Nancy Holt (New York: New York University Press, 1979), 220; Robert Morris, "Notes on Art as/and Land Reclamation," *October* 12 (April 1980): 98.

23. Dickinson, "Journey into Space: Interpretations of Landscape in Contemporary Art," 44–45; Morris, "Notes on Art as/and Land Reclamation," 98, 101–102.

24. ("Earth Art, with very few"): Michael Auping, "Michael Heizer: The Ecology and Economics of 'Earth Art,'" *Artweek,* June 18, 1977, 1. While the comparison to developers was specifically made in relation to James Turrell's *Roden Crater* (1979–)—which is still being carved out of a volcano—it also applied more generally to many of the movement's large-scale works; Ann Landi, "Moving Mountains, Walking on Water," *ARTnews,* June 2004, 87. Gary Shapiro, *Earthwards: Robert Smithson and Art After Babel* (Berkeley: University of California Press, 1997), 24; ("it's already destroyed"): "Discussions with Heizer, Oppenheim, Smithson," 67; ("the possibility of a direct"): Smithson, "Frederick Law Olmsted and the Dialectical Landscape," 65.

25. "Foreword," in *A Sense of Place: The Artist and the American Land,* by

Alan Gussow et al., vol. 2 (Omaha: Joslyn Art Museum, 1973), 7; Smithson, "Frederick Law Olmsted and the Dialectical Landscape," 65.

26. On Holt, see Alena Williams and Nancy Holt, *Nancy Holt: Sightlines* (Berkeley: University of California Press, 2011). Boettger, quoted in Landi, "Moving Mountains, Walking on Water," 87; Boettger, *Earthworks,* 118, 147–149; "Carl Andre: Interview," *Artforum,* Fall 1970, 23.

27. The Percent for Art program was a 1979 federal initiative that required 1 percent of public works capital improvement projects to be earmarked for public art. Other artists who participated in reclamation projects included Herbert Bayer, Michael Heizer, Nancy Holt, and Stan Dolega: see Bourdon, *Designing the Earth,* 225; David Allen Jones, *Earthworks: Land Reclamation as Sculpture—Final Report* (Kings County, Wash.: Kings County Arts Commission, Kings County Department of Public Works, United States Bureau of Mines, February 10, 1981), 4–5; Beardsley, *Earthworks and Beyond,* 94–97.

28. See Judith Russi Kirshner, "Non-Uments," *Artforum,* October 1985, 104.

29. See Liza Bear, "Gordon Matta-Clark: Splitting (The Humphrey Street Building): An Interview by Liza Bear," *Avalanche,* December 1974, 35.

30. Judith Russi Kirshner, "Interview with Gordon Matta-Clark, Museum of Contemporary Art, Chicago, February 13, 1978," in *Gordon Matta-Clark: Works and Collected Writings,* ed. Gloria Moure (Barcelona: Poligrafa, 2006), 330.

31. Gordon Matta-Clark, *Fresh Kill,* 16mm film, 1972, in Gordon Matta-Clark, *Program Three,* DVD video (New York: Electronic Arts Intermix, Inc., 2000).

32. Gordon Matta-Clark, "In Regard to the Many Condemned Buildings, from Notebook 1261, ca. 1970," in *Gordon Matta-Clark: Works and Collected Writings,* 73.

33. Gordon Matta-Clark, "Work with Abandoned Structures, ca. 1975, Typewritten Statement," in *Gordon Matta-Clark: Works and Collected Writings,* 141; Ned Smyth, "Gordon Matta-Clark," Remarks Presented at David Zwirner Gallery, New York, N.Y., 2004, *Artnet.com,* http://www.artnet.com/Magazine/FEATURES/smyth/smyth6-4-04.asp (accessed 1/30/12).

34. Gordon Matta-Clark, *Splitting,* Super 8mm film, 1974, in Gordon Matta-Clark, *Program Six,* DVD video (New York: Electronic Arts Intermix, Inc., 2000); Gordon Matta-Clark, "Splitting, 1973, Black Pen on Paper," in *Gordon Matta-Clark: Works and Collected Writings,* 161.

35. Gordon Matta-Clark, *Bingo/Ninths,* Super 8mm film, 1974, in Gordon Matta-Clark, *Program Six.*

36. In an intertitle Matta-Clark explains that the bulldozer (which was

actually a loader) arrived just an hour after the crew had finished their work. In fact, the work was incomplete when the crew initially appeared, and the artist was able to finish only by persuading them to commence work the next morning. See Matta-Clark, "Work with Abandoned Structures," 142. Gordon Matta-Clark, "The Earliest Cutout Works, Typewritten Statement, Undated," in *Gordon Matta-Clark: Works and Collected Writings,* 136–137.

37. See Kirshner, "Non-Uments," 103.

38. Liza Bear, "Interview with Gordon Matta-Clark, Antwerp, September 1977," in *Gordon Matta-Clark: Works and Collected Writings,* 251; Kirshner, "Interview with Gordon Matta-Clark," 330; Liza Bear, "Gordon Matta-Clark: Dilemmas, a Radio Interview by Liza Bear, March 1976," in *Gordon Matta-Clark: Works and Collected Writings,* 263.

39. Bear, "Gordon Matta-Clark: Dilemmas," 265; Gordon Matta-Clark, "Cannibalism Suburbia, from Notebook 1262, ca. 1969–71," in *Gordon Matta-Clark: Works and Collected Writings,* 75.

40. David Wall, "Gordon Matta-Clark's Building Dissections, an Interview by Donald Wall," *Arts Magazine,* May 1976, 74; Gordon Matta-Clark, "Proposal to the Workers of Sesto San Giovanni, Milan, 1975, Typewritten Statement," in *Gordon Matta-Clark: Works and Collected Writings,* 120; Bear, "Interview with Gordon Matta-Clark, Antwerp," 253.

41. The New York City loft building in which Matta-Clark grew up was also demolished: Bear, "Gordon Matta-Clark: Splitting (The Humphrey Street Building), an Interview by Liza Bear," 37; Kirshner, "Non-Uments," 104; Pamela M. Lee, *Object to Be Destroyed: The Work of Gordon Matta-Clark* (Cambridge: MIT Press, 2000), 6–9; ("So what I am"): Wall, "Gordon Matta-Clark's Building Dissections," 78.

42. Matta-Clark, "Work with Abandoned Structures," 142.

43. Bear, "Interview with Gordon Matta-Clark, Antwerp," 252; Matta-Clark, "Work with Abandoned Structures," 141.

44. Bear, "Gordon Matta-Clark: Splitting (The Humphrey Street Building), an Interview by Liza Bear," 34; Wall, "Gordon Matta-Clark's Building Dissections," 79; "Gordon Matta-Clark," *Artforum,* October 1978, 75.

45. ("More direct community involvement"): Gordon Matta-Clark, "Gordon Matta-Clark's Building Dissections, Typewritten Statement, Undated," in *Gordon Matta-Clark: Works and Collected Writings,* 132; Wall, "Gordon Matta-Clark's Building Dissections," 79.

46. Kirshner, "Interview with Gordon Matta-Clark," 330.

47. Karl Linn, "Landscape Architect in Service of Peace, Social Justice, Commons, and Community," interview by Lisa Rubens, transcript, 2003 and 2004, 118, Regional Oral History Office, the Bancroft Library, University of California, Berkeley, at http://digitalassets.lib.berkeley.edu/roho/ucb/text

/LinnBook.pdf; Marilyn Berlin Snell, “Down-to-Earth Visionary,” *Sierra,* June 2001, 28. See also Karl Linn, *Building Commons and Community* (Oakland, Calif.: New Village Press, 2007).

48. Linn, “Landscape Architect in Service of Peace, Social Justice, Commons, and Community,” 5, 37, 99, 115, 119, 188.

49. Linn also engaged with the urban renewal bureaucracy through meetings with figures like Ed Logue, in Boston; ibid., 94, 100, 153.

50. Ibid., 88, 90.

51. In Chicago, the offshoot group's aspirations expanded to also include housing and economic development; ibid., 118.

52. Linda Aguilar, “Between Clearance and Renewal: Curtis Place Park,” *HUD Challenge,* August 1972, 27.

53. Linn, “Landscape Architect in Service of Peace, Social Justice, Commons, and Community,” 137; Karl Linn and Urban Habitat Program, *From Rubble to Restoration: Sustainable Habitats Through Urban Agriculture* (San Francisco: Urban Habitat Program of Earth Island Institute, 1991).

54. Linn, “Landscape Architect in Service of Peace, Social Justice, Commons, and Community,” 97–98.

55. Ibid., 142, 157.

56. Dorothy Shinn, *Robert Smithson's Partially Buried Woodshed* (Kent, Ohio: Kent State University Art Gallery/Ohio Arts Council, 1990), http://www.robertsmithson.com/essays/pbw.pdf. See also John Fitzgerald O'Hara, “Kent State/May 4 and Postwar Memory,” *American Quarterly* 58 (June 2006): 301–328; Jennifer L. Roberts, *Mirror-Travels: Robert Smithson and History* (New Haven: Yale University Press, 2004).

57. See Lippard, *Six Years,* xiii.

58. As Smithson put it, “Sooner or later the artist is implicated and/or devoured by politics without even trying”: quoted in “The Artist and Politics: A Symposium,” *Artforum,* September 1970, 39.

Conclusion

1. Edward Abbey, *The Monkey Wrench Gang* (1975; New York: Perennial Classics, 2000), 85–86, 88, 90, 93–94.

2. Mischa Richter, [Untitled], *New Yorker,* May 1, 1971, cover.

3. Mary Anne Guitar, *Property Power: How to Keep the Bull-Dozer, the Power Line, and the Highwaymen Away from Your Door* (Garden City, N.Y.: Doubleday, 1972), 2; Stuart Chase, “Bombs, Babies, Bulldozers,” *Saturday Review,* January 26, 1963, 23.

4. Douglas Adams, *The Hitchhiker's Guide to the Galaxy* (New York: Har-

mony, 1979); Theodore Sturgeon, et al., "The Thing Called . . . Killdozer!," *Worlds Unknown* #6 (April 1974), Marvel Comics Group.

5. *Soylent Green,* directed by Richard Fleischer (Warner Brothers, 1973).

6. Harold Gilliam, "Beating Back the Bulldozers," *Saturday Review,* September 23, 1967, 67.

7. On protest against the bulldozer, see, for example, Raymond A. Mohl, "Stop the Road," *Journal of Urban History* 30, no. 5 (July 2004): 674–706; Adam Rome, *The Bulldozer in the Countryside: Suburban Sprawl and the Rise of American Environmentalism* (Cambridge: Cambridge University Press, 2001); Christopher C. Sellers, *Crabgrass Crucible: Suburban Nature and the Rise of Environmentalism in Twentieth-Century America* (Chapel Hill: University of North Carolina Press, 2012); Christopher Klemek, *The Transatlantic Collapse of Urban Renewal: Postwar Urbanism from New York to Berlin* (Chicago: University of Chicago Press, 2011); Samuel Zipp, *Manhattan Projects: The Rise and Fall of Urban Renewal in Cold War New York* (New York: Oxford University Press, 2010), 197–249.

8. Mrs. Bruce W. Klunder, "'My Husband Died for Democracy,'" *Ebony,* June 1964, 27–36; United Press International, "Bulldozer Kills Racial Protester: Cleveland Minister Crushed During Clash at School," *New York Times,* April 8, 1964, 1.

9. See, for example, "Apartheid = Squatters = Bulldozers = Despair," *Economist,* August 20, 1977.

10. Tom Gorman, "Earth First! Tactics in Fight to Save Planet Anger Some, Tickle Others," *Los Angeles Times,* August 14, 1988, sec. San Diego County, SD_A1, A8–A10; Ken Wells, "Earth First! Group Manages to Offend Nearly Everybody: Militant Environmentalists Are Either Perpetrators or Victims of a Bombing," *Wall Street Journal,* June 19, 1990, A1, A8; Michael A. Lerner, "The FBI vs. the Monkey Wrenchers: Dave Foreman and His Environmental Guerrillas Say They're Saving the Planet. The Government Calls Them Criminal Saboteurs," *Los Angeles Times Magazine,* April 15, 1990, 10–21; Dave Foreman, *Ecodefense: A Field Guide to Monkeywrenching* (Tucson: Earth First! Books, 1985).

11. Gorman, "Earth First! Tactics in Fight to Save Planet Anger Some, Tickle Others," SD_A1.

12. See Arthur H. Westing, *Arthur H. Westing: Pioneer on the Environmental Impact of War* (Heidelberg: Springer, 2013), 47–48; Eric C. Orlemann, *LeTourneau Earthmovers* (St. Paul, Minn.: Motorbooks International, 2001), 54–56. "New Device Rips Trees from Earth," *Hartford Courant,* October 6, 1957, 6C.

13. See Joshua B. Freeman, "Hardhats: Construction Workers, Manli-

ness, and the 1970 Pro-War Demonstrations," *Journal of Social History* 26, no. 4 (July 1993): 725–744.

14. Jane Jacobs, *The Death and Life of Great American Cities* (New York: Random House, 1961); Rachel Carson, *Silent Spring* (Greenwich, Conn.: Fawcett, 1962). See also David Kinkela, "The Ecological Landscapes of Jane Jacobs and Rachel Carson," *American Quarterly* 61, no. 4 (December 2009): 905–928; Rebecca Solnit, "Three Who Made a Revolution," *Nation,* April 3, 2006, 29–32.

15. On Controlled Demolition, Inc., the major American implosion company, see Jeff Byles, *Rubble: Unearthing the History of Demolition* (New York: Harmony, 2005), 69–104.

16. "Japanese Bulldozers Invade U.S. Market," *Business Week,* January 23, 1971, 48–49; Masahiro Sakane, "Winning in the New Workplace," in *Rediscovering Japanese Business Leadership: 15 Japanese Managers and the Companies They're Leading to New Growth,* ed. Yozo Hasegawa (Hoboken, N.J.: John Wiley, 2010), 145–146.

17. Human Rights Watch, "Razing Rafah: Mass Home Demolitions in the Gaza Strip," October 2004, http://www.hrw.org/reports/2004/rafah1004/rafah1004images.pdf; *George of the Jungle 2,* directed by David Grossman (Walt Disney Pictures, 2003); Laura Hodes, "Caterpillar Versus Disney: Why the First Amendment Does Not Protect the Use of the Bulldozer Company's Logo in George of the Jungle 2," FindLaw, October 23, 2003, http://writ.news.findlaw.com/commentary/20031023_hodes.html.

18. Stefan Al, "Introduction," in *Villages in the City: A Guide to South China's Informal Settlements,* ed. Stefan Al (Honolulu: Hong Kong University Press, University of Hawai'i Press, 2014), 1; Thomas J. Campanella, *The Concrete Dragon* (Princeton, N.J.: Princeton Architectural Press, 2011), 145–171. On Shanghai, see Qin Shao, *Shanghai Gone: Domicide and Defiance in a Chinese Megacity* (Lanham: Rowman and Littlefield, 2013).

19. See Leigh Gallagher, *The End of the Suburbs: Where the American Dream Is Moving* (New York: Portfolio Trade, 2013), 6; Smart Growth America, *Measuring Sprawl 2014,* April 2014, 5, http://www.smartgrowthamerica.org/documents/measuring-sprawl-2014.pdf; Nate Berg, "Urban vs. Suburban Growth in U.S. Metros," CityLab, June 29, 2012, http://www.theatlanticcities.com/neighborhoods/2012/06/urban-or-suburban-growth-us-metros/2419/; Nate Berg, "Exurbs, the Fastest Growing Areas in the U.S.," CityLab, July 19, 2012, http://www.theatlanticcities.com/neighborhoods/2012/07/exurbs-fastest-growing-areas-us/2636/.

20. Ken Belson, "Vacant Houses, Scourge of a Beaten-Down Buffalo," *New York Times,* September 13, 2007, sec. New York Region, B1, B6; Monica Davey, "Detroit Urged to Tear Down 40,000 Buildings," *New York Times,* May

27, 2014, A1, A15; Timothy Williams, "Blighted Cities Prefer Razing to Rebuilding," *New York Times,* November 12, 2013, A1, A15; Lindsey Rupp, "BofA Donates Then Demolishes Houses to Cut Glut of Foreclosures," *Bloomberg,* July 27, 2011, http://www.bloomberg.com/news/2011-07-27/bank-of-america-donates-then-demolishes-houses-to-get-rid-of-foreclosures.html; Janell Ross, "Cleveland to Use National Mortgage Settlement Money to Demolish Homes, Still Wants Banks to Pay," *Huffington Post,* February 19, 2012 (updated February 20, 2012), http://www.huffingtonpost.com/2012/02/19/cleveland-mortgage-settlement_n_1275896.html; Peter Y. Hong, "Housing Crunch Becomes Literal in Victorville," *Los Angeles Times,* May 5, 2009.

21. Sherri Duskey Rinker, *Goodnight, Goodnight Construction Site* (San Francisco: Chronicle, 2011); Candace Fleming, *Bulldozer's Big Day,* illus. Eric Rohmann (New York: Athenaeum, 2015); *Demolition City,* online game, available at multiple web addresses, including https://games.yahoo.com/game/demolition-city-flash.html (accessed 6/3/15); "Our History," Diggerland Home Page, http://www.diggerland.com/useful-info/history/ (accessed 6/3/15).

22. Building demolitions calculated based on U.S. Bureau of the Census, *Census of Housing, Components of Inventory Change,* 1959–1997. Farmland losses from U.S. Bureau of the Census, Census of Agriculture, tabulated in United States Department of Agriculture, National Agricultural Statistics Service, http://www.nass.usda.gov/Publications/Trends_in_U.S._Agriculture/Farm_Numbers/index.asp (accessed 3/1/2012).

Index

Page numbers in *italic* type indicate illustrations.

Abbey, Edward, *The Monkey Wrench Gang,* 287–288, 295
Adams, Douglas, *The Hitchhiker's Guide to the Galaxy,* 289
aerial photography, 56, *101, 102,* 107, *108,* 199, 255
affirmative action, 14, 152–153
African Americans, 16, 36, 170; in Army Corps of Engineers, 13, 32–33; bulldozer school for, 126; construction workers, 13–14, 32–33, 152–153, 282; protests by, 293–295, *294, 296;* urban renewal and, 154, 160, 164–165, 180, 212, 293–295, *294*
agriculture. *See* farms
airfields, 36, 61, 97, 186
Alaska, 75. *See also* Alcan Highway
Alcan Highway, 30–32, 43, 187, 188
Allis-Chalmers, 6, 62, 67, 78, 89, 90, 121, 188
American Road Builders' Association, 192
Anaheim (Calif.), 115, 133, 137, 303
Anaheim Hills (Calif.), 7, 115–117, 121, 126, 128, 129–138; fire damage, 136, 138; landform grading, 129–130, 133
anarchist environmentalists, 287–288. *See also* Earth First!
Andersen, Hans Christian, 236
Anderson, Martin, *The Federal Bulldozer,* 179
Andersson, K. S., "The Bulldozer—An Appreciation," 21, 23
Andre, Carl, 259–260
Andrew Dickson White Museum of Art (Ithaca, N.Y.), *Earth Art* show, 262–263
anthropomorphism, 8, 236–238, *237,* 289. See also *Benny the Bulldozer*
archaeology, 29–30
Architectural Forum (magazine), 57, 123
Argento, Teresa, 165, 166, 175
Army Corps of Engineers, 1–2, 7–8, 10, 139, 149, 186, 238; African American units, 13, 32–33; environmental issues and, 59–60; equipment, 21, 62, *62, 63;* formal founding of, 24; World War II activities, 23, 29–34, 41–43, 71, 74, 75, 187, 188
art, 265–266, 283–284. *See also* earthworks
Artforum (magazine), 253, *254*
artifacts, 29–30, 134–135, 217
Artzybasheff, Boris, 78, 81, *82,* 289
Aspen (Colo.) *Earth Mound,* 255
atomic bomb, 12, 62
automobiles, 103, 106, 133, 212

backhoes, 232, 252, 283, *284*
Baker Manufacturing Company, 67, 78, *80,* 90
Baltimore, 211
Bankers Trust, 188, *189*
Banks, Michael A., 242
Barnum Lumber Company, 156, 177

Bate, Norman, *Who Built the Dam?,* 228
Bayer, Herbert, *Earth Mound,* 255
Bechtel, Steve, 189, *189*
Bechtel Corporation, 188, *189*
Beck, Dave, *189*
Benny the Bulldozer (Hurd), 8, 221, *222,* 225, 236, 237, 249, 251
Best, C. L., 17
Black, Millard. *See* McClellan, Jack
Black Power, 13
blade attachments, 11, *22, 65,* 70–71, 90, *120,* 121, 127–128
Blatnik, John, 217
Bobcat Company, 90, *91,* 92, 120
Borglum, Gutzon, 253
Boston, 145, 211
Bourke-White, Margaret, *184*
bricks, 30, 169–171, 177, 279, 280, 282
Bronson, William, *How to Kill a Golden State,* 135
Browning, James, *The Busy Bulldozer,* 228, 231–232, 236–237
bucket loaders, 239
Buckeye equipment, 41
Bucyrus-Erie tractor equipment, 37, 67, 81, 90
Buda Company, 90
Buffalo (N.Y.), 173, 303
building rehabilitation, 176–177, 218, 243
building wrecking. *See* demolition; wrecking companies
built environment. *See* cities; suburban development; urbanization; urban renewal
bulldozer, 1–3, *4,* 6–18, *15,* 38–43, *45, 95,* 122–124, 127–128, 149, 193–194, 197; advertisements for, *80,* 81–82; anthropomorphization of, 8, 236–238, *237,* 289 (*see also* bulldozer books); blade design, 17, 67, 70–71, 121; capabilities of, 17; earthworks artists and, 251, 252, *260,* 268, 269, *273,* 276; feminization of, 45–46; hydraulic controls, 71, 119–120; iconic nature of, 3; masculine images and (*see* Bulldozer Man); metaphoric power of, 138–139, 146–147; negative images of, 5, 7–10, 288–293, *292,* 305; operation schools, 124, *125,* 126; postwar prominence of, 1–2, 6–7, 138–139, 298; progress and, 226, *227,* 228; symbolisms of, 9–10, 138, 293–303; taming of natural environment by, 31, 51, 52–60, 226; as tank attachment, 41–42, 70–72; toy models of, 239, *241;* Vietnam War and, 298–299; World War II and, 5, 21–22, 27–28, *35,* 38–52, *39, 45,* 62, *63,* 74–82. *See also* earthmoving
bulldozer books, 3, 8, 12, 16, 41, *219,* 221–238, *235,* 244, 252, 304; four common themes, 226–232; gender roles in, 231–232; negative depictions in, 246, 248–249, 300
Bulldozer Man, 23–24, 43–60, 121–122, 232, *233, 234,* 249, 255; as all-American ideal, 46–49, *47;* earthworks artists and, 255, 266; film portrayals of, 44–49, *45, 47;* masculinity of, 6–8, 14, 23, 44, 49, 230, 235–236, 249; as modern cowboy, 14, 44–45, 266
Bureau of Public Roads (BPR), 186, 187, 188, 190, 214
Burton, Virginia Lee, 223, 224, 243; *Katy and the Big Snow,* 236, *237; The Little House,* 229, 243; *Mike Mulligan and His Steam Shovel,* 228, 232, 243, 289

California, 7, 98–138; brush and wildfires, 135–136, 138; citrus industry, 104–107, 112 (*see also* orange orchard losses); earthmoving firms, 126; freeway revolts, 211, 293; grading loopholes, 137–138; grading regulation leadership, 136; grassroots protests, 293–295; hillside development, 114–134; interstate highways, 183, 186,

198–199, 203; land-clearing reforms, 113–114; Land Conservation Act (*see* Williamson Act); postwar moves to, 100; rights-of-way policy of, 202–203. *See also specific place names*
Cappel, Constance, and Raymond Montgomery, *Vermont Roadbuilder,* 230, 231, 232, *235,* 246
Carroll, Quentin, 43
Carson, Rachel, *Silent Spring,* 300
Caterpillar Military Engine Company, 37
Caterpillar Tractor Company, 1, 6, 17, 18, 23, 69, 70, 225, 287–288; advertisements, 78, 81; blade designs, 121; British plant, 89; bulldozer models, 2, *4, 17,* 29, 31, 42; copyright, 302; highway construction equipment, 191, *191;* Japanese competitor, 9, 301; model toys, 225, 239, *241,* 242–243; postwar reconversion, 88–90, 92, 108–109, 121–122; urban renewal equipment, 149, *150, 153;* wartime annual reports, 61, *62, 63,* 68, 69–70, 75; wartime employment numbers, 72–74; World War II equipment dominance, 67–68, 77
Central Valley (Calif.), 111
chain saws, 107, 252, 275
Chase, Stuart, "Bombs, Babies, Bulldozers," 289
Chavez Ravine project (Los Angeles), *15,* 15–16
children's books, 3, 8, 12, 16, 41, *219,* 221–238, 252, 304; activity suggestions, 230, 300; career-oriented, 230; counternarratives in, 243–250. *See also* bulldozer books; *specific authors and titles*
Chinese redevelopment plans, 302–303
cities: community design and, 278–286; current demolition programs, 303–304; earthworks and, 267–271; historical eviction and clearances, 5, 147–148; interstate highway construction and, 204–209; preservationists and, 177–180, 300; proactive environmental destruction and, 52; resurgence of, 203–204; U.S. growth (2011) of, 203; World War II heavy bombing of, 23, 53, 55–57, 59, 61, 279. *See also* urbanization; urban renewal
Civil Rights movement, 14
Clark Equipment Company, 92
Clay, Grady, 209–210
Clay, Lucius, and Clay Committee, 188, *189*
clearance. *See* culture of clearance; demolition; earthmoving; wrecking companies
Cleveland construction site protest, 294–295, *296*
Cleveland Tractor (Cletrac), 69
Cleveland Wrecking Company, 159
Coca-Cola, 49, *51*
Codding, Hugh, 98, 100
Colby, Jean Poindexter, *Tear Down to Build Up,* 226, *227,* 229, 230, 232
Cold War, 9, 99, 186, 192, 226, 230–231
Cole, Albert M., 146–147
Colman, Hila, *Andy's Landmark House,* 245
Colorado interstate highway construction, *190, 198,* 208–209, 212, 215
Colt, Sloan, *189, 201*
comic books, 289, 291, *292*
Commercial Contractors, 151, 159, 160–162
community-based projects, 279–282, 283
community design, 278–286
Community Development Block Grants (CDBGs), 180
compaction, 29, *30,* 127, 130, *132,* 137
compact skid-steer loader, 90, *91,* 92
conceptual art. *See* earthworks
conservation. *See* culture of conservation
construction equipment, 5, 8, 12, 17, 18, 23, 61–93; adverse portrayals of, 287–289, 291; anthropomorphization of, 8, 236–237, 289; children's toy models, 238–239, *241,* 249; earthworks art based on, 252, 255,

construction equipment (continued) 256, 257, 283; sales growth (1941–1948), 64; societal adverse view of, 289, 291; wartime, 61–82; World War II experience with, 61–82; written histories of, 76. *See also* *specific types*

construction workers, 9, 72–82; children's book portrayal of, 232, 235, 249; demolition and, 151–153; employment discrimination, 13–14, 32–33, 282; "hardhat" negative image of, 299–300; masculine image of, 230–232, 249; World War II recruitment of, 25–27, 298. *See also* Bulldozer Man

contour grading (sculpting), 128–129, *131*

Cooke, David C., *How Superhighways Are Made,* 225, 226, 229, 232, 235, 238

Cornell University, 262–263, 268

Corrie, Rachel, 295

cowboy image, 14, 44–45, 266

cranes, 11, 71, 72, 83, 139, 149, 154, 226–228, *227,* 239

crawler tractor, 17, 77, *79,* 88, 89, 108–109

creative destruction, 12, 259

Cubbedge, Robert, *Destroyers of America,* 135

culture of clearance, 1–18, 93, 103, 133, 297–305; challenges to, 243, 252–253, 259–261, 293, 299–300; as children's books theme, 225–238, 243–250, 304; international landscape and, 302–303; World War II record of, 27–37, 61, 298. *See also* demolition; earthmoving; urban renewal; wrecking companies

culture of conservation, 287–305

Dahl, Roald, *Fantastic Mr. Fox,* 243–244

Danner, Catherine, *Buster Bulldozer,* 8, *219,* *223,* 236–237

Dasmann, Raymond, *The Destruction of California,* 135

debris. *See* refuse disposal

Deering Harvester, 65

Defense Highway Act (1941), 200

De Maria, Walter, 9, 261, 266; "Art Yard," 251–252; *The Color Men Choose When They Attack the Earth,* 257–258; *Earth Room,* 283; manifesto, 285; *Munich Earth Room,* 258; *Three Continents,* 259, *260*

demolition, 3, 5–8, 92, 139–141, 144, 147–172, 173; accommodation of, 229–230; Chinese program of, 302–303; costs of, 151–152; downgrading of, 180, 300; earthworks and, 268–277; "fear and dust" of, 163–172; of houses, 15, *15,* 34–35, *35,* 148, 246, *247;* implosion method of, 301; as natural evolution paradigm, 226, 228; negative image of, 297; in nineteenth and early twentieth centuries, 147–148; public health hazards of, 168–170; Rust Belt cities and, 203–204; viewed as progress, 226, 228. *See also* urban renewal; wrecking companies

Demolition City (video game), 304

Detroit, 148; Blight Removal Task Force, 303–304

diesel engine, 66, 70, 88, 90

Dietrich, Marlene, 55–56

Diggerland (theme park), 304–305

dinosaurs, 226

dirt. *See* earthmoving; earthworks

Disney Company, 302

displacement. *See* interstate highways; suburban development; urban renewal

Doepke, Charles and Fred, 238–239, 242

Doepke Manufacturing Company. *See* Model Toys

Donohue, William T., 178

Dove, Vinton Walsh, 42

Dresden, bombing of, 53

drones, 12

Duchamp, Marcel, Readymades, 253

dumping sites, 155, 268

dump trucks, 239, *269*

Dwan, Virginia, 253, 261
Dwan Gallery, *Earth Works* exhibit, 57, 253, *254,* 256, 258, 268

Earth Art show (Cornell University), 262–263
earth-based sculpture. *See* earthworks
Earth Day, 216
Earth First!, 295, 297
earthmoving, 1–7, 13, 37, 77, 97–139, 188; allure of, 123–124; archaeological finds, 29–30; bulldozer importance to, 119, 122–124, 127–128; California soil loss from, 111–112, 116; children's books on, 220–223, 225, *226,* 227, 243–245, 249; collateral damage from, 31–32, 117, 134–139; cut-and-fill phase of, 127–128, *128;* decline in, 305; defense of, 117; earthworks and, 257–258; English theme park on, 304–305; mechanics of, 117–139; plastic surgery metaphor for, 82; social concerns over, 287–289; technological development and, 3, 6, 17–18, 62, 67, 114–139, *118, 119;* temporary uses for, 175; Tonka toys and, 239–250; unintended consequences of, 114–115, 133; World War II and, 6, 11, 21, *22,* 23, *28,* 28–29, 43, 103–107, 119. *See also* hillside development; interstate highways
earthworks, 3, 9, 14, 16, 250, 251–286, 300; ephemerality of works, 261; political ramifications of, 262; site-specific projects, 251–267
Earthworks: Land Reclamation and Sculpture project (Wash.), 266, *267*
Eastman, P. D., *Are You My Mother?,* 246
Edelstadt, Vera, *A Steam Shovel for Me!,* 231
Eisenhower, Dwight D., 71, 186, 187–188, *189*
Ellett, Roy, 43
eminent domain, 33, 238
Engineers Corps. *See* Army Corps of Engineers
England: earthmoving theme park, 304–305; London World War II Blitz rubble, 55, 56–57; Roman artifacts discovery, 30
Environment. *See* natural environment
Environmental Impact Report (1974), 134
environmental movement: children's books and, 245; critics of earthworks and, 266, 267; emergence of, 116, 216, 229, 243, 300; fictional anarchists and, 287–288; landmark loss and, 134; legislation for, 9, 133–134, 300–301; militant tactics and, 295, 297
Environmental Protection Agency, 168, 216
Environmental Quality Act (Calif., 1970), 133–134
erosion, 136
Euclid Company, 18, 121, 188; model dump trucks, 239
evolution, 226
excavation. *See* earthmoving
expressways, 183. *See also* freeways; interstate highways
exurban areas, 303
Eyerman, J. R., 46–49, *47, 48*

Fair Housing Act (1968), 14
fairy tales, 225, 226
farms: declining numbers of, 7, 103–117, 305; equipment for, 17, 65–66, 89, 90–91; highway right-of-way and, 202, 228; preservation legislation and, 100. *See also* orange orchard losses
Farm Security Administration (FSA), 36
Federal-Aid Highway Act (1916), 186
Federal-Aid Highway Act (1938), 186
Federal-Aid Highway Act (1944), 187, 200
Federal-Aid Highway Act (1956), 93, 188–190, 192, 199, 209
Federal-Aid Highway Act (1962), 213

Federal-Aid Highway Act (1973), 218, 300
Federal Highway Administration (FHWA), 200, 215
Federal Housing Authority (FHA), 99
feminists, 266
feminized machines, 45–46, 232, 236, *237*, 282, 289
Field, Weldon, 108, *109*, 109–110
Fighting Seabees (film), 44–46, *45*
films, 3, 5–6, 9, 23, 55–56, 57; Bulldozer Man portrayal, 44–46, *45*
fires, 9; California brush and wildfires, 135–136, 138; refuse burning, 109–110, 155; urban renewal, 160–162, 167
Fleming, Candace, *Bulldozer's Big Day,* 304
Florida, 106, 203–204
Foreign Affair (film), 55–56
Foreman, Dave, *Ecodefense: A Field Guide to Monkeywrenching,* 295, 297
forests. *See* trees
found objects, 258, 280
Fourth of July, 163, 221, *222*, 237
Frank G. Hough Company, 90
Freeman, Don, *Fly High, Fly Low,* 243
freeways, 105, 199, 206; revolts, 211, 293
Fresh Kill (film), 269–270
Friedrich, Heiner, 258, 261
Friends Neighborhood Guild (Philadelphia), 280
Friesen, Pete, 204
front-end loaders (shovel dozers), 18, 120, *184, 261*
Fugate, Douglas, 188–189, 199, 215
Functional City concept, 56
fungal disease, 111

Gans, Herbert, 180, 279; *The Urban Villagers,* 179
Garnett, William, 107, *108*, 111, 253
Gar Wood Industries, 67, 78
gender roles, 49, *50*, 231–232, 289. *See also* Bulldozer Man; feminized machines; masculinity; women
General Motors, 188
George of the Jungle 2 (film), 302
Germany: autobahn, 188; World War II and, 53, 55–56, 59, 81
G.I. Bill (1944), 6–7, 93, 98, 99–100
Gilliam, Harold, "Beat Back the Bulldozers," 293
Gil Wyner Company, 7–8
Glenwood Canyon project (Colo.), 215
Glueck, Grace, 255
golf course design, 128
grading, 32, 83, 107, *108*, 117, 119, 122, 126–130, *129;* California standards, 136; consequences of, 133–139; types of, *131*
Great Salt Lake, 259, *260*
Greene, Carla, *I Want to Be a Road-Builder,* 228, 230, *234*
Greer Earth Moving School (Ill.), 124, *125*, 126
Guadalcanal, 29, 46, 75
Guam, 29–30, 47, 87
Guitar, Mary Anne, *Property Power,* 289
Gussow, Alan, *A Sense of Place,* 265–266

Haase, Edward, 197, 208–210, *210*, 214
Halpin, Gerald, 98
Hard Core (film), 266
Hardy, Hugh, 117
Harwood, Pearl Augusta, *Mr. Bumba and the Orange Grove,* 244
Haussmann, Baron von, 147
Hawaii, 25, 33–34, 87
Heizer, Michael, 9, 252, 257, 262, 263, 265, 266, 267, 285; *Dissipate #2,* 258; *Double Negative,* 258–259, 260, 261; *Munich Depression,* 261; *Nine Nevada Depressions,* 258
highways. *See* Alcan Highway; expressways; freeways; interstate highways

Highway Trust Fund, 183, 218
hillside development, 114–117, 122, 126–139; cultural consequences of, 134–135; guidelines publication, 126–127. *See also* Anaheim Hills; grading
Hilton, Suzanne, *How Do They Get Rid of It?*, 244
Hine, Lewis, *Men at Work*, 231
historic preservation, 177–180, 216, 243, 246, 277, 300; children's books on, 245–246; legislation, 9–10
Hollywood Hills (Calif.), 136
Holt, Benjamin, 17, 41
Holt, Nancy, 266
Holt Company, 69
Home Builders Manual for Land Development, 126–127
homosexuals, 14, 165, 230–231
house moving, 178, 185, 204–208, *205*, 229, 243
housing: demolition, *15*, 15–16, 34–35, *35*, 148, 246, *247*; demolition statistics (1950–1980), 5; discrimination, 14, 295, *296*; federal programs, 99; market collapse (2008), 203; mass construction of, 103, 104 (*see also* tract housing); public housing, 15, 165, 166, 301; single-owner, 105–106; urban rehabilitation, 180. *See also* mortgages; suburban development
Housing Act (1937), 148
Housing Act (1944), 99
Housing Act (1949), 7, 93, 144–145
Housing Act (1954), 93, 176
Housing and Community Development Act (1974), 180, 300
Housing and Home Finance Administration, 146
Human Rights Watch, 302
Huntsville (Ala.), 145
Hurd, Edith Thacher, *Benny the Bulldozer*, 8, 221, *222*, 225, 236, 237, 249, 251
hydraulic technology, 5, 71–72, 119–120, 205

Internal Revenue Code (1954), 7, 100
International Harvester Company, 6, 21, *22*, 62, 64–67, *95*, 98, 225; ad campaigns, 78, *79*; crawler tractors, *65*; expanded product line, 89, 90; farm equipment, 65, 89; formation of, 65; postwar conversion of, 88–89; World War II operations, 65, 69, 71–76
interstate highways, 3, 10, 92, 93, 114, 121, 130, 145, 181, 182–218; aerial view of, *190*; bulldozer procedure for, 31, 181, 191; children's books on, 221, 222, 225, 226, 228, 230, 232, 235, 238, 244; collateral damage from, 213–214; costs escalation, 217–218; design of, 189–190, 192–193, 197–198; displacement compensation for, 204; displacements from, 8, 13, 164, 182–183, 197, 203–204, 210–213, 293; federal funding of, 7, 188–189; first formal concept of, 186–187; impact of, 208–218; land clearing for, 7, 183, 185, *191*, 191–192, 197–199, 238; landscape imprint of, *198*, 198–199, 216–218; as limited access, 187, 200, 202; as multilane, 187; network mileage of, 183; overpass clearance standards for, 192, 193; property speculators and, 202; protests against, 180–181, 293; public hearings on, 209–210, *210*; reorientation of, 300; right-of-way acquisition for, 17–18, 199–204, *201*, 207, 213, 218; scope of takings and displacements from, 210–211; standards and, 183; urban areas and, 180–181, 203–208
irrigation systems, 104, 111, 112
Israeli-Palestinian conflict, 295, 302
Ithaca (N.Y.) art sites, 261, 262–263

Jacobs, Jane, 211, 279; *The Death and Life of Great American Cities*, 179, 300

Japan: postwar bulldozer manufacture, 9, 301–303; World War II and, 25, 38–40, 53, 59, 75–76, 81
Jeffries, Willie, 160
Johnny's Machines (Little Golden Book), 246
Johnson, Philip, 122–123
Johnson Pit (Kings County, Wash.), 266–267, *267*

Kearney (Neb.), 36–37, 86–87
Keller, Louis and Cyril, 90
Kelley, Ben, *The Pavers and the Paved,* 213
Kent State University, 283, *284*
King, Stephen, 2
Klunder, Bruce, 295, *296*
Komatsu, 9, 301–302
Korean War, 87, 92

labor unions, 42, 146, 152, 188
Laderman, Mierle, 266
Lakewood (Calif.), 107–108, *108,* 111
land clearance. *See* earthmoving; interstate highways; suburban development; urban renewal
landfill, 269–270
landform grading, 129–130, *131*
landscape. *See* built environment; natural environment
landscape architects, 278–286
landscape-based art, 283–284. *See also* earthworks
landscape painting, 265–266
landscaping, 130, 132, 135
landslides, 7, 59, 136, 137–138
LaPlante-Choate Manufacturing Company, 17, 70, 90
Lassen, Ken, 177–178, *179*
lead-dust exposure, 168, 212
Leavitt, Helen, *Superhighway—Superhoax,* 213
Le Corbusier, 56
Ledo Road (Burma), 29, 33, 74
Lee, Dick, 140–145, *141, 142,* 152, 154, 161–162, *170,* 172–173, *173,* 175–178
Les McCoy & Sons, 7, 121–122, 126, 128
LeTourneau, 18, 62, 67, 70–72, 88–90, 119; "Goliath," 121
LeTourneau, R. G., 76
Levitt, William, 6, 97–98, 110
Levittown, 6, 97–98, 110
Life magazine photo-essays, 8, 28–29, *51,* 58, 75, 100, *125,* 135, 191, *196,* 205, *205;* bulldozer operator, 46–49, *47;* urban renewal, 140, *141, 142,* 145–146, 147, 172–173, *173,* 176, 177; World War II, 32, 36
Lillard, Richard, *Eden in Jeopardy,* 135
Linn, Karl, 278–282, 285; *From Rubble to Restoration,* 282
literary works, 9, 44, 49, 51, 287–293. *See also* children's books
Little Golden Books, 223, 246
Logue, Edward, 144–145
London Blitz, 55, 56–57
Long, Richard, 262
Long Island (N.Y.) Levittown, 6, 97–98, 110
Los Angeles, 105, 115, 135, 295, 303; house moving, 204–205, *205,* 206; razing of houses, *15,* 15–16
Los Angeles County, 109, 115–116
Louis' Lunch (New Haven), 177–178, 179, *179*

MacArthur, Douglas, 23
Macaulay, David, 223–224; *Unbuilding,* 248–249
MacCornack, Walter, 56–57
MacDonald, Thomas H., 187
MacIntyre, Elisabeth, *The Affable, Amiable Bulldozer Man,* 231, *233*
Madalla, David, 262–263
malls. *See* shopping centers
manual labor, 13, 146, 152, 160, 231

Marvel comics, 289, 291, *292*
masculinity, 6–8, 14, 23, 44–45, 49, 230–232, 235–236, 249, 266
Massachusetts Institute of Technology, 56, 281
Matta-Clark, Gordon, 9, 252, 267–278, 283, 285; *Bingo/Ninths,* 272, *273,* 276; *Circus* or *Caribbean Orange,* 273–274; *Conical Intersect,* 273, *274; Day's End,* 272–273; *Fresh Kill,* 269–270, 276; *Office Baroque,* 273; *Splitting, 271,* 271–272
McClellan, Jack, and Millard Black, *What a Highway!,* 244
McCormick, Cyrus, 65
McCoy, Les. *See* Les McCoy & Sons
McManus, Francis, 169, 170
McWilliams, Margo, and Patricia Reisdorf, *Let's Go to Build a Highway,* 229, 230, 236
Meader, Stephen, *Bulldozer,* 1–3, 8, 242
Melon Neighborhood Commons (Phila.), 280, 282
Mendieta, Ana, 266
Mexican Americans, *15,* 15–16
Michener, James, 49–50; *Tales of the South Pacific,* 49, 51, 57
Middle East, 203, 295, 302
"middle landscape" (rural-urban), 104
Milwaukee, 148
minimum wage, 152
minorities, 8, 9, 12, 13, *15,* 152–153, 162, 164–165, 211. *See also* African Americans
Miss, Mary, 266
Mitchell, Lucy Sprague, 225–226; *Here and Now Story Book,* 225
Model Toys, 238–240, *241*
Mono Island Pillbox episode, 38–40, *39*
Montgomery, Raymond. *See* Cappel, Constance
Moore, Anne Carroll, 225
Moreel, Benjamin, 25
Morison, Samuel Eliot, 29
Morris, Robert, 252, 264–265, 285; *Earthwork,* 258, 266–267, *267*
Morrison-Knudsen, 24, 25, 33–34
mortgages: federal subsidies, 99–100, 103; foreclosures, 304; gender discrimination, 14; veterans, 6–7, 99–100
Moses, Robert, *4,* 144, 211
Mound Metalcraft Company. *See* Tonka Toys
mountain cropping, 116, 117–139, *118*
Mount Rushmore National Monument, 253
Mowbray, A. Q., *Road to Ruin,* 213
Muhonen, Walter, 103–104, 107
Mumford, Lewis, 193
Museum of Contemporary Art (Chicago), 273–274, 278
Museum of Modern Art (NYC), *Family of Man* exhibit, 56

NAACP (National Association for the Advancement of Colored People), 32
Napa Valley (Calif.), 256
National Association of Home Builders, 114–115, 126–127
National Association of Wrecking Contractors, 148
National Defense Education Act (1958), 226
National Demolition Association, 148
National Environmental Policy Act (1960), 300–301
National Environmental Policy Act (1969), 216, 217
National Historic Preservation Act (1966), 180, 216, 300
National Interregional Highway Committee, 187
National School of Heavy Earthmoving Operation (Charlotte, N.C.), 124
National System of Interstate and Defense Highways, 185, 187
Native American relics, 134

natural disasters, 106. *See also* fires; landslides
natural environment, 5–8, 168–170, 263–264, 285; American historical conquest of, *4*, 5, 6, 13; bulldozer taming of, 31, 51, 52–60, 228; conservation ethic and, 287–305; as enemy to be conquered, 24; highway construction impact on, 185, 197–198, 208–217; hillside grading consequences and, 133–139; land clearance and, 103, 110, 113–114, 116–117; land reclamation and, 104, 263–286; suburban development impact on, 106, 110, 111, 265; Vietnam War damage to, 298–299; World War II scars on, 57–59. *See also* environmental movement
Naval Construction Battalions. *See* Seabees
Neighborhood Renewal Corps, 280–281
Newark (N.J.), community design and, 281–282
New Deal construction projects, 24
New Haven (Conn.) urban renewal, 7–8, 140–178, *141, 142, 157, 170;* assessments of, 172–173, *173,* 176–178; bidding for, 149, 151, 159; displacements from, 164–171, 173, 293; empty spaces from, *173,* 173–174, *174;* first six demolition contracts, 151; mixed emotions about, 175–176; neighborhood boundaries and, *153,* 154; Oak Street Connector and, *170,* 175; preservation and, 177–178, *179;* riots (late 1960s) and, 170; Shoninger fire and, 160–163, 167
New Haven Coliseum, 301
New Haven Redevelopment Agency, 155, 163, 173, 178, 180
New Jersey Turnpike, 183
New York City: Matta-Clark site-specific works, 269–273, 278; Moses projects, 144, 211; urban renewal bidders, 148; West Village mixed-use, 179
New Yorker (magazine), cover cartoon, 288–289, *290*
New York State Thruway, 182–183, *184,* 193, 195, *196,* 197, 216
Nichols, Herbert L., *Moving the Earth* (workbook), 127
Nimitz, Chester, 46
Nixon, Richard M., 226
Noble, Charles M., 183
Noguchi, Isamu, *This Tortured Earth,* 253, 255
noise pollution, 212, 217
Normandy landing (1944), 41–42
North Dakota air base, 186

Oakland (Calif.), 175
office complexes, 98
Ohio interstate highway, 183, 187
Okinawa, 42–43
Operating Engineers (union), 145, 152
Oppenheim, Dennis, 256, 260, 262, 263, 264, 285; *Accumulation Cut,* 261; *Contour Lines Scribed in Swamp Grass,* 256; *Directed Seeding—Cancelled Crop,* 256; *Landslide,* 256
Orange County (Calif.), 7, 100–131, *102;* as fastest-growing U.S. county, 100. *See also* Anaheim; Anaheim Hills
orange orchard losses, 7, 100, 103–113, 244, 293; equipment used for, 107–109; selective salvage and, 110–111

Palestinians, 295, 302
Palma Contracting Company, *95*
Paris, nineteenth-century remaking of, 147
parking lots, 8, *174,* 175
patriotism, 6, 8, 44, 77, 78, 82, 237, 298
Peace Corps, 280
Pearl Harbor attack (1941), 25
Pennsylvania Turnpike, 183
Pershing Map, 186
pesticides, 106
Philadelphia: community development, 279–281, 282; Society Hill, 180

Philadelphia College of Art, 279–280
Philippines, 59, 86, 188
photogrammetry. *See* aerial photography
photography, 3, 5–6, 9, 16, 231; Earth First! utilization of, 297; of wartime urban rubble, 55, 56. *See also* aerial photography; *Life* magazine photo-essays
Pisani, Adolph, 29
Pitman, Ray, 98
Pittsburgh urban renewal, 144
Pivetti, Chuck, 193, 194
political conservatism, 103
political dissent, 275–276, 283, 295
Pollock, Jackson, 255
pollution: highway, 212, 213, 217, 293; suburban development, 106, 293; urban demolition, 168–169
preservation, landscape, 110, 111, 113. *See also* historic preservation
progress, 13, *227;* children's books on, 226, 228, 230; demolition-based, 226, 228; earthworks artists' critique of, 259; questioned definitions of, 297
property speculation, 202
property takings, 212; compensation for, 212–213
protest: bulldozer as symbol of, 9–10, 293–302; bulldozer used as weapon against, 295, *296;* suburban development, 100, 293, 295; urban renewal, 180–181; Vietnam War, 283, 299
Pruitt-Igoe housing complex (Saint Louis), 301
public housing, 15, 165, 166, 301
Public Works Historical Society, 185

race: of bulldozer operator, 14, 44, 124, 126; of urban renewal workers, 152–153. *See also* African Americans
race riots, 9, 170–171, 281
racial discrimination, 13–14, 32–33
Rauschenberg, Robert, 255
reaper, horse-drawn, 65
recreation areas, 114, 117
recycling, 244, 249
Red Hill (Hawaii), 33–34
refuse disposal, 154, 155–156, *170,* 268; by burning, 109–110, 155
Reisdorf, Patricia. *See* McWilliams, Margo
relocation: interstate highway construction, 183, 185, 203, 213, 218; urban renewal, 8, 9, 13, 162, 164–172, 183, 213; World War II, 35–36
Revere Copper and Brass, 83, *83,* 139, 172
Revolutionary War, 24
Reybold, Eugene, 21, 23
Richter, Mischa, 289, *290*
Rinker, Sherri Duskey, *Goodnight, Goodnight Construction Site,* 304
riots, 9, 170–171, 281
Ritter, James R., 25
Riverside County (Calif.), 109
R. J. Reynolds Tobacco Company, 40
Roberts, Bill, *189*
rock removal, 128, 191
rock slide, *137*
Rodgers Hydraulic Inc., 81, *84*
Rome Plows, 298
Roosevelt, Franklin, 38, 186, 187
Rowe, Lucius, 143–144
rubble clearance, 23, 30, 53, 55, *56, 63,* 70, 268
rural areas: highway right-of-way, 200, 202, 208; house moving from, 229, 243; as new frontier, 14; wartime ruins, 58–59. *See also* farms
Russell Grader Manufacturing Company, 17
Rust Belt, 203–204

Saint Louis (Mo.), 173, 301
salvage, 149–151, 154, 159, 177; children's book on, 244; sculptures made from, 282

San Diego, 202
San Francisco, 211, *294*
San Penasquitos Canyon (Calif.), 295
Santa Ana Canyon (Calif.) Improvement Association, 133
Santa Clara County (Calif.), 113
Scenic Easement Deed Act (1959), 113
Schor, Horst J., 129–130, *131*
science education, 226
science fiction, 289, 291
scrapers, 3, 6, 11, 61, 67, 90, 119–120, *120*, 121, 127–128, 130, 138, 191, 239; capabilities of, 119; first high-speed motor, 18
Scull, Robert, 261
Seabee Museum (Calif.), 46
Seabees, 6, 23–52, *35, 39, 50*, 71, 99, 187; construction experience of, 97, 98, 208, 238; creation of, 25–27; environmental damage and, 58–59; fictional accounts of, 49–52; glamour of, 298; heavy equipment deployment by, 61–62, 75; heroism of, 38, 43, 75; high pay of, 26; Hollywood portrayal of, 44, *45;* infrastructure building by, 27–37, *28;* Korean War reactivation of, 87; media attention to, 40–49, *47, 48;* motivations for joining, 26–27; postwar and, 86–87, 97, 186; projects statistics, 23
Seattle-Tacoma airport, 266–267, *267*
Sebald, W. G., 56
Selby, Glenn, 46, *47*, 48, *48*, 49
Sellers, Robert H., 47–49, *48*
Servicemen's Readjustment Act (1944). *See* G.I. Bill
Sharp, Willoughby, 262
Sherman tanks, 71, 72
Sherrod, Robert, 58
Shoemaker, William J., 42
Shoninger Building (New Haven), 160–163
shopping centers, 97, 98, 100
shovel dozers. *See* front-end loaders
shovels, 11, 252
Shumate, Charles, 200, 215–216
Skelly, James R. *See* Zim, Herbert S.
slum clearance. *See* urban renewal
Smart Growth America, 203
Smithson, Robert, 9, 252, 253, 256–263, 265, 267; as earthworks voice, 285; *Franklin Non-site*, 258; *Major Displacement (Cayuga Salt Mine)*, 269; non-sites, 262, 263; *Partially Buried Woodshed*, 283, *284; Spiral Jetty*, 259, *260*, 261
Smyth, Ned, 270
social concerns, 185, 275–282, *284*, 285, 287–288, 293, 299, 304
soil. *See* earthmoving
Solomon Islands, 29, *30*, 34, 36
Sonoma County (Calif.), 98
South Pacific (musical), 57–58
Soviet Union, 192, 226
Soylent Green (film), 291
Sparkman, John, 145
Sputnik, 192, 226
state highway departments, 194–200, 204
steam shovel, 5; bulldozer as successor to, 17–18; children's books on, 228, 232, 243, 282, 289; as dragon, 289, *290;* earth-based conceptual art and, 251; feminization of, 232, 289; Tonka toy, 239
Steichen, Edward, *Family of Man* exhibit, 56
Stevenson, Adlai, 143
Stilwell Road. *See* Ledo Road
strip-mining sites, 263, 264
stump removal, 108, *109*, 127, 128
Sturgeon, Theodore, et al., "The Thing Called . . . Killdozer," 289, 291, *292*
suburban development, 3, 5–7, 10, 12, 13, 92, 97–139; adverse environmental effects of, 106, 110, 111, 265; backlash against, 111–133, 293; California reforms, 113–

114; critics of, 135, 275; hillside, 114–117; Matta-Clark building cuts, *271,* 271–272, 275; negative consequences of, 135–139; protests against, 100, 293, 295; scale of, 115; sprawl and, 9; tree selective retention for, 111; urban growth (2011) vs., 303; World War II and, 23, 28–29, 36, 61. *See also* tract housing
Surface Mine and Reclamation Act (1977), 263
surveying methods, 199

Talbot, Toby, *My House Is Your House,* 245–246, *247*
Tallamy, Bertram, 183, 195, *196,* 197, 216
tankdozers, 41–42, 70–72
tank models, 41, 88
Tashlin, Frank, *The Bear That Wasn't,* 229–230
Tassone, Aurelio, 38–41, *39,* 42, 43, 75
taxes, 7, 76, 92, 106, 113; excise elimination, 92; postwar code changes, 100
tax incentives, 5
Teamsters Union, 146, 188
Tibbetts, Abbott, McCarthy, and Stratton, 256–257
Tokyo, firebombing of, 53
Tonka Toys, 8, 225; product line, 239–240, 249
topographic maps, 127, 199
torque converters, 120
toys, 8, 225, 238–243, 249
tract housing, *4,* 6, 97–98, 110, 112, 293; repetitive design of, 135
tractors, 3, 11, 61, *62, 63,* 66; attachment versatility, 67, 71–72, 81–82, 108–109, *109, 120,* 128; blade-fronted crawler (*see* bulldozer); bulldozer blades, 11, *22, 65,* 70; California land clearance and, 107–108; children's book about, 230, 232, 244–245; crawler, 17, 77, *79,* 88, 89, 108–109; development of, 17; earth-based sculptural art and, 252; hydraulic controls, 71; models, 66, 67, 240, *241;* model toys, 239, *241;* postwar production of, 6; sheepskin rollers, *132;* wartime production of, 66, 67, 75
traffic volume, 133, 134
tree cutter, 107
treedozer, 108, 191, 298–299
tree removal, 107–111, 127. *See also* orange orchard losses; stump removal
trees: anti-logging protests, 295, 297; damage to, 59; salvaging of, 130, 132
trucks, 66, 149, 239, *269;* military convoys of, 186; toys and models (*see* Tonka Toys); transcontinental convoys of, 186, 188
Tupper, Margo, *No Place to Hide,* 135
Turnbull, Charles, 38, 39, 40
Turner, Frank, 188, *189*
Turner Construction, 24
Tysons Corner (Va.), 98

Ukeles, Mierle Laderman, 266
Uniform Relocation Assistance and Real Property Acquisition Policies Act (1970), 213
unions. *See* labor unions
United Nations Relief and Rehabilitation Administration (UNRRA), 86
University of Pennsylvania, 279
Urban Development Action Grants, 180
urbanization: factors in, 5–6, 12; social consequences of, 177; soil loss to, 111–112. *See also* cities
urban renewal, 3, 5, 7–8, 10, 13–16, 53, 81–82, 92, 140–181, *141, 142,* 275, 301; alternatives to, 179–181; bulldozer role in, 7–8, 83, 138–139, 146–147, 149, 154; collateral damage from, 147, 160–162, 212; common visual perception of, 146–147;

urban renewal (continued)
community-based projects and, 279–281; displacement from, 164–171, 173, 212–213, 282; documentation of, 146; federal official end (1974) to, 180, 300, 304; fires and, 160–163; interim solutions, 174–175; Matta-Clark critique of, 275, 276, 285; metaphors for, 57, 82, *84,* 172; minorities and, 13, *15,* 15–16, 180, 212; as new frontier, 14; popular rejection of, 178–181, 293, 295; protests against, 293–295, *294;* rehabilitation and renewal vs. clearance in, 176–177, 218, 243, 279–282; relocation from, 8, 9, 13, 162, 164–172, 183, 213; resident upsets from, 165–172, 212, 245–246; sadness of, 145; statistics, 144; vacant lots left by, 8, *174,* 175, 281; workers' wages, 152. *See also* New Haven (Conn.) urban renewal

vacant city lots, 8, 280, 281–282, 304
Ventura County (Calif.), 112
vertical integration, 89–90
vest-pocket parks, 282
veterans, 185–186, 195; G.I. Bill (1944), 6–7, 93, 98, 99–100
Victory Ordnance Plant (Ill.), 73
video games, 304
Vietnam War, 9, 46, 258, 262, 288, 300; antiwar protests, 283, 299; South Vietnamese environmental damage, 298–299

Wagner, Martin, 57
walnut trees, 110
War Production Board, 65, 73, 77
Washington, D.C.: community involvement, 280–281, 282; urban renewal, 144–145
water trucks, 130
WAVES (Women Accepted for Volunteer Emergency Service), 49, *50*
Wayne, John, 44, *45,* 46
Wells, Malcolm, 135
white-collar workers, 235–236
Whyte, William, 135
Wickwire Spencer Steel Company, 53, *54,* 55; advertising, 81, *82*
Wilder, Billy, 55–56
Wilderness Acts (1964, 1970), 216
wildfires, 135–136
wildlife, 59, 114, 243–244
Williamson Act (Calif., 1965), 113–114, 116
women: children's book publishing dominance of, 224, 232; construction jobs and, 12, 13, 14; earthworks genre and, 166; mortgage discrimination against, 14; wartime employment of, 12, 49, *50,* 73. *See also* feminized machines; gender roles
Woolley, Catherine, "Butch the Bulldozer," 244
Wooster Square (New Haven), 165, 175, 176–177
World War I, 24, 25, 56, 64, 67, 69; surplus equipment donations, 82–83; truck convoys, 186
World War II, 5, 6, 16, *19,* 21–71, 253, 279, 292; ads mocking Axis, *80,* 81, *82;* advanced technologies and, 61–65, *62, 63,* 298; bombing of cities, 23, 53, 55–57, 59, 61, 279; construction industry growth, 64–68, 238, 289; construction projects, *20,* 23–51, *28, 30, 39,* 58, 61, 75; construction training, 72–82, 97, 98; as culture of clearance context, 27–31, 67, 298; D-Day landing (1944), 41–42; as engineer's war, 23; environmental destruction from, 52–60; equipment technology advancements and, 68–72, 75–76; evictions of natives, *30,* 34–35; Greatest Generation and, 195; heavy equipment, 18, 61–71; machinery, 11–12, *29,* 119; social consequences of, 12; surplus equipment disposal, 82–87; Viet-

nam War vs., 299. *See also* Army Corps of Engineers; Seabees; veterans
wrecking. *See* demolition
wrecking balls, 72, 139, 140–141, *141,* 268
wrecking companies, 3, 5, 7–8, 147–170, 268; challenges faced by, 154–170; children's bulldozer books and, 225; fee calculation, 149–152; New Haven urban renewal and, 152, 154, 177; pre- and postwar differences, 149; wartime experience and, 149
Wright, Frank Lloyd, 56

Zim, Herbert S., and James R. Skelly: *Hoists, Cranes, and Derricks,* 236; *Tractors,* 230, 232, 244–245
zoning, 113